SPIRIT OF ENTERPRISE
THE 1984 ROLEX AWARDS

Foreword by
Michio Nagai
Senior Adviser to the Rector, United Nations University
Former Minister of Education, Science and Culture, Japan

Preface by
André J. Heiniger
Managing Director, Montres Rolex S.A.

Aurum Press  London
AURUM

Photographs and art not credited below were submitted by the entrants.

Page 190   G. Mazza Musée d'Histoire naturelle, Paris
Page 230   Shiro Shirahata – Japan mail order
Page 302   University of Agriculture Tokyo – Japan mail order
Page 350   C. de Klemm – Jacana Visage – Jacana
Page 430   Ms. Julia Horrocks

This book was produced in Switzerland by Rolex
Production coordinator: Georges Naef

Published and distributed in the United Kingdom and Commonwealth
by Aurum Press Limited, 33 Museum Street, London WC I.

ISBN 0-906053-951

Printed in Switzerland by S.R.O. S.A., Geneva.

CONTENTS

Rolex Laureates are indicated by a gold crown (👑)

Honourable Mention winners are indicated by a black crown (👑)

FOREWORD

The "Spirit of Enterprise" can be summed up as an incessant human striving for a better life.

Human history is an endless relay race in which the baton of life is handed over from one generation to another. It is a continuing challenge to transmit life to future generations. So long as this chain of life stays viable on earth, the 'spirit of enterprise' will continue to enlighten humanity.

The life of an individual is short and transient, and one's sphere of activities is limited by both temporal and spatial boundaries. Within this short span of life, only so much can be inherited from the past and imparted to the future. Thus, the most significant mission entrusted to each human being is to act as a transmitter by selecting from the enormous historical and cultural inheritances of the past the richest assets of mankind to be passed down to future generations.

We are approaching the dawn of the 21st century. Technological advancements have brought tremendous progress in industrializing a very small part of this world. But, on the other hand, the greater part of the world is still unable to free its peoples from the threats of poverty, hunger and depletion of natural resources, as well as the natural and cultural disruption of the environment. Our globe is now equipped with the potential capabilities of self-destruction. We are also exposed to the fear of war. Today, the 'spirit of enterprise' is needed more than ever before to eliminate the source of fear and to bring prosperity to the poorer majority suffering daily want.

The Selection Committee of The Rolex Awards for Enterprise — 1984 addressed one important issue throughout the selection process: What is the "Spirit of Enterprise" that is most urgently needed today? It was a great pleasure and a privilege for me to be a member of the Selection Committee to select the Rolex Awards Laureates and Honourable Mention winners from a flood of excellent projects. Individuals were invited to submit their ideas within any one of the three broad categories of competition: Applied Sciences and Invention, Exploration and Discovery, and The Environment. Virtually all the projects fully exhibited human ingenuity and thought. They also provided a key to understanding the 'spirit of enterprise' essential to ensure a brighter future.

What is pressingly needed today is the perception to shape humanity into a peaceful community. If the wisdom, foresight and dedication revealed by the Rolex Awards Laureates and Honourable Mention winners, as well as the other candidates, many of whose projects appear in this book, are fully employed to create a better world, then we can be optimistic for the future freedom, peace and happiness of all mankind.

Tokyo
April, 1984

Michio Nagai
Senior Adviser to the Rector, United Nations University
Former Minister of Education, Science and Culture, Japan

THE ROLEX AWARDS FOR ENTERPRISE 1984

MEMBERS OF THE SELECTION COMMITTEE

PREFACE

Once again, I have the enjoyable privilege of welcoming readers to the company of a very special group of people; men and women from around the world whose individual efforts and ambitions have earned our respect, gratitude, and recognition.

If you are among the tens of thousands of readers who have delved into one or both of the two previous books on the Rolex Awards for Enterprise (the 1976-77 and 1981 books), little explanation is necessary, but you will be pleased to know that this third book contains even more candidates than before.

If not, you will, with this book, enter into the fascinating world of The Rolex Awards for Enterprise, an undertaking that we, at Montres Rolex S.A., believe is of fundamental importance in our world; the encouragement of the "spirit of enterprise" in individuals around the globe.

For us at Rolex, there is a long — and very highly valued — tradition of the "spirit of enterprise" in our pursuit of quality. We know it has been a basic, necessary factor in the many innovations Rolex has brought to the world of watchmaking. We therefore applaud this "spirit of enterprise" as an integral part of the quest for excellence in all those efforts that contribute to a better quality of life in our world.

Thus, in 1976, on the 50th Anniversary of one of our "firsts", the invention and patenting of the Rolex Oyster, acclaimed in 1926 as the world's first waterproof watch, it was natural to celebrate this milestone with a tribute to the "spirit of enterprise": The Rolex Awards for Enterprise, which were inaugurated in 1976 and granted for the first time in 1978. The response was large, worldwide, and exciting. The Awards were presented again in 1981, with similarly rewarding results. At the beginning of The Rolex Awards for Enterprise 1984, we realized that we had embarked upon a grand and long-term venture, but by this time, we knew that our way of seeking to encourage the "spirit of enterprise" was an international catalyst, affecting many people.

Candidates, whose projects had appeared in the two previous books, had written to tell us of new worlds opening to them, of offers of help and advice from others with similar interests. Equally, institutions with interests in the three broad areas of the Rolex Awards had asked to be included in the distribution of information about our next Awards.

In the early days of The Rolex Awards for Enterprise, it had been our hope that books on the Awards would provide additional assistance and encouragement to the "spirit of enterprise" in individuals and that these books would bring enterprising projects and individuals to a wider audience than that the candidates might reach with their own resources. We met, and are meeting, our objective, as readers around the world respond to the Rolex Awards candidates and extend support, interest and encouragement to them.

We trust that you will find this book a spur to your own "spirit of enterprise" as well.

When the Rolex Awards for Enterprise 1984 were given at the Awards

Presentation Ceremony on April 26, 1984 in Geneva, it was the third time that five Rolex Laureates had come to us as our guests, been given cheques for 50,000 Swiss francs and specially engraved gold Rolex Oyster Day-Date Chronometers. It was the third time that some two-dozen Honourable Mention winners had been advised that they would receive their steel and gold Rolex Oysters in local ceremonies around the world. And it was the third time we announced the publication of a book with details on the winning, and many other, projects. The Rolex Awards for Enterprise are our way of trying to help improve the quality of life in our world, and to share our own belief in the value of the "spirit of enterprise".

It was at the 1984 Awards Ceremony that I announced the next Rolex Awards for Enterprise, to be given in Geneva in 1987, and I hope you will be joining us in them.

Geneva   André J. Heiniger
May 1984   *Managing Director, Montres Rolex S.A.*

INTRODUCTION

Some two hundred years ago, Edmund Burke (1729-1797) wrote: "All that is necessary for the forces of evil to win the world is for enough good men to do nothing." Today, in our 'global village' where ideas move with much greater speed and to much wider audiences, his message still rings just as clearly and just as validly. And good men – and good women – are with us, as hoped for, and as much needed as they have always been.

Two hundred twenty six of them are here, in this book. People who are doing *something*, not "nothing". Each, in his or her own way, striving to achieve an idea that will make our world a better place for us and our descendants. Some have goals that are exquisitely refined, focusing on the further extension of knowledge in some highly specialized area. Others pursue objectives with global perspectives and potential. Whichever the case, these individuals have been and are exhibiting the exhilarating human quality of 'enterprise'.

On April 26, 1984, five of these outstanding 'enterprisers' were brought to Geneva, as the guests of Montres Rolex S.A., to be given their Rolex Laureate prizes at an Awards Ceremony. Each received a cheque for Swiss francs 50,000, a specially inscribed gold Rolex Oyster Day-Date Chronometer and scrolls attesting their Awards. Elsewhere, around the world, another twenty-six Honourable Mention winners were feted in local ceremonies and presented with steel and gold Rolex Oysters and scrolls.

In addition to the official recognition and prizes given to the Laureates and Honourable Mention winners, and as with the two previous Rolex Awards, Montres Rolex S.A. wished to extend further encouragement and assistance to the winners and to many of the other enterprising candidates through publication of information about them and their projects in this book.

As the details about the candidates and their projects were submitted on the 19-page long Official Application Forms for the Awards, descriptions have been edited down to briefer chapter form, to enable inclusion of more candidate's projects than in the two previous Rolex Awards books.

The division of the book into project categories is done for reasons of easier broad general reference. Numbers of projects per category represent the approximate proportion of projects submitted per category, rather than the division of winning projects.

It is hoped that you will be able to add your own encouragement to one or more of the enterprising men and women in this book. And that the book may stimulate your own "Spirit of Enterprise". The Rolex Awards for Enterprise will be granted again in 1987.

APPLIED SCIENCES
AND INVENTION

The projects appearing in this section were submitted in competition under the "Applied Sciences and Invention" category, which was defined in The Rolex Awards for Enterprise — 1984 Official Application Form as follows:

Projects in this category will be concerned primarily with science or technology and should seek to achieve innovative steps forward in research, experimentation or application.

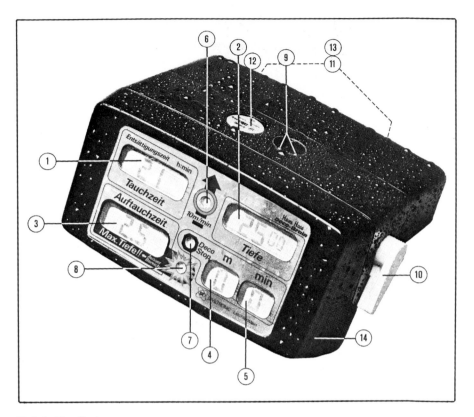

Technical Specifications

1. Air pressure display in millibar (indicated for 10 sec. after unit is switched on)
    — Actual dive time (during the dive)
    — Completed total dive time (during 5 seconds per minute for the duration of the surface interval)
    — Actual desaturation time (during 55 seconds per minute for the duration of the surface interval)
2. Actual dive depth (Accuracy: 0.5 m (1.6 ft))
    — Maximum attained depth as a threshold value (together with the total dive time during 5 seconds per minute for the duratin of the surface interval)
3. Actual ascent time (its calculation considers the ascent rate of 10m/min. and the sum of all decompression stops)
    — If display 8 illuminates ("Out-of-range" mode) simultaneously, the maximum attained depth is displayed
4. The next decompression depth
5. Decompression time in minutes which must be observed at the respective decompression depth
    (4+5) Blinking zeros provide 2 min. warning before the nondecompression limit is exceeded
6. Red LED blinks if the maximum ascent rate (10 m (33 ft)/min) is exceeded; continuous LED illumination in the event of a 20m/min ascent rate.
7. Red LED begins blinking 1.5 meters (5 ft) prior to the first decompression level and blinks as long as the countdown duration (displays 4 + 5)
8. Red Out-of-Range LED blinks in 0.5 second intervals when the diver: Dives deeper than 70m; dives longer than 255 minutes; exceeds maximum permissible saturation limit values due to several repetitive dives; dives in mountain lakes at elevations greater than 3500 m (11,500 ft); ascends to the surface from a depth of 7 m J(23 ft) without completing decompression; switches on the Deco-Brain® underwater or switches it off and on again under water. In the "Out-of-Range" mode, displays 4 and 5 are extinguished, displays Nos. 1 and 2 continue normal operation while display 3 indicates the maximum attained depth. Consequently, the diver may now complete the dive according to his own estimates.
9. Charging connection consisting of two gold-plated contacts directly exposed in the water which do not require additional seals
10. Power switch which can only be activated by "pulling and turning" to prevent accidental unit switch-off under water. As an added feature, switch design is such that the diver must swith the unit on before fastening it to his arm.
11. Flexible straps for fastening unit to forearm or BC
12. Piezoresistive pressure sensor
13. Sealed mounting screws
14. Impact-resistant plastic housing with replaceable power pack. The housing has no openings for switches or cables; the electronics module and power pack are self-contained, airtight units.
Top of the line CMOS 8 bit microprocessor with 10 kByte ROM for assembler program and data base.

The World's First Solid State Personal Diving Computer

Jürgen Hermann

Rennhof 546, 9493 Mauren, Fürstentum, Liechtenstein

Liechtensteiner, born July 24, 1955. Company president. Engineering degree in Electronic Measuring and Control Engineering from Interstate College of Technology, NTB, Switzerland.

As a solo entrepreneur, active sports diver and electronics engineer, I have developed the Hans Hass *"Deco-Brain"®* — the world's first solid-state *dive computer*, after 3 1/2 man-years in collaboration with R. Vogler. Through this invention, for which worldwide patents are pending, the dream of unencumbered diving has finally become a reality. For the first time in the history of diving, a microcomputer accompanies the diver, giving the diver much greater diving safety and pleasure. The Hans Hass Deco-Brain® calculates the nitrogen saturation and desaturation of eight different tissues and indicates the dive time, dive depth, and ascent conditions in an easily readable countdown display. The Deco-Brain® warns the diver if his ascent rate is too rapid and of incorrect actions, thus thoroughly preventing improvised decompression plans with tables, watches and depth gauges which have cost the lives of many divers. Through the test reports of a well-known diving instructor and articles in leading diving publications and technical journals, I was successful in obtaining financing from prominent companies in Switzerland and Liechtenstein — and among others, underwater researcher Professor Hans Hass — for production and distribution of the Deco-Brain®. DIVETRONIC Instruments AG was established in Schaan, Liechtenstein in October of 1982, and we began with series production of the first 10,000 dive computer units. The Hans Hass Deco-Brain® will be available from retailers worldwide in April 1983, and diving will be revolutionized for the better.

How did the idea for this invention, which satisfies a genuine need in the diving community and closes a market gap, originate?

As a graduate of the Interstate College of Technology, NTB in Buchs, Switzerland, and with in-depth work in microprocessors and more than 10 years of diving experience, I was adequately prepared to launch development of the "diver computer" project, which lasted 3 1/2 years.

The incentive for my invention orginated with my knowledge of diving hazards and the awareness that diving accidents such as that suffered by my brother in the Mediterranean in 1974 could be prevented if dive data were calculated by a computer that would not be affected by depth, cold or darkness, and that would function impeccably even if the diver's concentration were affected by nitrogen narcosis.

I am deeply grateful to my colleague, R. Vogler, for his active collaboration, as well as for the permission by the NTB to use part of my dive computer development for my thesis.

The following is a brief survey of the development of the dive computer.

Apr '80 — Start of thesis "The Dive Computer" at the NTB.

Nov '80 — Approximately 1/6 of dive computer development concluded; graduation with engineering degree.

Apr '81 — Transfer of operation of injection molding plant and establishment of HERMAPLAST, myself as president, for continuing dive computer development.

Sep '81 — Exhibition of the first 6 prototypes at the SPOGNA, Cologne. The dive computer creates furors in the diving literature and trade journals. "A New Era in Diving." Initial marketing work and studies. Six-month test phase in the ocean, lakes and decompression chambers.

Oct '82 — Based on the enthusiastic response regarding the function and method of operation of the world's first dive computer, as well as numerous requests for sales rights from many western countries, it was possible to obtain financial backing for series production and distribution from oceanic researcher Prof. Hans Hass and others. As a result, we were able to establish DIVETRONIC Instruments AG in October, 1982.

Dec '82 — Exhibition of the improved dive computer model at the BOOT 1982 in Düsseldorf. Acceptance of the first orders from divers and retailers. Strategic support with test reports and advertisements in diving publications. Additional tests in the ocean, lakes and decompression chambers by the Royal Navy, the Deep Diving Research Laboratory in Zurich, Switzerland, diverse diving publications, diving instructors and by us.

Jan '83 — Production of the first 100 series units.

Apr '83 — Worldwide delivery of the first 2000 series units.

General Comments Regarding Safe Diving

Every diver must be aware of ascent conditions to prevent decompression sickness. To this end, the diver must observe a safe ascent rate and specific decompression stops at different depths during ascent. Decompression stops must be observed depending on the maximum dive time (bottom time) and depth. Throughout the history of diving, ascent conditions were always determined "by hand", i.e., with a diver's watch, depth gauge and dive tables.

The diver often faces difficulties when using dive tables to determine decompression stops. In addition, the determination of decompression stops in relation to optimum decompression is highly inaccurate since convenient use is a trade-off for precise dive profile determination.

Dive tables must:

1. Be easy to use.

2. Have a decompression plan for each depth/time profile with the required degree of safety.

Both of these demands are contradictory, since simple table use requires a small number of table value input values which, in turn, precludes precise determination of dive criteria. Consequently, the precisely required decompression conditions for each dive profile cannot be determined. Therefore, all tables are based on the compromise that a small number of table input values are at the expense of a minimum required decompression time, and only in extreme cases does the diver attain the precisely required decompression; in all other cases the diver's decompression time is much too long. This is precisely the reason why use of the tables is so hazardous since, in the most common cases, the diver is induced to shorten the decompression time according to his own estimates. Improvisation of this nature can lead to fatal results. The Hans Hass Deco-Brain® totally eliminates hazardous, improvised decompression since only the dive computer is capable of recording and processing the multitude of dive data for optimum decompression calculations on every dive, and providing the diver with an easily readable countdown.

A Few Comments Regarding Dive Physiology
The diver's regulator provides a pressure balance when diving with compressed air. This means that the air the diver breathes is under ambient pressure. At increasing depths, the diver breathes air at high pressure, and this results in an increased quantity of air dissolved in the diver's blood. Different tissues are saturated with the different gases from which air is composed at different rates, according to specified saturation factors, i.e., desaturation. However, if the ambient pressure is rapidly reduced resulting from too rapid an ascent, the air in the blood and the tissues cannot be discharged fast enough. This is particularly the case with nitrogen, since the excess of oxygen and carbon dioxide is substantially less. This is due to the fact that oxygen is consumed by the tissues to a large extent, and carbon dioxide leaves the body faster than other gases due to its high diffusion rate. If the nitrogen is subject to a rapid pressure reduction, bubbles can form, resulting in decompression sickness which can be avoided by gradual gas discharge from the blood and tissues so that bubble formation does not occur.

The factor of gas bubble formation which may lead to bodily injury results when nitrogen gas bubbles form in the blood vessels and become trapped in the capillaries. Here, they prevent blood and oxygen supply to the surrounding tissues. When a blood vessel is blocked (embolism), the respective tissues are no longer nourished and die.

To avoid decompression sickness, the diver may not exceed a safe ascent rate of 10m/min., and must observe specific decompression stops during ascent, depending upon his saturation and the ambient air pressure on the surface. Saturation and desaturation are also affected by different factors, the major ones of which are given below:
— Period spent at the dive location before the dive.
— Previous dives within 12 hours.
— Course of the dive depth/time profile, exertion under water.
— Tissue composition of the diver (obesity or athletic build, man or woman).
— Composition of respiration gas.
All of these varying, and to some extent continually changing, factors are considered by the calculation algorithms of the Hans Hass Deco-Brain®, thus ensuring safe diving.

The first "Sun Pump" in Pakistan, with the Pakistan Agricultural Research Council. Pump mechanism is floating in the water under the buoys.

Micro Solar Technology for the Third World

♛ Stephen Allison

Honourable Mention – The Rolex Awards for Enterprise – 1984
31 Amies Street, London SW11, England

*Canadian, born August 9th, 1935. Self-employed Consulting Engineer.
Educated in Canada and U.S.A.; Dr. Eng. (Water Resources) from
University of California in 1966.*

BACKGROUND

Almost all the poorer countries of this world have two things in common;
conventional forms of energy are expensive, and sunshine is abundant. Many
efforts have been made to develop appropriate (affordable/simple/reliable)
alternative energies for these regions.

It has now become apparent that the widespread adoption of these technologies has been constrained by the fact that most of the devices proposed so far
have lacked the characteristics required to stimulate the interest of entrepreneurs who would inject the dedication and sustained effort essential to the
success of any *enterprise*, including technology transfer.

The aim of this project is thus to develop, and transfer effectively to the less
developed countries of the world, a series of small items of solar powered
equipment that will contribute to improvement in the welfare of the people of
those countries.

Specifically, these are devices that:
   a) Meet critical perceived needs of the rural poor.
   b) Draw on the abundant energy available from the sun, and,
   c) Are developed with the explicit recognition that for any technology to
spread effectively in the Free World, it must have attributes which make it the
basis for a sensible, reasonably profitable business.

THE CONCEPT

Over 50 million of the world's poorest farm families exist on less than 50
million hectares (125 million acres) in the fertile alluvial valleys, deltas and
coastal plains of the Third World. Most of these areas experience extended dry
seasons during which cropping is impossible without irrigation. Adequate
irrigation, however, is frequently either unavailable, or beyond the financial
means of a small farmer. The use of diesel powered pumps for irrigation is
limited by the fact that the smallest of these have the capacity to irrigate well
over ten acres of land, and many farms are nowhere near as large as this.

Our SEI Sun Pumps were developed to meet the needs of these small farms, as solar-powered water pumping systems which can dramatically alter the irrigation options open to these farmer. The units are specifically designed to supply irrigation water to farms of 1 to 4 hectares, depending on lift and cropping pattern, where conventional methods are either inapppropriate or becoming prohibitively expensive. In addition, SEI Sun Pumps can be used for stock watering in remote locations and as a component of village and other rural water-supply systems.

DESCRIPTION

The Sun Pump available today is the SEI-250M model; a lightweight, self-contained system designed to deliver 2.5 liters/second at a lift of 5 meters. The unit comprises a 250 watt photovoltaic array, submersible pumpset and a portable mounting frame. The entire apparatus can be easily moved from house to field, or from one water source to another. The simple operation and low maintenance of the Sun Pump make it particularly well suited for use in rural areas and in those places not served by an electric grid.

OPERATION, MAINTENANCE AND COMPONENTS

The SEI-250 Sun Pump is quiet, safe and simple to operate. The user merely places the array in the proper sun-facing direction, lowers the pumpset into a well or canal, and switches the unit on. If he does not intend to move the Sun Pump daily, he leaves the switch on; the unit will automatically turn on and off as the sun rises and sets. Additional daily discharge is obtainable by repositioning the array manually three or four times a day to face the sun.

Routine maintenance involves occasional water rinsing to remove any accumulated dust or dirt. The array is designed to a trouble-free lifetime of 15 years, while the pumpset is designed for 15,000 hours of maintenance-free operation.

The photovoltaic array is rugged, highly reliable, and easily maintained. Its outer surface is protected by high transmittance tempered glass.

The pumpset is submersible, with high efficiency variable voltage and variable speed. Its nominal operation is 3,000 RPM at 60 Volts DC. Brushless, permanent magnet design eliminates the need to unseal the motor for periodic maintenance. The vertical axis, single-stage centrifugal pump is close-coupled to the motor. It has a high impact plastic impeller and glass reinforced plastic casing. The pump is optimized for high-flow, low-head, canal and open well conditions. The sturdy frame supports the array, allows proper orientation to the sun and facilitates transport of the unit.

PROGRESS TO DATE

Since this 'project' was launched in 1978, it has succeeded in developing a series of solar powered "micro pumps", used primarily for irrigation. A total of over 200 of these units are now in operation in 25 countries around the world. These pumps have functioned well and served to confirm the validity of the basic concept of portable, solar-powered pumping systems.

PRESENT STATUS

In 1978, conventional energy costs were rising and the cost of photovoltaic cells were dropping in a fashion which indicated strongly that the point of economic competitiveness would be reached in 1983/84. Since then, world oil prices have stabilized and may even decline in real terms, while the pace of technological development and the growth of world markets for "Solar Equipment" has been slowed by changes in the political and economic environment.

Thus, while the technology has been demonstrated successfully, the point of real economic competitives has been delayed, perhaps until the early 1990's.

The dissemination of the technology, however, has been well launched. Whereas the number of solar pumps in use around the world was less than 100 in 1979, the total is now well over 500. A dozen manufacturers have developed pumps and the market continues to develop, albeit too slowly to support many small entrepreneurs for the moment.

FUTURE DEVELOPMENTS

While we wait for oil prices to start to climb again and for further decline in solar cell prices, another product has been identified which meets the criteria set forth above and is economically competitive (and thus marketable) now. This is a solar recharger for the batteries used in the billions to power transistor radios and flashlights in the unelectrified villages of the world. A small firm to assemble these systems has been established and product development initiated. Testing will take place in the first half of 1983, and first shipments overseas are plannned for August/September. Beyond this, there are a considerable number of small solar devices, including such handheld agricultural implements as weeders and sprayers, awaiting development.

The high-protein, third instar of *Papilio polytes romulus*; a potential help in solving Third World malnutrition problems.

Fighting Malnutrition in Tropical Asia with "Natural Protein Capsules"

♛ Parakum Maitipe

Honourable Mention — The Rolex Awards for Enterprise — 1984
'Pasadika', Maitipe, Galle, Sri Lanka.

Sri Lankan, born June 27, 1955. Scientific Officer at National Aquatic Resources Agency. Educated in Sri Lanka and Austria; B.Sc. from Faculty of Natural Sciences, University of Colombo

Lack of animal protein in the human diet is a serious cause of malnutrition in many parts of the world. Conventional sources of such animal protein are normally high on the food chain, and thus too costly in energy terms to be afforded by many people who need animal protein in their diets. The search for more inexpensively available animal protein is one that concerns those of us who work in scientific areas of research into better nutrition and food production. In my institution, research is carried out to exploit the marine, brackishwater and freshwater resources in Sri Lanka. I am one of the scientists who work under the Inland Fisheries Project, which carries on research in our man-made reservoirs with the hope of improving freshwater fish production, for purposes of increasing our food supply.

This general direction of research has led me to the examination of numerous elements in the food chain, and to explore those opportunities for producing animal protein rapidly and efficiently.

One area of promise is the exploitation of certain insect larvae, long known to be edible, but never broadly popularized as human food. The reasons for the non-utilization of this potential food source probably include the lack of knowledge required to distinguish the palatable species from the venomous, and lack of experience in their mass (commercially profitable) production and processing of readily acceptable product forms. Such obstacles are not necessarily impossible to overcome, and in view of the need for animal protein for human consumption, I have investigated the opportunity posed by achieving solutions to these problems.

The larvae of *Papilio polytes* (Insecta; Lepidoptera — a tropical butterfly) grow up to an optimal 'harvesting' size within 17 days, if kept at 28°C and fed on certain alkaloid rich leaves.

In the context of food production, these tiny 'natural machines' are exceptionally fast at the job of converting plant protein into animal protein. In the early stages of the development of these small creatures, they have no other natural function apart from feeding for growth, so there is virtually no energy wasted on the functions of reproduction and locomotion; a rare degree of 'manufacturing efficiency' indeed.

A further example of 'energy efficiency' is the fact that the larvae feed on primary producers — plants — and therefore the loss of energy is at only one link along the food chain.

Since the animals are very small, the food conversion rate is very fast, which adds yet again to the efficiency factor. The food conversion rate — the number of kilograms of plant material translated into the number of kilograms of animal material — takes a shorter period, in 17 days, than that of any other popular animal food production duration.

With such initial efficiencies evident, it can only be considered as a bonus that this source of animal protein should present itself so easily as a product when it reaches an appropriate stage for use. At the fourth instar (those periods of insect development between the natural molts that accompany growth), these larvae can literally be 'harvested' as a "natural protein capsule", ready for consumption; there are neither bones nor skin requiring discard. When taken, the larvae can be consumed by humans (or readily used as a feed for fish — an area of interest in my work — or poultry). No skinning or slicing, no other preparation requirements.

Work to date indicates that these larvae are relatively easy to handle and cultivate. They are non-venomous, non-cannibalistic, and no larval parasites have been encountered.

The program of work ahead is to pursue a number of objectives.

One of the tasks is to investigate a variety of acceptable plants to determine the most appropriate for use as food for the larvae. It must be easy to cultivate, economical and optimal for the larvae. Observations so far show that different plant leaves containing different alkaloids are accepted by the larvae, but that differing growth rates of the larvae can be achieved depending upon plant selection. For example, citrus leaves are a far better food base than are 'woodapple' leaves.

Secondly, further work must be done to determine optimal conditions for larval development on a large scale. It has been found that excessive humidity increases mortality, indicating that humidity control will be important in large-scale growing operations. Further, it was found that the larvae are sensitive to low frequency sounds or vibrations, with the result that their feeding rate was retarded.

It will be necessary to identify necessary precautions against possible diseases, as well as remedies, in case of infectious disease conditions. This is applicable to the food plants as well as to the larvae.

One of the difficulties encountered is the preliminary work necessary for the preparation of the larvae 'cultivating process'. It becomes laborious, with present techniques, if large scale production is concerned. Even a single batch of butterfly eggs hatches within a span of two days, and will provide optimal initial larvae within the span of seven days. This means that selective initial harvesting must be carried out daily for seven days with each batch of eggs. How best to do this will require further research. Final harvesting, of course, must be done at the right moment or the larvae will use up their energy stores, prior to pupation.

The nutritional and calorific values of the larvae "protein capsules" need to be further assessed by detailed chemical surveys. These should be repeated with whatever 'processed food' form they may be given, in later steps of experimentation. Quality control and hygienic tests will also be needed.

Quick methods of processing or preservation are essential because live larvae use up their energy stores quickly after harvesting, and dead larvae are highly vulnerable to microbial attacks. Based on our experience, quick processing seems to be the more rewarding path to follow, since preservation techniques would mean additional costs.

The mass production of *Papilio polytes* eggs, the protection of eggs from egg parasites, and control of hatching needs to be studied further. Rearing of the adult butterfly and aided fertilization has been found to be feasible.

Research on this one species, already shown to be promising as a potential food source, will provide baseline information for research into other, possibly more suitable species from the insect world. The development of such new sources of animal protein could provide significant help in the fight against malnutrition.

The "Cyclo-Crane" at mast in August, 1982. A fuel-efficient, versatile competitor to the helicopter.

The Cyclo-Crane: Lighter-Than-Air Hybrid Aircraft for Ultra-heavy Lift Operations

Arthur G. Crimmins

4105 Blimp Boulevard, Tillamook County, Tillamook, Oregon 97141, U.S.A.

American, born January 4, 1931. President and Chief Executive Officer for Aerolift, Inc. Educated in U.S.A., in Electrical Engineering/Business Studies.

The proposed Cyclo-Crane is a hybrid aircraft utilizing aerostatic lift from a helium filled centerbody to support all structural weight, plus 50% of the slingload specification. The balance of the slingload support and thrust for control and translation is supplied by a system of airfoils that rotate in hover and become aligned with the direction of flight when Cyclo-Crane reaches its maximum designed forward speed.

The Cyclo-Crane can be used for worldwide logging from remote areas, transport of complete factory-built housing or movement of existing houses, for offshore surveillance of coastal waters, in salvage and mine sweeping, oil drilling, the application of agricultural chemicals and in passenger transportation.

A principal element of the Cyclo-Crane design is the aerostat. The aerostat (balloon/envelope) uses standard, modern balloon materials, i.e., a dacron fiber impregnated with urethane, for helium retention. The aerostat includes a rip-stop network that limits any envelope damage to the width of a single panel. Repair of such damage is relatively simple, and would not present any safety problem. While the history of aerostatic vehicles in general (blimps, etc.) does not indicate that envelope failure is a significant problem, the use of this rip-stop technique is intended to minimize the possibility of loss of aerostatic lift. As helium is intended as the lifting medium, no fire hazard exists.

The aerodynamic elements of the Cyclo-Crane (wings, blades, tail surfaces) use completely standard aircraft fabrication techniques. These surfaces and the center structure benefit from the basic weight insensitivity of the Cyclo-Crane concept. All components can be designed with safety factors far in excess of normal aircraft practices. However, this is a new design that has not been fully flight tested. Actual flight testing may require design changes.

As all the structural weight is supported by the buoyancy of the centerbody, it is not necessary to design for minimum structural weight. Thus, the criteria for selection of components, or systems, e.g., fore and aft bearings, hydraulic systems, etc., are safety, low maintenance cost and low acquisition cost rather than the normal aircraft design concern for low weight.

This design philosophy must be fully appreciated to understand the design, construction and operational cost estimates. The design effort uses a very high safety factor that accepts weight penalties in return for a strong structure that can be fabricated from low cost components using simple construction techniques. Maintenance costs are likewise low (compared with aircraft experience) due to use of parts and systems far more massive and durable than those normally used in airframe manufacture.

The Cyclo-Crane utilizes standard aircraft engines that power what is essentially a tip-driven, rotor configuration. Vertical takeoff and landing capabilities are available with any two of the four engines operating.

The design objective of the Cyclo-Crane is to create an aircraft capable of expanding the vertical takeoff and landing capabilities of helicopters into areas of higher slingloads, at a very much reduced cost of operations. This hybrid aircraft will utilize far less power than equivalent helicopters, but will offer similar maneuverability. The anticipated performance characteristics will be achieved by virtue of the design philosophy, wherein all structural weight, plus 50% of the slingload, is carried by highly cost effective aerostatic methods, while the active power systems (aircraft engines) are only required to generate lift forces equal to 50% of the slingload and thrust forces for control and translation.

The two-ton slingload Cyclo-Crane now under test is powered by four 150 hp Avco-Lycoming piston aircraft engines, installed with overall power requirements of only 93 horsepower per engine. Fixed pitch, wooden, four-bladed propellers are used, and operate at the single design point of 60 mph airspeed over the wing, which provides a forward speed of the vehicle in the area of 40 mph, fully loaded.

The aerodynamics of the Cyclo-Crane, although very large in size, utilize light aircraft design and construction practices. This is true because the flight speeds of all the aerodynamic elements of the Cyclo-Crane do not exceed 100 mmph. These speeds are well within the technology of aircraft that were built during World War I and shortly thereafter.

The pilot controls the thrust vectors of the airfoils via systems analogous to cyclic and collective helicopter controls. Analytical studies (supported by wind tunnel data) by Princeton University indicate that the Cyclo-Crane has the same controllability as a helicopter (of equal slingload specifications) under gust conditions.

The vehicle can be safely moored in winds up to 80 mph by mast mooring techniques similar to successful airship mooring methods, and may be designed to float hundreds of feet high on a single line tether. When major storms are predicted, standard aircraft procedures would be followed, and the vehicle flown out of the area of danger. With a full slingload of fuel, the Cyclo-Crane can travel over 3,000 miles. Thus, relocation to avoid storms, or for ferry purposes, can be accomplished easily.

The largest commercially available helicopter (the S-64) has a maximum slingload of 12 tons (minimum fuel on board) and averages about eight tons in crane type missions such as logging. The S-64 sells for over $7,000,000. It also costs well over $3,000 per hour to operate (1,500 hours per year).

A 16-ton slingload Cyclo-Crane will sell for approximately $2,500,000, with operation costs of about $747 per hour (1,500 hours per year). The 16-ton Cyclo-Crane specification is with eight hours of fuel on board. As the Cyclo-Crane has a built-in overload capacity of 20% in hover, the average slingload carried should be roughly twice the helicopter average. As the actual operational speeds of the vehicles will be similar, i.e., the helicopter rarely reaches its maximum speed with a full slingload, a process improvement factor of about eight should be within reach.

The Cyclo-Crane concept is valid for vehicles up to 75 tons.

The project began in 1978, with the first prototype being completed and tested in 1982. Following damage to the vehicle in October 1982, its reconstruction process is estimated to be 15 months, from February 1983 to May 1984.

As of September 8, 1981, 13 patent applications had been filed in various countries; six have issued patents as of the date of this application.

Enzyme Replacement Therapy in Porphyrias: A Counter to Lead Poisoning?

Alcira María del Carmen Batlle
Viamonte 1881 10'A', 1056 Buenos Aires, Argentina

Argentine, born January 5, 1937. Professor of Biochemistry at the School of Sciences, University of Buenos Aires, Head of National Research Institute on Porphyrins and Porphyrias. Educated in Argentina and U.K.; Ph.D. in Science (cum laude) from University College, University of London in 1965.

Enzyme replacement therapy (ERT), for patients with various disorders in which a single enzyme is deficient or missing, has been attempted in some instances. To date, however, this promising therapeutic maneuver is often regarded as a "futuristic medical tour de force", which has not been achieved as yet.

The majority of porphyrias are hereditary disturbances of porphyrin metabolism. A specific enzymatic defect, genetic or acquired, generally accompanied by a secondary increase in the activity of aminolevulinic acid synthetases $(ALA-S)$, is the basic cause of all these diseases, which explains the diversity of biochemical patterns of accumulation and excretion of porphyrins and their precursors that characterize each type of porphyria. We have been investigating the application of ERT in porphyrias, an approach we believe has not been tried before.

In order to successfully apply ERT, certain criteria must be fulfilled: 1) a suitable carrier for the enzyme, which protects both the enzyme from degradations and inactivation, and the host from unwanted pharmacological and immunological effects, needs to be developed; 2) a source of enzyme should be selected, human in origin if planning is for use in human therapy; 3) very high purity and stability must result from the preparation of the enzyme; 4) the enzyme should be delivered to the tissue in which it is to function, and; 5) the enzyme should interact with its substrate long enough to render clinical benefits.

Enzyme covalently bound to poly-aldehyde dextrans, or enzyme entrapped in liposomes or in erythrocyte hosts, can be good candidates as vectors for ERT.

In selecting the porphyria to start trying ERT, we have chosen lead intoxication because of many considerations. It is a disorder which occurs with very high frequency; consequently, it is easy to find suitable test subjects and the number of people who might potentially benefit would be large. It is also one of the most advantageous experimental animal models. An important biochemical abnormality in lead poisoning, the increased concentration of

ALA, is very easily detected in urine as a consequence of a significant inhibition of ALA—Dehydratase (ALA—D) by the metal; measurement of erythrocyte ALA—D activity is the most sensitive indicator of lead exposure.

As the first step in this project, a procedure for obtaining a highly purified and stable preparation of of ALA—D from human blood was developed.

The second step was to find the carrier which best satisfied our requirements. First attempts to obtain an active preparation of ALS—D and other enzymes immobilized to microspheres of chemically modified Sephadex were unsuccessful. Two other approaches were then followed: Entrapment of enzymes in a) liposomes, b) erythrocyte ghosts.

With entrapment in liposomes, encapsulation of enzymes in artifical lipid vesicles and their delivery to cells has been a most promising development in the field of ERT. (The term liposome was coined to describe these spherules, which resemble natural membranes and are concentric, bi—layered structures formed on the addition of an aqueous phase to dry phospholipids.) Different types of neutral and negative, large, small and multilamellar liposomes were prepared and conditions for entrapping ALA—D were studied. Of all preparations, best entrapment values for this enzyme were consistently obtained with negatively charged large multilamellar vesicles. Concerning this phase of the project, we have so far achieved the step of obtaining an active preparation of ALA—D loaded liposomes for their use as an enzyme carrier in ERT, and the fate and effects of these both 'in vitro' and 'in vivo' must still be studied. With entrapment in erythrocyte ghosts, the possibility of using loaded erythrocytes in the treatment of lead intoxication is a fascinating one, posing a number of potential advantages. Among these, the patient's (or animal's) own blood may be used, and the resealed erythrocytes can circulate as long as normal erythrocytes — which allows enzymic reactions to be carried out within the blood stream. The potential use of ghosts as carriers for ALA-D was investigated, and the optimum conditons for the preparation of the erythrocyte ghosts and encapsulation of ALA-D were determined. A very good yield procedure for entrapment of highly purified human blood ALA-D, into both normal and defective erythrocytes, was developed.

In a clinical trial, a patient was treated with only one course of highly purified human blood ALA-D entrapped in autologous erythrocyte ghosts, given intravenously. No ill effects were observed during or after infusion; there was an immediate increase (observed 1 hour later) in the the patient's erythrocyte ALA-D activity, reaching maximum and nearly normal level two days later, with recovered activity kept practically levelled off for months. This biochemical and clinical improvement shows that this novel therapy has been beneficial, and can be safely and successfully used in the treatment of human lead poisoning. We are pursuing further studies in the belief that ERT can be successfully used in the treatment of many porphyrias.

"By the Year 1990": Newsletter for Immunization Programs in Developing Countries

Duncan Guthrie

Wildhanger, Amberley, West Sussex BN18 9NR, England

British, born October 1, 1911. Director (part-time) of Child-to-Child Programme, Institute of Child Health, London University. Educated in U.K.; honorary degrees from three universities.

"By the Year 1990" is a project to encourage the involvement of the lay populations of Developing Countries in poliomyelitis immunization programmes. Without this added manpower, it will not be possible to achieve immunization of more than a small portion of the child population.

Poliomyelitis has now been virtually eradicated from the industrialised countries. The same is by no means true of the developing countries, the so-called Third World, where the incidence of crippling as a result of poliomyelitis continues to increase, a matter of very real shame to the privileged nations of the "North" (in the context of the Brant Commissions's North-South concept), a continuing health disaster in the less developed countries and a potential health hazard to the populations of the world generally.

Small-pox has been totally eradicated throughout the world and, although the nature of the poliovirus differs from that of the variola virus and the techniques for protection from the two infections are quite distinct, the small-pox story must surely give great encouragement to anybody considering the poliomyelitis situation world-wide.

Further encouragement comes from the knowledge that there are already two types of vaccine conferring protection against polio; the killed (or inactivated) vaccine that is injected, and the live vaccine that is administered orally.

In the world's more developed regions, the child population has roughly stabilised. In the less-developed regions, however, the child population figures are rising steadily, to an estimated 1,680,000,000 by the year 2000. Each of these children must be offered immunization against poliomyelitis.

The cost, in cash and man-power, of achieving this degree of coverage must be examined. Looking at costs, we see, for instance, that in 1976 in India each state was spending around U.S.$1.00 per year per head of population on total health care. Maharashtra spent $1.60, rather more than most. However, 80 per cent of this expenditure was in three cities, and only 4½ percent was spent in the villages. It was calculated that only two cents per year were spent on the health care of each villager — far less than would make immunization a possibility.

Neither Maharashtra nor India is by any means unique.

Limited funds, and immense numbers of children *needing immunization*. This is the measure of the problem of immunizing the world's children against poliomyelitis.

When I reached the official age for retirement in 1976, I was clearly still in excellent health, with only a minor war wound on the debit side and the prospect of a good number of active years ahead of me. For the previous 25 years, I had been the director of the National Fund for Research into Crippling Diseases, originally the National Fund for Poliomyelitis Research, which I myself had established in 1952. Starting with no funds at all, by the time I retired the National Fund was raising from the public, and dispensing for research, almost £1 million per annum.

In view of this, and of the fact that over the previous quarter of a century I had acquired considerable experience both of disablement and non-governmental organizations, I set up an un-endowed and non-profit organization, the Disabilities Studies Unit (DSU) which was accepted by and registered with the Charities Commission, so benefitting from certain tax exemptions and other fiscal privileges.

The DSU has carried out a number of projects, but has no permanent staff. Research workers and secretarial assistance are employed on an ad hoc basis, as and when needed. The day-to-day administration of the DSU is normally carried out by myself.

I decided to use the DSU to promote a campaign for encouraging involvement in immunization programmes on the part of the lay populations throughout the world, or at least in those countries in which poliomyelitis still exists. It was clear from what has already been said that existing medical services in these countries are inadequate for this purpose and that an additional in-put from lay or untrained people of good will is essential. The campaign has been entitled "By the Year 1990", that being the target year set by WHO for the availability of immunization to every child in the world.

The first step has been to publish a newsletter, also called "By the Year 1990"; the first edition is now available. This newsletter is the core of the programme and the central means of communication with the programme's supporters, as well as fulfilling a 'missionary' role to the as yet uninvolved. Its information about vaccination, the cold chain, etc., will help to give knowledge, confidence and enthusiasm to the lay populations, and lead to the spread of popular concern for an effective control of poliomyelitis. When funds are available, publications will translated into appropriate languages.

It will have been noted that I myself am over 70 years of age. Although my health is excellent, it is necessary to be realistic and accept that it may be that I shall not reach the year 1990 myself. Steps will be taken at an early date to ensure the continuation of the project until the target of "immunization available to all children" has been achieved.

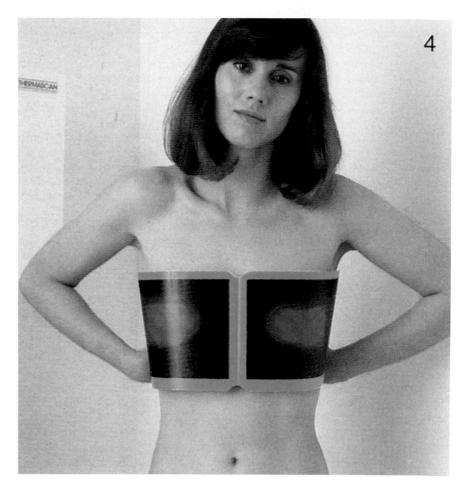

The "THERMASCAN" at work: quick, inexpensive, non-invasive detection of heat abnormalities that may signal breast cancer.

Early Detection of Breast Cancer with a New Wide-range Thermographic Foil

Aldo Colombo

Via Frua 14, Milano, Italy 20123

Italian, born March 5, 1926. Managing Director of FINPAT ITALIA srl. Educated as an economist.

I have developed, in Italy, a new, safe, simple device to detect the presence of heat producing breast diseases at their earliest stage. The device is a flexible strip of mylar foil which women simply wrap around their breasts. It contains encapsulated liquid crystals which produce color thermal patterns of the breast. It is a *breakthrough* product for all women in high risk categories who find self-palpation difficult and X-rays too risky.

The breast is the site of the most common cancer in women today, accounting for about 28% of all cancers in American women. It is estimated that women have a one-in-eleven chance of developing breast cancer in their lifetime. This year, over 110,000 American women will develop the disease. Breast cancer is the leading cause of death in women 39-54 years of age. Incidence of the disease has been increasing by 1% per year for the past 40 years, while the death rate from breast cancer has remained stationary. Of great concern is the fact that there has been a steep rise in the occurrence of breast cancer in women under 40 years.

The stage of cancer at the time of diagnosis is a vitally important factor in the prognosis of patient survival. About one-third of breast cancer is presented to the physician in too advanced a stage to expect a possible cure. In the U.S.A., only about 25% of women with breast cancer are alive and cured 10 years after diagnosis. Over 90% of malignant lesions are initially detected by the patient, usually by accident and too late (Stage II-Stage III, Nodal Involvement).

Yet no definitive cure or prevention technique exists. The major means available today to possibly effect a decrease in the incidence of the disease, and an increase in prognosis, is early detection. Two factors are required in any early detection program: 1) an aware patient population, and 2) a routine breast screening program.

There are a number of risk factors generally agreed upon as contributing to breast cancer; age, previous history (prior mastectomy), family history of cancer (especially premenopausal occurrence of breast cancer), ductal hyperplasia, lump, localized pain, or nipple discharge, previous benign breast biopsy, early menarche or late menopause, nulliparity, increased age at birth of first child, increased body size, large breasts, Jewish or European extraction, radiation exposure and a previous abnormal thermogram.

The major aim of a screening program should not be differential diagnosis; but rather the detection of breast abnormality. The THERMASCAN/B.T.D device is intended to serve as a valuable adjunct to diagnostic techniques such as palpation and baseline mammography. Furthermore, in young, average-risk patients for whom mammography would not be advised, thermography is both safe (non-radiative, non-invasive) and valuable since it can discover even tiny heat producing cancers missed by physical examination. THERMASCAN/B.T.D. is never to be considered as diagnostic for cancer; clinical judgement should always prevail. The technique, by detecting areas of abnormal breast heat and showing asymmetrical patterns between the breasts, does warn of the possible existence of various breast pathologies, one of which could be cancer.

A recent report in the literature has emphasized that biopsy on the basis of a suspect thermogram is not warranted, and that mammographic or clinical evidence is necessary for confirmation and localization. In the absence of confirmatory findings, the patient with a suspect thermogram evaluation is considered "at risk" until the positive finding reverts to normal or an adequate explanation becomes apparent.

Though the device and technique has focused its application and usefulness on the primary care physician, it is the first product of its kind that women can use at home to get an early warning of breast abnormalities long before they reach the critical stage.

Known as B.T.D. (Breast Thermo Detector) in Europe, and THERMAS-CAN in the U.S.A., the device consists of a polyvinylchloride support in a suitable size and form to be applied to the breast. It incorporates 2-3 layers of microencapsulated liquid crystals, and a layer of a special black printer's ink. The entire side of the device that contacts the skin is protected with a special transparent varnish. The construction of the device permits accurate coverage of the temperature range of 29°C (+/− 0.5°C) to 35°C (+/− 0.5°C).

The device is intended to supplement self-palpation and self-examination by women and to function as an early warning sign of breast abnormalities. Studies have shown its effectiveness in Italy, Spain, Germany, Greece and Japan. Additional studies are on-going in these countries.

The device has been tested in the U.S.A. in order to evaluate its accuracy as a heat sensitive device, and to compare it with existing screening methods for the detection of breast anomalies. Use in practice is by applying it to the breast; the test takes less than one minute to perform.

The device is safe, not toxic, not irritating, not invasive, and it does not emit dangerous radiation. When the device is held firmly across the breasts, the liquid crystals react to the thermal activity of the breasts and display an instantaneous color pattern which stabilizes in 15 seconds. These colors and their symmetrical or asymmetrical appearance form the basis for the THERMASCAN interpretation.

Encapsulated Liquid Crystals (E.L.C.), of which the device consists, have the property of changing colors according to the different temperatures of the breast area. The cooler areas will be dark brown colored, medium hot areas will be green to pale blue colored, and very hot areas will display an intense blue color. The aim of the absolutely harmless thermographic-self-examination with this device is to reveal possible temperature variations of the breast. Breast pathology is indicated by asymmetrical images between the 2 breasts (i.e., bright blue spots showing a high temperature area in one breast only).

Thermal differences, corresponding to color differences on the device, detected in any part of the breast are abnormal signs that have to be evaluated by the physician.

The device is not a diagnostic for self-diagnosis of breast anomalies. It is up to physicians to diagnose possible illness of the breasts of women, and women should periodically have their doctors check their breast condition, or contact them in case of doubts when effecting self-examination.

The objective of the device is to make women aware of the problem of breast anomalies, so that they will be able to understand them without baseless fears. Self-examination and palpation are of fundamental importance for an early detection program in the fight against breast cancer.

Currently sold to doctors, the device aims to detect the presence of heat producing breast diseases at the earliest stage. Approval by the United States Food and Drug Administration is the result of a 24-month, $1,250,000 clinical research effort. Ten extensive clinical studies, involving over 5,000 women, were conducted at leading medical institutions in the U.S.A. The studies include a long-term study under way at Georgetown University, under the direction of Dr. Patrick Byrne. In this study, women with previous breast cancer are using the device at home as a monthly monitor to check for possible cancer recurrence. In the other studies, 60-90% of the patients with breast cancer showed an abnormal color pattern. In fact, some of the THERMAS-CAN-detected cancers were not discovered by other diagnostic measures. One study suggests that THERMASCAN can detect lesions at the earliest possible stage, i.e., as small as 2.5 mm, the size of a pea. All the studies were performed according to exact scientific protocols and F.D.A. guidelines.

To assist the spreading of this useful detecting device, instruction booklets for women have been published in the following languages: Italian, English, Greek and Spanish.

Easily Interpreted "Heat Pictures"

Stanley James French

3 Woodlands Avenue, Lugarno, New South Wales 2210, Australia

Australian, born February 28, 1941. Self-employed furnace consultant and infrared surveyor. B.Sc. in Chemical Engineering from University of New South Wales, 1962.

The project aims to take the normal output picture from an industrial thermal imaging camera and inexpensively convert it to a form that can be more readily used in industrial, commercial and domestic applications.

This will be done by applying the principles of computer image processing of multi-spectral satellite photographs to the normal thermal picture which is obtained from a single waveband. In this case, the thermal picture will be combined with a normal video picture.

The technique has already been shown to be feasible using a simple but laborious photographic method. It will allow the essential heat information from the low resolution thermal camera to be superimposed on a higher resolution normal picture, effectively lifting the spatial resolution of the thermal information and allowing easy interpretation, even by non-specialists.

Industrial thermal imaging cameras normally have a video type picture output in black and white and shades of grey, with the lightest shades representing highest temperatures and the darkest shades the lowest temperatures. Because of the complexity of the electronics and the high amplification involved, the picture quality is greatly inferior to that of a conventional video camera. To this is added the difficulty in interpretation of these pictures by non-skilled people used to seeing objects according to differences in their colour rather than differences in their temperature.

This project will simplify the interpretation of thermal pictures by taking a normal video picture as seen by the eye in black and white, and overlaying this with various colors representing the thermal information to form a composite picture. The colors, the temperatures ranges and the number of temperatures represented could all be selected to give the optimum combination of ease of interpretation and maximum thermal information.

The experimental photographs included with this application demonstrate part of the range of uses for the method, from close-up fine detail to broad panoramic information, and instant identification of areas of interest. These photographs were made using a thermal imaging camera and inexpensive video equipment with multiple exposure photographic processing. The results achieve the desired quality but are more suited to laboratory research applications than to industrial applications because of their cost and the need for close control. For example, the production of a photograph with 7 temperature levels (7 colors plus black and white) requires 9 negatives and 8 separate exposures, all in perfect alignment.

The computer processes used for satellite image processing provide a more promising answer. These techniques routinely combine several images from different wavebands to enhance the information in any one waveband and simplify its interpretation. Unfortunately, such equipment is sophisticated and expensive, and of much higher resolution than can be utilised by a simple industrial thermal imaging camera. The benefit of computer processing is the speed of the image manipulation and the automatic alignment of the images, resulting in lower costs and wider potential applications for the technique.

The resolution of satellite images is about 3,200 x 2,400 picture elements, for a normal monochrome video camera about 700 x 500, and from the thermal camera about 300 x 300. Suitable advanced microcomputers currently available offer resolutions of about 750 x 450 picture elements for graphics applications and offer special graphics software for image manipulation. Speed of the computer's operation will be an important consideration in its selection.

With appropriate computer hardware in hand, it is proposed to develop the process along the following lines:

a)   Read in the thermal picture via the video digitiser.

b)   Read in the normal picture via the video digitiser.

c)   Display both images on the screen together, by either displaying alternate lines of each image, or a similar method such as alternate frames.

d)   Determine the amount of movement required to cause the images to coincide and convert this to a format compatible with the graphics commands available, such as translation (vertical and horizontal movement), rotation (angular movement) and zoom (reduction or magnification).

e)   Modify one image only, and re-display.

f)   Repeat "c", "d" and "e", if required.

g)   Copy the image. In the first instance, this will continue to be a photographic process by first photographing the normal picture on the monitor screen, and consecutively, in color, those parts of the thermal picture representing increasing temperature (higher brightness). Direct color printing of the images on a printer may be a further development.

Whereas current industrial thermal imaging is expensive, and used mainly by large companies operating very valuable equipment, this new technique would provide much lower cost surveys. It would enable hitherto impractical energy conservation studies and pinpointing of heating and cooling loss areas in buildings, the spotting of potential fire hazards in banks of electrical switchgear, location of dangerous heat buildups in coal mine stockpiles and overburden dumps, identification of needed repairs to furnaces in the petroleum, petrochemical and metallurgical industries, and a wide variety of uses in other areas, such as advertising, medical research and even "pop art".

A Smart Buoy to Stop Water-Borne Vector Diseases

Robert Francis Lavack

Dybensgade 3, DK-1071 Copenhagen K, Denmark

Canadian, born October 10, 1924. Pilot/geologist, managing director of companies in Denmark and U.K. Educated in Canada and U.S.A., B.Sc. in Geology in 1961.

The Ecological Control Unit (ECU) is a programmed, water-borne, larvaecide/pesticide dispensing buoy, designed to be positioned on a pylon/tripod for shallow streams/waters, or as a floating or submerged buoy in deep streams/waters. It was designed for a 12-month operating cycle, and can be re-positioned after servicing. It is a cost effective alternative to the use of helicopters and fixed-wing aircraft, permitting a five-fold expansion of areas now being treated within the same budget, in specific areas of public health, e.g., WHO Onchocerciasis Control Programme.

The ECU is a simple, rugged, electronic/mechanical tool, similar to the radio controlled marine mines or anchored meteorological reporting buoys that have been in use for many years. Using proven "off-the-shelf" electronic and mechanical components, the on-board 'computer' package handles a limited number of functions; 1) Emergency Location Transmitter (ELT) action, 2) Water velocity/Water sounding, 3) Monitoring/Test Sequence: larvaecide tank level, metering tank level, battery state, CO_2 state, buoyancy control, seasonal larvaecide parameters, component malfunctions and anchoring/mounting state, and 4) Larvaecide/Pesticide Application Cycle. The ECU concept is designed for a 12-month on-station cycle. The larvaecide tank holds 600 liters of chemical and can be increased or decreased in capacity for other control programmes. The unit under discussion here was designed serve the programme requirements of Onchocerciasis, Bilhartzia/Schistosomiasis, Malaria and related water-borne, vector diseases.

The ELT, common to those used in most light aircraft, transmits signals on a monitored HF/VHF designated frequency when directed by the computer/programmer control, and in emergency situations. ELT transmissions will continue over a minimum period of 24 hours and will transmit an identity code for each ECU unit, as well as providing a radio homing signal for helicopter-borne maintenance crews. In the case of submerged units, the ELT will only function when the ECU has been surfaced through activation of the buoyancy control. In general operation, the ELT will only operate to signal the end of the application 'on station' cycle and the need for replacement servicing.

Water velocity/Water sounding data are collected through a mini-doppler unit, and fed into the computer/programmer control to determine an accurate water volume-larvaecide ratio. If calculations fall outside the min/max seasonal water volume parameters stored in the control memory, the reading will be ignored and the cycle repeated. If, after three such readings, the error persists, a seasonal water volume average will be used by control to make the immediate larvaecide application. When the data fall within the parameters, a larvaecide application will be made in the normal manner using that reading. Failure of the data input sequence will be treated as a malfunction and cause control to initiate ELT action for servicing.

Monitoring/Test Sequences for all systems are designated at set intervals to ensure ECU capability for its prime function. With so few moving parts within the electronic/mechanical package, battery power drain for these sequences is minimal. The monitoring battery system is separate from the operational Ni-Cad battery package, and has a normal life of 24 months. When a malfunction is encountered during a monitoring sequence, a three-cycle interrogation is repeated. If the malfunction persists, ELT emergency servicing action is initiated.

Larvaecide/Pesticide Application is a data programmed/programme-override function. Doppler water volume date must produce a reading to fit the high-low seasonal water volume parameters. If everything fits, the application cycle follows the doppler override path. If not, the programmed seasonal average is used after the normal repeat test cycles. Both methods activate chemical application. This is followed in the non-doppler application by ELT action. The application sequence is:

— Electronic timer starts cycle.
— Fill valve opens, filling metering vessel by gravity.
— Fill valve closes after fixed time, metering vessel discharges into river at 0,2 bars CO_2 pressure.
— Preceding two steps may be repeated "n" times to discharge larger doses of larvaecide.
— Cycle closes.

ECU larvaecide/pesticide application is more exact than that done using aircraft. The aircraft method relies mainly on seasonal records, plus "eyeballing" the river bank to get a rough idea of the water volume. An estimation is then made by the crew and the larvaecide dumped into the river. The natural tendency is to overdose, which could affect the river and surrounding flora and fauna.

Detailed analysis of costings for the present WHO/OCP aerial application requirements (10 helicopters, 3 fixed wing aircraft, required crews and flying times, etc.) gives a total aircraft application cost for the programme in West Africa of U.S.$10,368,000 per year. The ECU program (900 ECU's at $2,500 each for 'in position' costs, plus attrition estimated at 10% annually, plus annual aviation support requirements) is estimated to cost $2,011,800 per year over a three year programme, or about 20% of the current WHO/OCP. What is important is that the ECU concept is a cost-effective approach to vector disease programmes that will permit a 5-fold expansion of the present WHO/OCP control area within the same budget.

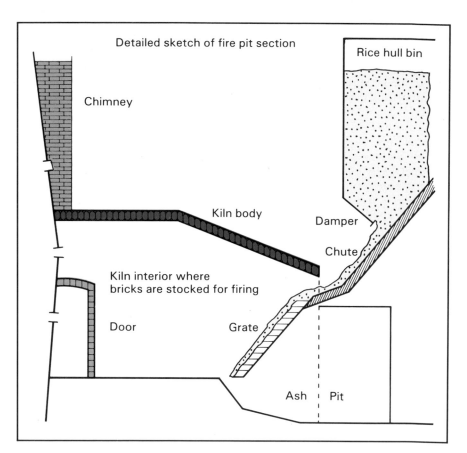

Part of the schematic design of a new kiln for making needed building bricks —and fired by waste product rice hulls.

Using Waste Rice Hulls for Firing Building Industry Bricks

♕ Lazaro D. Pineda, Jr.

Honourable Mention – The Rolex Awards for Enterprise – 1984
Tagum, Davao, Republic of Philippines

*Filipino, born July 25, 1932, Market Administrator, Tagum Public Market,
Suaybaguio District La Fortuna, Tagum, Davao, Philippines. Educated
in Philippines; pre-medicine at University of Mindanao, B.A. from
International Harvardian University, 1965.*

The firing of bricks and tiles for use in the building industry requires large
amounts of fuel to obtain and sustain the high temperatures needed to produce
good quality end products. In the industrialized, more developed countries,
relatively strong economics are able to use fossil fuels in the firing of these basic,
necessary building materials. Lesser developed countries, lacking the financial
resources for all but the most critical uses of fossil fuels, routinely turn to the use
of wood, which is fast disappearing in once tree-laden regions. This practice
has resulted in costs rising rapidly, and does not help in the conservation of our
trees and forests.

My project/experiment is to use the large amounts of rice hull available
practically all over my country, the Republic of the Philippines, for the firing of
bricks and tiles. The successful implementation of the idea would solve many
problems and will result in a number of advantages:

1) It would contribute to the solving of the problem of disposing of the
   tremendous amounts of rice hull, which is the waste product of rice
   production. Here in the Philippines, we average 2.5 rice harvests per
   year; the waste rice hulls are a serious environmental problem.

2) The process, correctly managed, will provide free fuel for the firing of
   bricks and tiles, which are badly needed building supplies and which
   could be made much less expensively by this system.

3) Widespread use of the rice hulls for firing bricks and tiles at the village
   level will contribute significantly to saving our trees and forests from
   being used as fuel.

4) In another economic saving, the mineral rich ashes left after the burning
   process will provide free soil additives, thus lessening the need for
   energy-costly commercial products.

As a major rice-consuming and producing nation, the Philippines has many rice processing plants whose adjacent lands are filled with piles of rice hulls, quite resembling mountains of rich, brown desert.

We know that rice hull, being a carbonaceous material, is at least in theory a good source of energy. The problem is how to convert this huge amount of trash into a viable source of energy, for industries that can be adapted to its use. Certain major industries will probably never be able to use rice hulls efficiently, but there are others where size of the manufacturing process may be suited to the requirements of local markets, and where rice hulls as fuel could be of particular advantage.

The specific industry that interests me is that of brick manufacturing. We have an acute housing shortage in my country, and the inexpensive, local manufacture of bricks and tiles is a critically needed process. The brick-making industry, as it is presently constituted, uses expensive fuel oil for firing the kilns, which in turn produces a prohibitively high-priced product. Wood is also used to fire the kilns to produce bricks, but it is also fast becoming expensive and scarce, and its continuing use threatens our forests.

As an initial experiment, in my town of Tagum, I first constructed two commercial bakery ovens, based on my design ideas for the efficient burning of rice hulls, to gain practical experience with the combustion needs of this potential new fuel. Each of the two customers involved have been delighted with the results, as they have been able to save a few thousand pesos each month in fuel costs by taking advantage of the readily available — and free — rice hulls. We have found that the average baking temperature produced is in between 350°F and 450°F, achieved with one small fire pit fed with the rice hulls.

With experimentation, we learned that rice hull would not burn readily unless it is properly fed into the 'fire box', across and down a grate inclined at about a 45° angle. Below the grate, there is a need for an ash pit, from which the mineral rich ash can be regularly recovered for later use in soil enrichment. The heat and energy generated by the burning rice hull depends upon the amount of rice hull burned (the rate at which it is fed across the grate) and the efficiency of the insulating brick of the kiln itself.

Being myself a brickmaker led me to the idea of harnessing rice hulls as fuel for brick kilns. The goal now is to design a structure with an ample overhead storage bin to contain sufficient rice hulls to maintain a steady flow of rice hulls across a specially made grate, thus achieving fairly fast burning of rice hull to obtain the high temperatures required for the production of high quality brick and tiles.

With a novel design of a kiln, equipped with a certain number of specially designed fire-boxes to burn rice hulls fast enough to achieve the desired temperature of around 1,000°C, I honestly believe rice hull will be the ticket for cheap bricks. Though other attempts have been made to use rice hulls as a form of fuel, virtually all of them commence with the energy-costly conversion of rice hull into briquets. Such an expensive alternative does not achieve the goal of bringing a fuel-efficient manufacturing process to the local level, where it is needed.

The next steps are the designing and building of various shapes of unconventional kilns to get optimum results with the use of our vast amounts of rice hull, for bricks that would be cheap because of free fuel. Additionally, work needs to be done on devising simple methods of kiln building and firing that can be learned easily by non-technical people without expensive and lengthy training.

Establishing a Salt-Damage-Proof Farm for Arid Lands

Yasuo Tsuruta

22-2, 4-chome, Sanno, Ota-ku, Tokyo 143, Japan

Japanese, born October 30, 1930. Councillor, TOHO Research Institute, Tokyo. Educated in Japan; Bachelor of Architecture from Waseda University, 1955, qualified 1st Class architectural engineer.

In the case of agriculture in arid lands, classic irrigation methods may well have been the cause of present day desertification. The collapse of the once-fertile Mesopotamian civilization may well have been accelerated by the twin damages to soil brought about by conventional irrigation practices:

1)   Salt contained in irrigation water can be concentrated by the evaporation of water from the surface of the soil, and it can be deposited within the soil.

2)   Irrigation water infiltrated too deep into the soil brings about unneeded and wasteful waterlogging.

This project involves a unique infiltrating technique, using revolutionary irrigating concepts entirely different from normal irrigation use, and is being built in cooperation with the government of China's Heilongjiang Province, in the Nun-Ziang plain situated in the north eastern area of China, where drought and salt damage have for so long prevailed. This project's goal is to establish a desert greening farm, which can overcome not only the drought so long present in the area, but also to correct the salt damage to the soil.

The two desertification phenomena mentioned above are inevitably present in the case of conventional irrigation methods, which supply water by the application of motive power (gravity).

In contrast to irrigation methods employing gravity power, invented in Mesopotamia about 4,000 years ago, the Infiltrating Irrigation Technology (IIT) does not bring about the two phenomena described above, as it is designed to use and amplify the absorbing function of the soil itself, and allows the salt to concentrate naturally around an emitter. In this, the technology differs entirely from the conventional irrigation concept, which allows the salt dissolved in irrigation water to move together with the irrigation water. In fact, under this unique technology, the largest part of the soil absorbs the water only, as has been proved clearly by experiments carried on for one year.

The technique uses a number of different structures to accomplish this highly efficient, virtually salt-free water delivery to growing plants.

First, a 'constant water level' tank is filled (water level within being controlled by a buoy device). Tubing carries the water to "emitters" (small inverted pot-shaped structures located near to-be-watered plants), where it exits from water supply vessels within the emitters and is absorbed by a porous resin in direct contact with the ground. At this point, with the effect of gravity minimized, the absorptive capacity of the soil (plus the root system of the plant) creates a steady, and agriculturally correct draw upon the moistened soil, draining more moisture from the emitter as required. Therefore, gravity is not required to let water move from the constant water level tank; in fact, the influence of gravity is viewed as a rather undesirable factor leading to water wastage.

The moistened area prepared in the soil by the application of Infiltration Irrigation Technology is therefore able to remain quite constant, with the soil absorbing what it 'needs' and yielding up little or no water to 'loss'. In our experiments, we have shown that even when the surface temperature of the soil is raised beyond 100°C, evaporation of water within the moistened area does not occur, due to the protecting influence of the dried surface of the soil. Such evaporation as does occur is limited to a small area immediately around the emitter, thus minimizing one of the major problems of conventional irrigation, that of surface salt deposits left to leach into the soil.

When irrigation is done by IIT, surface activity of the grains of soil is amplified sharply, and salt contained in the water irrigated concentrates around the emitting surface of the emitter, being absorbed into the grains of soil, leaving the majority of the moistened area at a neutral pH.

An additional benefit of IIT is that its technique maintains proper soil temperatures for growing plants. Our experiments have shown that the moistened soil remains at stabilized temperatures ideal for the growth of roots, thus heightening the water absorbing nature of the soil. With the soil automatically absorbing the correct amount of moisture from the emitter, soil temperature is kept at constant levels. This stability is important from an agricultural viewpoint. As there is no interval between irrigating operations, such as is found with conventional irrigation methods, there is no stress placed on the plants due to variations in the temperature of the soil, thus achieving the ideal irrigation for the growth of plants.

As water irrigated by Infiltration Irrigation Technology remains within the soil without evaporation or loss, and since the plants absorb the water quite effectively, minimal amounts of water are needed. One hundred liters is enough to let one tomato grow perfectly. Therefore, it can be said that this technology can achieve outstanding water savings.

Having proven this technology in our laboratory procedures, this project is now moving to establish a pilot plant designed for the resurrection of Nun-Ziang district, China, which has suffered severely from salt damage and permanent drought, aiming at the same time toward the greening of the desert.

A New Radiopharmaceutical for Study of Normal / Pathological Bone Metabolism

Frank Paul Castronovo, Jr.
26 Lloyd Street, Winchester, MA01890, U.S.A.

*American, born January 2, 1940. Research Scientist. Educated in U. S. A.;
Ph.D in Radiological Sciences (Radiopharmacology), from The Johns
Hopkins University, in 1970.*

A new radiopharmaceutical named radioiodinated Phosphonate (phenyl-methylene hydroxy)bis, or *I-0PA, has been developed. This agent is uniquely suitable for short term (acute) and long term (chronic) investigations of the skeleton and other sites of calcium metabolism in both normal and pathologic states.

The objective of this project is to investigate the biologic behavior of this new bone-seeking radiopharmaceutical in suitable animal models, and to determine its clinical potential. *I-0PA rapidly concentrates onto bone and sites of calcium deposition after parenteral administration in animals, and can be labeled with several isotopes of iodine, e.g., I-123 ($T\frac{1}{2}p=13h$), I-131 ($T\frac{1}{2}p=8.08d$), and I-125 ($T\frac{1}{2}p=60d$). The potential of each will be investigated. The labeling characteristics and biodistribution of *I-0PA will be studied. The biodistribution investigations will involve in vivo studies to determine its toxicity (both chemical and radiation-induced) pharmacokinetics, biologic distribution and ultimate fate in normal and pathologic animal models.

These studies will eventually lead to the development of procedures for the utilization of *I-0PA as a radiopharmaceutical for a variety of pathologies related to calcium metabolism.

Our laboratory was instrumental in the research and development of the skeletal seeking agent Tc-99m, labeled 1-hydroxy ethylidene-1, 1-disodium phosphonate (Tc-99m-HEDSPA), and its ultimate clinical use as a radiodiagnostic agent for metabolic and metastic bone disease. This labeled complex is a prime example whereby the phosphonate moiety serves as a "handle" to translocate the Tc-99m to bone. This so-called "handle" and "carrier" concept was applied to the present molecule by addition of a phenyl group to the phosphonate moiety, which allowed for halogen labeling, most notably radioiodine.

The affinity of skeletal seeking radioactive tracers is thought generally to depend on two uptake mechanisms: (a) reaction with inorganic compounds by exchange or sorption; (b) reaction with organic components. The gamma camera-computer interface has allowed for investigations which image radiopharmaceutical distribution (qualitative) and determine regional tracer concentration (quantitative) statically or over time. Such studies have been performed with the Tc-99m phosphonates. The Tc-99m agents, however, can only measure the acute or input phase of bony metabolism because of the short physical half life of the tracer (6.06h). By labeling the bone with a long-lived radionuclide such as I-125 (T½p=60d), many measurements can be made over many months after only one injection, thus providing an index of bone resorption.

The clinical importance of the Tc-99m-HEDSPA complex as a skeletal seeking radiopharmaceutical has been well documented. The ease by which the phosphonate moiety of the HEDSPA molecule bound the radiometal Tc-99m prompted us to investigate the possibility of combining this chemical group with other compounds. One such compound combined the phosphonate moiety with a phenyl group to yield phosphonate (phenyl-methylene hydroxy)bis, or 0PA.

This agent was easily labeled with Tc-99m and, after i.v. injection in rats, produced excellent bone concentration. Its rate of clearance from the blood stream, however, was less than that of the conventional bone agent, Tc-99m HEDSPA.

While not an ideal Tc-99m carrier, the 0PA proved to be most noteworthy as a carrier of radioiodine, more specifically I-125 with its physical half-life of 60 days. With this long half life, one could label bone and follow I-125 0PA's skeletal release as a function of time.

For the first time, an agent was available to study the "output" function of bony metabolism (resorption) in conjunction with a gamma camera-computer interface for qualitative and quantitative analyses.

Whole body retention studies in mice with I-125 0PA showed a tri-exponential release pattern, with the longest component comprising 33% of the dose with a biologic half life of 962 days. Preliminary experiments with I-125 0PA have demonstrated the enormous potential of this radiopharmaceutical for measuring bony turnover in normal and in animal models associated with altered skeletal metabolism. Previous investigations have not been able to document this information with a gamma camera-computer interface.

The development of a labeled phosphonate which can provide both acute (several hour post administration) and chronic (months after administration) measurements of bony metabolism is especially attractive. With such, an iodine isotope could be chosen based on the clinical indication for which the radiopharmaceutical is indicated.

There is also a rationale for using a combination of these agents to obtain a complete understanding of bony changes. Initial injections with I-123 0PA, followed by I-125 0PA, could provide "base-line scans". Following treatment, a further scan several months later would measure the rate of release of this agent from the metastic sites. The dual study will enable the clinician to differentiate between reactive bone due to healing and reactive bone due to the spread to tumor. In both cases, a positive I-123 0PA scan would result. The I-125 0PA, however, would be released from the bone associated with the spreading tumor more rapidly than its release from healing bone. A radiopharmaceutical method for measuring the effect of tumor therapy would, therefore, be established.

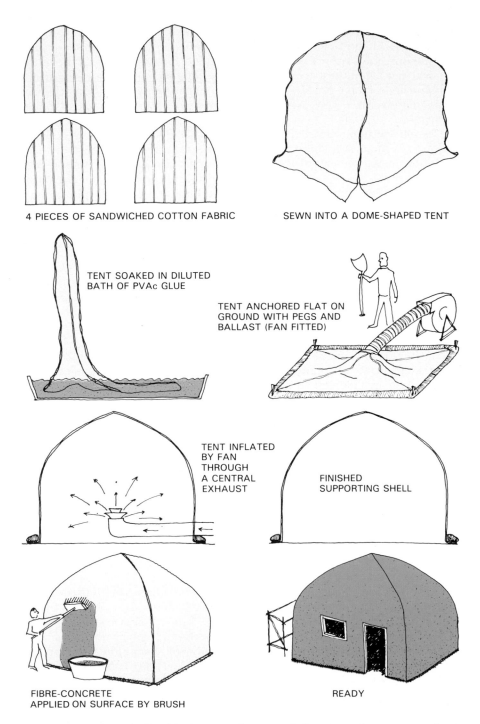

4 PIECES OF SANDWICHED COTTON FABRIC

SEWN INTO A DOME-SHAPED TENT

TENT SOAKED IN DILUTED BATH OF PVAc GLUE

TENT ANCHORED FLAT ON GROUND WITH PEGS AND BALLAST (FAN FITTED)

TENT INFLATED BY FAN THROUGH A CENTRAL EXHAUST

FINISHED SUPPORTING SHELL

FIBRE-CONCRETE APPLIED ON SURFACE BY BRUSH

READY

The 'permanent' inflatable house: cheap and fast (one day construction) for local production in emergency situations.

Inexpensive, Inflatable "Permanent" Housing – An Answer to Catastrophes

♛ Karl Bjarke Ryberg

Honourable Mention – The Rolex Awards for Enterprise – 1984
Pl. 173 Nordanå, S-23200 Arlöv, Sweden

Swedish, born April 7, 1949. Consulting architect. Educated in Sweden and U.K.; Architect's exam, Lund Institute of Technology, 1971; studies in Chinese language, psychology and psychotherapeutics.

The need for quickly available and cheap housing is very acute in areas where catastrophe (earthquakes, floods, fires, etc.) occurs, as well as in those areas where poverty precludes adequate housing for people. When disaster strikes, time is often a critical factor in providing needed shelter for people who are suddenly homeless and exposed to the elements. Erection of a shelter should thus ideally take not more than a day, including the manufacture of the necessary prefabricated elements.

These considerations rule out most types of traditional housing as solutions. While the emergency use of tents is a conventional answer in many situations, the temporary and insufficient nature of tent housing does not meet the very real longer term needs of suddenly homeless populations. The only 'instant structures' presently available to meet such needs are the pneumatic ones. In developing countries, however, one cannot reasonably plan to use sophisticated balloons of plastic supported by high air pressure systems. It is necessary to create a new method that uses high technology only at the minimum required levels and which, for the rest, can rely on local materials and skills for the completion of the needed shelter.

As an architect, I have worked for several years on an idea that solves these problems in a radically different way. The step-by-step description of this methodology is as follows:

1) A dome-shaped tent is sewn from three sandwiched layers of the cheapest cotton fabric available. Templates for cutting the cotton to appropriate sizes are easily made, stored and transported to the area in need. The cotton material is chosen for its wide availability, or, in exceptional circumstances, the ease with which it can be brought to the site. Stitching the cotton in three cross-grained layers provides the minimum necessary fabric strength for the subsequent steps in the process, and requires no special skills.

2) The layered and sewn tent is then soaked in a shallow bath of waterproof PVAc glue, which renders the fabric airtight. This shallow bath 'dip' may be centrally located at the site, and requires no special skills to maintain at proper levels.

3)   The tent is then anchored at its chosen site (on level ground or a concrete plate) by means of pegs and earth ballast around its edges.

4)   An air hose is inserted through a hole in the side of the tent and fitted to a nozzle in the middle of the tent, and a fan is used for pumping air into the structure.

5)   In a matter of a few seconds, the fabric structure becomes fully inflated and is left to dry for an hour or two until rigid.

6)   When dry, the air compression hose is disconnected, and the surprisingly strong shell remains as a self-supporting structure.

7)   The outside of the dome is thereafter covered with a layer of fibre reinforced concrete, plaster, lime or mud, as time allows. In case of urgency, this can be postponed.

8)   Openings are cut, and framings for doors and windows mounted as needed or desired.

9)   Finishing touches include optional weatherproofing on the outside, paintwork on the inside, fittings of doors and windowpanes, etc.

This approach to emergency housing offers a number of advantages, including the following:

1)   The foundation 'plate' can be very simple, and in most cases need only to consist of fairly level ground.

2)   No supporting scaffolding is required, thus doing away with a major logistics concern where large-scale building is required.

3)   The laminated layers serve as a cheap and efficient fibre reinforcement of the resulting structure.

4)   Comparatively low air pressure is required to inflate and sustain the tent during its drying phase, thus needing only minimal power and equipment at the site.

5)   The rigidity of the glued and dried textile shell solves the usual problem of deformation when heavy concrete is applied.

6)   The basic tent can be locally sewn from (normally) locally produced fabric.

7)   The method is rapid and easily taught as a self-help project.

8)   It allows for a great variety of housing shapes and configurations.

9)   It provides a relatively high degree of permanency and comfort as compared to military tents commonly used for emergency housing.

10)   A 3 x 3 meters shelter will nominally cost less than $50.00.

The idea of creating rigid shells by means of air support is, as such, not new. Architects Bini in Italy and Heifetz in Israel have developed methods for covering inflated plastic balloons with reinforcement mesh and concrete, thus casting domes in a simple manner. The British firm Frankenstein has used polyurethane foam to rapidly cover a bubble on both sides, but the use of this highly inflammable product cannot be recommended. These techniques all have the advantages of pneumatic structures, but their greatest disadvantages are:

—   the weight of the concrete deforms the inflated membrane, which later causes cracking of the finished shell,

—   very high pressures are needed (between 400 and 1000 mm water pressure) which calls for far stronger foundation plates than structurally necessary,

— the variety of shapes is limited to shallow domes of preferably circular outline,
— the reinforcement mesh is quite sophisticated and thus expensive.

I have discussed the technical details of my methodology with experts on both mechanical ventilation and glue chemistry. According to their judgements, this new method is realistic and feasible. My idea has also been presented in sketch form to engineers at the Industrial Operations Branch of UNIDO in Vienna for assessment. In their opinion, it could quite well be used for UN projects, once the technology has been perfected. The preliminary plan is to construct a model village at Rorkee, India, as a pilot project in which a total of 50 housing units will be constructed.

To be able to rapidly provide a decent little house at low cost will be of tremendous importance, particularly for developing countries. It is something of a humanitarian duty for us to make sure everybody has at least a minimum of clothing, food and housing.

In order to test and refine the idea, a great number of experiments and calculations are necessary. I have already been making a multitude of scale models during the last year, and thus know a great deal of the practical details involved. This is a time-consuming task, however, and conducting the full-size experiments will require additional machinery and equipment.

I expect the Rorkee project in India to start in spring of 1984, pending the outcome of present experiments and agreements. The proposed duration of the pilot project is anticipated to take three months.

Fighting Food Poisoning with the "Thermo-Controller"

Luis Fernando Knaack de Castilho
Rua Conrado Niemeyer, 12 ap. 101, 22021 Rio de Janeiro RJ, Brazil

Brazilian, born June 8, 1949. Doctor of Medicine (Pediatrician). Educated in Brazil; Post-Graduate Medical School, P.U.C., in 1974.

My project offers a solution to the problem and danger of human food poisoning as a result of eating foods that have been improperly refrigerated; a very serious public health problem.

The solution is very simple, it is cheap, and is based on a small plastic container, easily incorporated into food packages that require continuous refrigeration, that advises the consumer of any irregularity in refrigeration. In this way, foods damaged by improper refrigeration (and vaccines rendered inefficient) can be rejected. By means of permanently altering the original color (neutral) to an intense red, the device reveals with precision if the recommended temperature for the product has been maintained, or, if not maintained, that the quality of the product has been damaged.

The device, called "Thermo Controller", can be used to indicate virtually any desired temperature needed for proper conservation through the simple alteration of its contents. The final objective of its development and utilization on a large scale is to lower the number of cases of food poisoning, with its particularly dangerous consequences to young children and the aged.

In the unlikely, but possible case of accidental ingestion of the device's solutions, there is no danger because the only components of the solutions are water, gelatine, salt and food coloring.

The "Thermo Controller" is made of a plastic capsule, one part transparent and the other opaque. Each part contains a solution of the same chemical composition, differing only in that one is colorless and the other has had food coloring added to it.

These solutions have a point of fusibility immediately above the maximum temperatures indicated for the product to be controlled. In manufacture, the two solutions are frozen, or jelled, individually in casings similar to those of medicine capsules. The solution without color is contained in the transparent part of the capsule, and the colored solution is contained in the opaque part of the capsule. The two halves of the capsule, maintained at the proper temperature, are joined and the completed capsule is placed in a plastic blister sealed on the food package.

Once the capsule is mounted, with the components solidified, it should remain always at, or below, the temperature indicated.

In the case where temperature rises above the control temperature indicated, the two solutions melt and fuse, never again separating, and thus show the non-continuity of the indicated temperature.

The main ingredient for the "Thermo Controller" is water, used pure, if the control temperature desired is −1°C. If lower temperatures are indicated, to the water is simply added the amount of NaCl (Sodium Chloride) necessary to obtain a lower point of freezing. The concentration of NaCl must be precise, since it an anti-freeze, to ensure the controlled temperature.

The reason for using water is that most foodstuffs to be protected by refrigeration also contain a large percentage of water. Thus, the temperature changes observed in the interior of the "Thermo Controller" will be similar to the changes occurring in the interior of the package being controlled.

The food coloring used to color one of the solutions is soluble in water, and used in soft drinks approved by health authorities all over the world. In the improbable event that the contents of the "Thermo Controller" should leak into the food being controlled, there is absolutely no possibility of toxicity.

In case the product to be controlled requires a temperature higher than 0°C, powdered gelatine is added to the water in the same manner as for cooking purposes. Depending upon the quality and quantity of the gelatine used, the solution jells to a more or less firm consistency, calculated precisely to the point of fusing if the indicated temperature is exceeded.

The "Thermo Controller" should be small in size in order to be incorporated with the most varied types and sizes of packages. It also should be made in such a way that it cannot be reopened once filled with the two solutions, thus guaranteeing its inviolability. After numerous experiments, the size and shape I find most indicated is the medicinal capsule, made of two cylindrical halves, one fitting into the other, one transparent and the other opaque (white). Once filled, this capsule can be enclosed in a transparent plastic bubble, with aluminium below, allowing visibility and ensuring inviolability.

This remarkably simple, yet effective, device will offer a significant degree of public health protection when put into broad distribution.

Laser Studios — Bringing Holography Out of the Lab

Robert James Gourlay
Apartado 1423, Palma de Mallorca, Spain

Spanish/Canadian, born May 5, 1947. Author, radio commentator, freelance researcher in applied holography. Educated in Canada and U.S.A.

Until now, Holography (the art of making three-dimensional images on film using a laser beam) has been reserved only for those with an extensive scientific background and education. It is also a very expensive art, where lasers, lenses and expensive optical tables are needed to produce a hologram. For this reason, the common man cannot take advantage of this very useful technology for business, personal or artistic purposes.

My project involves opening the doors of a "Laser Studio" to the public, where a person can walk in with any reasonable object to be holographed, and have it done quickly, efficiently and at a cost that all can afford. Using my lab-studio layout, almost any hologram "demand" can be met with speed, efficiency and at a reasonable price.

To make holography even more popular, my project involves the application of this science in the fields of "studio photography", publicity and advertising, and, generally speaking, creating new uses and applications for this technology.

My dream is that someday, these "Laser Studios" will be as popular as any medium sized photo studio in cities and towns worldwide.

To make the "Laser Studios" a practical reality, the following techniques and technologies have been developed by me:

1. A Low-Cost, Efficient Optical Table

Apart from the laser itself, by far the most expensive component of a holography laboratory is the stable, anti-vibration optical table upon which the holograms are made. By drastically reducing the cost of this item, the per-hologram cost is also reduced as well as the laboratory investment capital. (As an example, the Newport Research Corporation of California, the world's leading manufacturer of specialized optical tables for laser research, quotes a price of U.S.$10,281 for the model RS 414-18 table top, plus U.S.$3,120 for the model XL4A-28 isolation system — the table 'legs' — for a total of U.S.$13,401, for ONE table system!)

My invention consists of taking advantage of the incredibly high anti-vibration characteristics of washed silica sand. Some amateur holography labs use ONE table of sand supported by rubber inner tubes. My system consists of a DOUBLE isolation set-up that results in a much better vibration-and-stress-free environment that meets, if not exceeds, the characteristics of the system offered by Newport Research. The important and most obvious advantage of this technique is its price: U.S.$200.00! A simple, effective solution for the kind of 4' x 8' table system needed in my Laser Studio project.

2. A Mass Production Laboratory Layout

To further reduce the per-hologram cost and increase speed and efficiency in making holograms, these laser studios will feature eight "cubicles", or rooms, each containing a complete laser and table system prepared for the production of specific types of holograms; for example, one cubicle would be for the making of cylindrical holograms, another for two-beam transmission holograms, one for focused image white light holograms, and another for reflection white light holograms, one for holographic stereogram portraits and another for the making of rainbow, white light holograms.

Each cubicle is separated from the other, and each has its own developing lab, lighting system and storage area. The largest cubicle would feature a screen and a more powerful laser for experimentation and innovation work, as well as studies on the projection of these three-dimensional images. Such a room would also be used for "special orders", or for use by students who have no such facilities in their school or college.

By reducing the costs of holograms drastically, a whole new world of three-dimensional photography opens up for ordinary use.

In the publicity area, in toys and games, in jewellery, in the home, and in many other areas, holography can create new products, ideas and benefits.

To make this project a reality, I have done everything possible to attract international attention to this worthwhile endeavor. With the proper financial backing, this project could become an industry of international stature and importance. The use of these studios for import−export of holograms, for experimentation, for artistic work and for general use by the public make this project worthwhile from many different points of view.

I have toured Spain doing demonstrations, lectured and written articles for various newspapers and magazines on the subject. I have been interviewed on numerous Spanish radio stations, and appeared on Spanish television, explaining my work and showing examples of this fascinating science.

Please do not consider this worthy project merely as a new means to "make money". With the world economy as it is, any major project that is not economically feasible has a very slim chance of becoming reality, no matter how worthwhile it is. What I propose is simply an efficient, economically practical way to open the fabulous world of holography to everyone..., for study, art, business or pleasure!

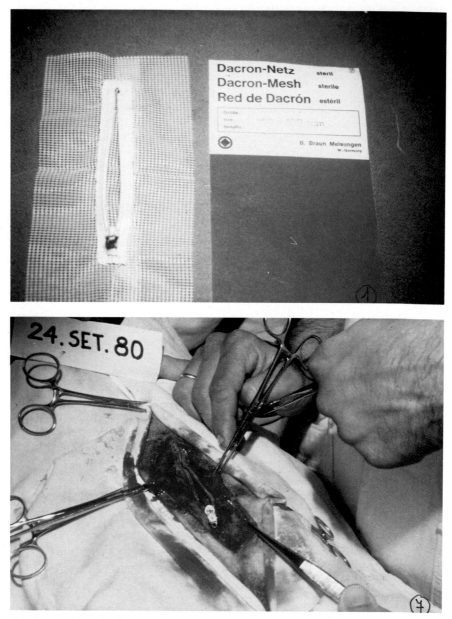

A piece of surgical dacron mesh, fitted with a nylon zipper *(top)*, enables surgeons to make repeated, less traumatic openings of the abdomen *(bottom)*.

A Novel Surgical "Zipper" for Use in Laparotomy Operations

José Luis Badano Repetto
Antonio Costa 3380 bis, Montevideo, Uruguay

Uruguayan, born February 10, 1916. Chief Surgeon of the Oncological Policlinic, Hospital Pasteur, Montevideo, Uruguay. Educated in Uruguay; Independent Researcher affiliated with National Cancer Institute, Bethesda, Maryland, U.S.A.

Surgical operations known as laparotomies involve cutting into the abdominal wall in order to accomplish medical objectives within the system. The nature of such operations is such that there may be a subsequent need to re-open the area for further work, or for further examination; a procedure that is traumatic for the system, and made more difficult in the presence of previous scar tissue. Easier means of repeated access to the internal organs would be a benefit for both patients and surgeons.

Description
This project presents the results of experimental work which shows the feasibility of employing a dacron mesh with zipper added, in the surgical operation known as a laparotomy, to gain easier and less traumatic access to the internal organs.

Synopsis
Using this technique and procedure, our laboratory opened and closed an abdomen six times on different dates. Overall tolerance of the subject throughout these operations was considered very good. When, after sufficient documentation had been recorded, it was decided to end the experiment, zipper resection was made.

Based on the findings of this experimentation, the author suggests that this new technique might be used as a potentially useful procedure for surgery on human beings, under certain circumstances.

Material Used
Given our objective of seeking a convenient means of reopening surgical wounds with minimum effort, we devised a piece of dacron mesh (of the familiar kind ordinarily and currently used in surgery) measuring 150 mm x 100 mm, and fitted with a nylon zipper through its mid-section.

A flap of the same dacron material was placed under the zipper as a buffer, serving to protect the anatomic elements over which it laid, and thus allowing the opening and closing of the zipper without difficulty. Following normal surgical practice, the entire assembly was first sterilized with gaseous ethylene oxide.

Experimentation
The subject of the experiments was a 20 kilograms adult female dog. A routine infraumbilical median laparotomy was performed under general anaesthesia (sodium pentothal). The aponeurotic surface of both recti was well exposed, and the abdominal cavity was fully opened in order to be thoroughly explored.

Following a normal examination, the dacron mesh device was sutured to the aponeurosis (anterior face) by means of separate stitches, using number 2.00 silk. The mesh device was placed so that the zipper completely "fit" the incision (median).

We found that this operation was facilitated if it done with the zipper of the mesh device in an open state. Even so, we found that it was necessary to open and close the zipper several times during the course of the operation, in order to keep exact parallelism between both rows of the zipper's teeth. In this way, we were able to ensure that it would continue to function correctly in its future role.

Controls and Evolution
In order to evaluate the immediate reaction to the mesh device, following initial implant of the zipper, opening and closing controls of the operatory wound were performed after 24, 48 and 72 hours. We experienced no technical problems in these steps.

Twelve days later, a sufficient time for initial healing, the wound was opened once again, and it was observed that there were slack adhesions under the rows of teeth of the zipper. The functioning of the zipper itself continued in correct fashion. Pursuant to this opening, the abdominal cavity was again explored.

A month later, the abdomen was opened again for the last time, once more without any technical problems. As the experiment had been a demonstration of the feasibility of the idea (tolerance of the body to the device, functioning of the system over a significant implant time span, etc.), it was decided to end the experiment. The whole zipper was resected and the mesh was left, a month and 12 days after putting it in place.

Late evolution is considered good.

Conclusions
We have proved the feasibility of using normal surgical dacron mesh with an integrated nylon zipper, as a practical means of re-opening laparotomy wounds.

We experienced no practical problems during any of the above operations, which were repeated 24, 48, 72 hours, 12 and 30 days later. The actual zipper portion of the device was finally resected a month and 12 days after the original operation, as it was decided to end the experiment.

Based on our experiences in this experiment, we now believe that the use of surgical dacron mesh with an integrated nylon zipper can be useful in human clinics. This may be especially true in some pathologies requiring urgent surgery, but we do not discard the possibility of using it in other kinds of surgery.

This project was begun in 1980, at the Faculty of Veterinary. Completion of our work in this area depends on the ability to produce the mesh at levels satisfactory for the market level and requirements.

The goal of the project is to introduce the dacron mesh/zipper device as an element in the surgical equipment of hospitals, and to have it available for prescription in urgent surgery situations, as a useful and beneficial tool.

To date, I have not received any aid in the development of the device.

Raising Much Needed Livestock from Test Tubes

Margaret Ongkeko Abrera

Km. 69, Buckville Subdivision, Bgy. Buck Estate, Alfonso, Cavite, Philippines

Filipino, born September 5, 1957. Owner/manager of goat upgrade breeding ranch, and newspaper columnist. Educated in Philippines; Cum Laude B.A. from St. Scholastica's College, in 1978; winner of Philippines' Young Ambassador award.

With the new technology of superovulation and embryo transfer, we propose to increase the size and weight of the average Philippine native goat by 550%, and its milk production by 300%, in ⅓ of the time now required. For breeding stock dispersal programs, we propose to produce, monthly, 696 pure breed European goats in only 22 months from project implementation. All would be born in the Philippines. All would be from native, 20 kg dams. All would be naturally acclimated. After proof of viability, this same basic procedure could be used for cows, water buffalo and pigs at a fraction of today's upgrade breeding cost.

Due to its low cost (U.S.$15-20 each), we will consider only the Philippine native goat, the most common source of milk, meat and leather for the Filipino peasant farmer. These goats average an adult weight of only 12 kgs, and their milk production is less than one litre per day. In 1973, there were more than 1.25 million of these native goats counted in the Philippines. On the other hand, European goats, referred to as pure breeds, will weigh an average of 65 kgs each, and yield 3 liters of milk per day.

Alarmed, the Philippine government moved to correct this obvious disparity. At great cost, they imported pure breed bucks and implemented dispersal programs. However, under even the most ideal conditions, and with no loss factors for false siring, sterility, abortion, disease, mortality or chance of sex (we are assuming that all upgrade offspring are healthy females), it would take 68 months of constant upgrade breeding using the services of 4 different pure breed bucks (at a cost of U.S.$750 each) to produce one single 12-month-old doe equal in size, weight and milk production to its European counterpart. It would take 14 years to accomplish a similar result with native cows and water buffalo.

Proposed Alternative. We wish to import 300 pure breed embryos and transfer them to 200 native recipients (allowing for ⅓ perishing in transit). The native recipients will be simple carriers and will not contribute to the genetic characteristics of the embryo. After this 1st and only importation of embryos, we will never have to import embryos, or for that matter, pure breeds again.

Seventeen months after intial embryo transfer, the female offspring of the transfer would be mature enough to become embryo donors. These donors would be superovulated with gonadotropins. Five days later, donors would be artificially inseminated with the sperm of pure breed bucks. Six days after that, donors would be flushed of their embryos. These embryos would then be transferred monthly, to a new group of native doe recipients. This whole procedure can be described in six steps:

1. Superovulation of Donor Goat. Donor does will be treated with gonadotropin and prostaglandin to accelerate estrus, followed by ovulation about ½ day later, and fertilization a few hours later.

2. **Artificial Insemination of Donor Goat.** Five days after gonadotropin treatment, donors will be artificially inseminated with the sperm of pure breed bucks.

3. **Non-Surgical Recovery of Embryos from Donor Goats.** Embryos move from the oviduct to the uterus 4 to 5 days after estrus (3-4 days after ovulation). Embryos can then be recovered from the uterus 6 days after the beginning of estrus. To do this non-surgically, an 18-gauge Foley catheter is inserted through the cervix into the uterus. The embryos are flushed out and flow into large graduated cylinders. Embryos are then stored in small containers in preparation for transfer.

4. **Non-Surgical Transfer of Embryos to Recipients.** Within 24 hours of recovery, embryos must be transferred to a recipient, previously screened for suitability (large, a previous normal healthy birth, disease free, etc). After selection, recipients will be confined to quarantine. Because recipients must be at the same stage of the estrus cycle as the donor, recipients will be treated with prostaglandin F. Actual embryo transfer will accomplished with the same tool used for artificial insemination.

5. **Verification of Pregnancy of Recipient.** Positive pregnant dams would be segregated and would receive supplemental feeds, vitamins and continuous medical care.

6. **Kidding 150 Days After Embryo Transfer.** Upon successful embryo transfer and the resultant kidding five months later of 120 pure breed kids, we plan to showcase and document the live results of this innovative, time-saving breeding procedure; pure-breed European goats normally costing 7,500 Philippine pesos, born from Philippine native dams normally costing 200 Philippine pesos.

The results of this project would be immediate. The impact of a live pure breed coming from a native doe — dramatic. The savings — tremendous. Breeding stock of goats in this country could increase by more than 8,000 pure breeds yearly. Most important of all, it would prove to many 3rd World nations that the method works, and could therefore be used for the more costly livestock animals as well.

We have erected 360-square meters of the most modern goat houses in the Philippines, along with two quarantines and a laboratory for parasitical disease detection. Convinced that superovulation and embryo transfer is the wave of the future, we are converting the direction of our 300 hectare goat breeding ranch toward the future.

Modifying Gasoline Engines to Save Fuel, Decrease Pollution

Pasquale Borracci

Via Benedetto da Foiano, No. 1, 50125 Firenze, Italy

Italian, born June 28, 1888. President of Italmeccanica S.p.A. Educated in Italy; Industrial Engineering at Polytechnic Institute of Turin.

This invention concerns the modification of an engine to achieve both fuel savings and reduction in polluting emissions. The modified engine has been tested at the Istituto Motori (Engines Institute) of the University of Pisa, according to the C.E.E. rules for town traffic. A 27% fuel saving has been reached for town traffic, and a 20% saving for extra-urban roads. There is a significant reduction of harmful emissions at the exhaust; during the idle running, the maximum CO percent allowed by the C.E.C. rules (3.5%, a poisonous level) is reduced to 0.12%, for a lowering of 97%.

The modification involves a rotating distributor synchronized with the camshaft, which initially introduces clean air and then a mixture of rich fuel. The results have shown that the two fluids remain quite separate during the compression stage; with this stratification, the cylinders are half-filled by the two fluids at reduced loads, thus maintaining the compression stroke in a fashion somewhat similar to that of the Diesel cycle.

The stratified charge is one of the improvements, allowing the filling of the cylinders with a higher quantity of air.

Solutions tried until now have not achieved practically successful results. This invention, however, has achieved valid endorsements, and relevant patents have been granted in the most important countries (U.S.A., Germany, Great Britain) where findings are submitted to strict examinations.

It is important to point out that this device requires no change in the geometry of gasoline engines. It can be mounted on any existing engine, and requires only that a rotating distributor in synchronization with the cam-shaft be inserted into the induction manifold.

The modified engine of a FIAT 126 has been thoroughly tested on the Florence-Siena highway, in comparison with a standard engine of the same age and with equal mileage.

The running tests prescribed for the Europe Cycle have been repeated several times on a roll-bench, completely equipped for the analyses of gases and measurement of fuel consumption.

The results concerning the exhaust gases during the hot-run (the most important for the use of cars) are stated in the following table:

Harmful Exhaust Gases	Standard Engine	Modified Engine
CO	grs. 65	grs. 65
HC	grs. 6	grs. 6
NOx	grs. 8.5	grs. 5.5

The analysis of the exhaust gases has proved to be very important at the idle-running of the engines. The standard engine has shown 3.5% for CO, corresponding to the highest value allowed by C.E.E. international rules; on the modified engine, CO has been almost cancelled, because its percent has been reduced to 0.12%. This value cannot be found in any gasoline car running anywhere in the world.

It is necessary to point out this result, because emissions of CO at idle-running are a cause mortality, even at the 'allowable' percentage of 3,5%. It is well known that, in city traffic, much engine operation takes place in 'idling situations'.

Fuel savings in town running are also significant, considering that all cars run 70%, on average, of their life in towns. Over 100 km distances, petrol consumption of the standard car has been 9.59 liters, while the modified car has only required 7.07 liters, a savings of 27%.

Furthermore, the report of the Engineers of the Istituto Motori stated considerable fuel savings in 'free road running', as follows:

Over Distance Constant Speeds	Standard FIAT 126 lt/100 km	km/lt.	Modified FIAT 126 lt/100 km	km/lt.
60 km/hour	4.8	20.83	4.02	24.87
90 km/hour	6.6	15.15	5.4	18.52
100 km/hour	7.5	13.15	7.31	13.68

The fuel consumptions at the speeds of 60 and 90 km/hr show a savings of about 20% for the modified car. At 100 km/hr, fuel consumption is quite the same because the two engines run with the same filling, i.e., at the maximum compression ratio.

The above data refer to results obtained by a State Institute qualified for such research analyses. In order to avoid any doubt, support from theoretic analyses has also been required, and is explained here.

Measurement of the quantity of air introduced into the two engines has been made at the rpm levels giving the most pertinent results, namely at engine speeds of 1,500 and 2,000 r.p.m., that of normal town-traffic. In terms of air intake (cubic meters/hour), The standard engine sucks in 7,43 cubic meters at 1,500 r.p.m. and 9.80 cubic meters at 2,000 r.p.m. The modified engine sucks in 13.04 and 15.78 cubic meters at the same r.p.m. levels.

Based on thermo-dynamic formulas, the modified engine has a theoretical efficiency advantage of 35% versus the standard engine. Such a result cannot be reached in practice, because the polytropics are not adiabatic. However, it makes fully admissible the 27% saving of fuel verified by the Institute tests.

In summary, the Borracci solution, which requires only the addition of an air intake and a rotating distributor, allows a 27% fuel saving in city driving.

This solution should interest all engine manufacturers, as well as the various Institutes concerned with petrol policy.

ሄዜ፥ሯሯ፥በሮክ፥ዞነዮ፥ሐ፥ጣነቀ፥ጎሬ
ዖን፥ይቄቍቒ፥ከርስቍክ፥ጣሬ
ሐይጣዮት፥ወጽይቅጄ፥ትወሐወ
ሁ፥ከዋክብ፥በዒይርክ፥ጣጡቅ
ውዎንበስብ፥ይወናት፥በዒሬ
ብ፥ወበርቀ፥ስበ፥ለሁርዔ፥በጄ
ማዒ፥ፍቍሬ፥ከብ፥ሐጣ፥ጎስ
ክ፥ስቍ፥ሐጣ፥ጥ፥ወዘወ
፥ሖሬ፥ብራሐሬ፥ኢይ፥ክርስ
ቍክ፥ጣጣዒ፥ጎወ፥ሰስ፥ዘበ

Enlargement (about 30 times) of a single character of the Ethiopic alphabet
(top). An example of ancient Ethiopic calligraphy, typical of Ethiopian
manuscripts going back several centuries (bottom).

Analysing, Developing and Producing a "New" Ethiopic Alphabet by Computer

John Shattuck Mason

104 Neal Drive, Bristol, Tennessee 37620, U.S.A.

American, born January 13, 1945. Theology teacher and missionary. Educated in U.S.A., Lebanon, Canada and Ethiopia; M.Div, ThM from Westminster Theological Seminary, Philadelphia, Pennsylvania, U.S.A.

The goal of the project is to develop a system for producing high quality printed material in the characteristic Ethiopic alphabet, initially for use in the Tigrinya language.

Necessary or defining characteristics of the system will be: superb quality, ease of operation, low cost and speed. Components of quality are technical, rational and aesthetic features. The project thus involves analysis of the Ethiopic alphabet, determining appropriate forms of letters, and coupling the alphabet to a keyboard and printer, to yield camera-ready copy, *within set requirements for cost and appearance.*

The idea is to make it possible for Tigrinya-speaking people to preserve, expand, and develop their culture in literature.

The Ethiopic alphabet serves as the way of writing several Ethiopian languages, including Amharic and Tigrinya. These alphabets have an aggregate total of more than ten million speakers. The "alphabet" is technically a *syllabary:* it has at least 226 different characters, each representing a particular syllable. No convenient or inexpensive way of writing in this syllabary exists today; there is only handwriting and conventional (!) metal type. Yet these languages have a literary culture which goes back over a thousand years. Especially in the context of the international situation, in which great numbers of the speakers of these languages have become displaced persons or refugees, the importance of facilitating the culture of literacy is easily grasped. An economical, efficient, and satisfactory means of producing written materials in this alphabet is the immediate goal of the project.

The first step is to assemble samples of handwritten and printed materials. These can be of several types: cursive (everyday) handwriting, calligraphic handwriting, typefaces (only two fonts are in use today), typewriter (one font only), placards and signs. These materials must be studied for logical structure, appearance and legibility. Logical analysis means determining what is the inner form or essence of each letter; that is, what factors make a particular character what it is. The defining norm becomes the basis for the development of letter styles. That defining norm will be discovered in two ways: first, by independent analysis of existing materials (described above), and second, by information and feedback from persons whose first language (or mother tongue) is one that employs Ethiopic characters. It requires to be known which form most "looks like" a certain letter, to a native reader. As letter style is developed in accord with the defining norms, they will be checked by ordinary readers for accuracy, legibility, and appearance. Acceptability is clearly dependent on accurate perception of the inner, defining norms.

The project is committed to evolving two styles of writing as primary goals. The first will be derived from the calligraphic handwriting that has been in use for centuries and which is the basis for all type fonts, modern and historic. The second style will be what is technically known (in the United States) as "gothic"; that is, a style in which all the strokes are of the same width, as if written with a ball-point pen. Such alphabets are commonly perceived as "modern". Typically, not only the variation of width is eliminated in such alphabets, but the small marks (called serifs) found at the beginning and ending of strokes are dispensed with. Such an alphabet has the advantage of having an intrinsic capability for great clarity of form. A second advantage is that such forms can imitate writing done with ordinary pens and pencils, and vice-versa: no special instrument is needed to reproduce their fundamental form. This last aspect is valuable for teaching literacy.

The goal of visible quality will be met by utilizing new forms of printing. The use of dot-matrix printers is envisioned at present. These printers construct characters by using assemblies of dots. The Latin alphabet can be done with a matrix of only 7 x 10 dots, although visual quality is poor; a greater number of dots is desirable. Characters in Ethiopic produced on a 50 x 50 matrix and photo-reduced to printing size would be smooth to the eye. Office copying machines have reducing capabilities which bring down the cost of this method of constructing small size letters to affordable levels.

Visual design of an alphabet-font is incomplete without proper coupling to a keyboard. The basic keyboard to be adopted is the conventional keyboard of 50 or 60 keys arranged in four rows of about 12 keys each. The reason for adopting such a keyboard is its availability, low cost and familiarity. The keys obviously need to be redefined to accommodate the Ethiopic syllabary. In addition to the 266 letters, there are 20 numerals and a few punctuation marks. Two successive strokes would suffice to define each character; in fact, on a keyboard of 50 keys, there are 2,500 different combinations of two keystrokes.

The problem to be addressed is to program the keystrokes for maximum simplicity and efficiency. The relationship of particular keystrokes to the character they designate should be clear. For efficiency, the order of strokes should be arranged so that the index and second finger do most of the work, and so that there is alternation from one hand to the other, or at least from one finger to another. This makes for increased speed. Working it out properly means studying written material to determine the frequency and order of the characters. Because the conventional English keyboard (QWERTYUIOP) is not particularly efficient in these respects, attempts have been made to replace it. But, despite its indadequacies, the traditional order of letters has proved impossible to dislodge in the English-speaking world. The opportunity of giving attention to this aspect of keyboard design comes very rarely, indeed perhaps only once!

The main benefit of the system as now envisioned is to facilitate the production of literary materials, both time-bound (newspapers and magazines) and permanent (books).

A second benefit is that it would become possible to file and alphabetize Ethiopic words automatically. This will greatly facilitate producing some badly-needed books: dictionaries (e.g., Tigrinya-Tigrinya, Tigrinya-English, English-Tigrinya), and a concordance to the Bible. These works, not presently in existence, will serve the aim of assisting to preserve and extend the culture of this language group.

It would also become possible to transliterate automatically, from Latin characters to Ethiopic, and vice-versa. Certain works (like grammars) which have been published in transliteration only could be economically reprinted in the right alphabet.

Such applications can be expected to encourage literacy in the general public in ways not feasible now.

A New Solar Chimney Design to Harness Energy from the Atmosphere

Mario Alberto Castillo

Tucuman 187, Santiago del Estero 4200, Argentina

Argentine, born September 26, 1928. Engineering Director in Ministry of Public Works. Educated in Argentina; Civil Engineering at Universidad Nacional de Buenos Aires (1953), Mechanical Engineering at Universidad Nacional de Cordoba (1956).

Among the systems for exploiting solar energy, wind power plants are playing an important role in the power supply scene. Unfortunately, wind energy is only available when Nature choses to provide it, and when it does only a limited range of wind speed is useful. Wind energy conversion could be classified as a "Random Intermittent" source of power.

The adoption of a new system, the "Very Tall Natural Draft Chimney", assisted by solar radiation, can provide momentum to a mass of air in order to extract energy from its movement. The lack of this system's dependence on natural occurrence of wind makes it seem a very attractive development.

Extracting energy from the mass of air flowing upwards inside a chimney is not a new concept. Natural draft chimneys have been known for years, as well as the calculations concerning the upward flow of warm air.

An installation using this concept, called the "Solar Chimney" is about to go 'on load' on location at Manzanares Ciudad Real in Spain. The plant is Project Number 4249 of the German Federal Ministry of Research and Technology. This pilot plant has a chimney diameter of ten meters and stands two hundred metres high. It is made of corrugated steel sheets, supported by steel rods.

There is also another project developing under the name of "Aeroelectric Tower", also based on the use of a chimney, but having the air mass flowing downward and driving conventional air turbine generators. The technical data concerning this project are awesome: a tower (chimney) 1.5 miles high and nine hundred feet in diameter is being considered, in order to generate 2,500 Mw at peak, with a construction cost of two billion dollars.

Although these projects differ in their technical approach, each system basically depends on the performance of its chimney.

For the "Solar Chimney" project, a 200-meter high prototype tower was chosen for initial experiments, but 900-meter high towers are under study that should be able to produce 1,000 Mw, or about 400 times more power than the biggest windmill operating today (the U.S.A's NASA/Department of Energy "MOD-2", built by Boeing, and rated at 2.5 Mw).

Experience to date clearly indicates that increases in height of the towers provides increases in the efficient production of energy. The desired "Very High Towers", however, pose very large technological and financial challenges.

We believe that a "soft" structure approach to chimney building, instead of the conventional "rigid" one, could meet these challenges successfully. We define a "soft" structure as one having a light, articulated frame suspended from a conventional captive balloon, using an outer skin to enclose space, and made of modern plastic-woven fabric. It would be a structure that would have neither the weight nor the stiffness that make ordinary "rigid" buildings stable.

This kind of construction does not need heavy footing, nor heavy structural elements, and thus avoids shear and compression stresses. Basically, it would be a tension structure, yielding to external forces, such as the prevailing winds found at high altitudes. It would be moored to the ground at the base, and stretched vertically by the buoyancy forces exerted by the captive balloon, which would be tethered to the ground by cable.

The articulated frame, hanging from the balloon, will keep the fabric sleeve from collapsing due to internal or external forces. A series of ring frames would be centered around the mooring cable.

To increase the natural draft of the chimney, the outer fabric skin should be black to maximize absorption of solar radiation in order to heat the inside air column. A substantial increase of generating power can be achieved by adopting the use of existing, patented devices such as the "Sun Powered Generator Cone", or the "Solar Chimney", both based on non-concentrated solar collectors to raise the temperature of the incoming air at the peripheral air inlet. The old and proven concept of the greenhouse thermal collector should prove to be the best answer to heat the incoming air at the base of the chimney.

An energy conversion plant using every day materials and technology requiring no water or wind whatsover, with the inherent simplicity of the vertical axis air turbine, being able to produce very large amounts of power at a very moderate cost, can bring a new era in the power scene.

In remote areas of industrialised nations, or in developing countries where purchase of electricity or engine fuel is simply unavailable, having no wind potential, our proposal, the "Very Tall Natural Draft Chimney, Assisted by Solar Radiation Thermal Collector", may very well be the only viable alternative.

Developing New Adaptogens for Healthier "Old Age"

Maurice Mmaduakolam Iwu

Division of Medicinal Chemistry, Pharmacy, Ohio State University,
Columbus, Ohio 43201, U.S.A.

Nigerian, born April 21, 1950. Fulbright Research Scholar; laboratory studies of medicinal plants for the treatment of diabetes, liver toxicity and cancer. Educated in England; Master of Pharmacy (1976), Ph.D. (1978) from University of Bradford, United Kingdom.

Asian and Far Eastern peoples have for many centuries exploited the ability of certain herbs to improve stamina and resistance and to help individuals reach optimum mental and physical performance without any of the side-effects associated with the so-called 'pep-drugs'.

The international scientific community was astonished by the report (NEW SCIENTIST, 21 Aug 80) that Soviet athletes, cosmonauts and workers were using extracts from the shrub *Eleuterococcus* to improve their mental and physical capability. In the article, Fulder stressed that the Soviet athletes at the 1980 Olympic games used the drug to increase their performance, and that Soviet cosmonauts Vladimir Lyakhov and Valey Ryumin took 4 ml of eleuterococcus every morning of their record breaking stay in the Salyut 6/Soyuz 32 space station. Such plants, known as adaptogens, macrobiotics or somantensics, improve human vitality, eliminate stress and sustain life even in the most arduous conditions.

In a comparative study, I have identified five plants with adaptogenic properties. These plants extended the average life span of laboratory animals, increased the stamina of rats forced to exercise to exhaustion, and prolonged the swimming time of rodents by 50%. On humans, these plants enhanced the physio-chemical profile of aged humans (60+ years), improved scores on bicycle ergonometer tests and enhanced physical performance, coding, radio telegraphy and other tests that determine psychophysical performance. The plants at the doses tested did not produce excitation, jitteriness or insomnia, while greatly improving human stamina, motivation, coordination and concentration.

This project is therefore addressed to the development of the adaptogenic properties of *Garcinia, Uapaca, Sphenocentrum, Icacinia* and *Bridelia,* and to the isolation and chemical/pharmacological study of their active principle components. The most active compounds would be patented and developed as a medicinal agent — not for the treatment of diseases but for maintaining optimum health — especially for the aged.

Senescence, or ageing, represents a phase in which the adaptability to external and internal stresses decrease, the homeostatic mechanisms deteriorate, and susceptibility to diseases increases. With the relative rise in the human life span over the past 20 years, it has become necessary to improve and sustain the quality of life of the aged with plant products that exert restorative action on various organs of the body.

A detailed study of ageing will reveal that death occurs at some point during this phase not because all the functions reach zero level, but because of one or more diseases or afflictions which affect certain organs so seriously during this period of declining functional ability that recovery is not possible.

Modern medicine undoubtedly has made spectacular progress in the understanding of pathology and pharmacology, but not much is known about the science of health. The Russians and many Eastern countries have devoted time and finances to the study of this new class of drugs. Work at the Institute of Marine Biology at Vladivostok, U.S.S.R. on the three known adaptogens (2-benzyl-benzimidazole, ginseng and eleuterococcus) has established that, irrespective of the trend of preceding pathological changes, these drugs exert a normalizing effect, improve dark adaptation of the eye, visual acuity, color and auditory perception; all these properties would be advantageous not only to athletes, soldiers and manual workers, but also to the aged.

Adaptogens present a dilemma to classical pharmacologists since the development of such drugs calls for a plethora of biological tests and new approaches to drug designs. According to Prof. I.I. Brekhman, head of department at Vladivostok's IofMB, there are three criteria of a true adaptogen; a) The adaptogen should be completely innocuous, and should cause minimal changes in physiological function of the organism, b) The adaptogenic action should be non-specific, i.e., should enhance the organism's resistance to the adverse effects of a wide range of physical, chemical and biological factors, and c) The adaptogen should, irrespective of the diseased state and the prevalent physiological condition, exert a normalizing effect.

Extracts of *Garcinia, Uapaca, Sphenocentrum, Icacinia* and *Bridelia* satisfied these criteria. *Garcinia kola*, or bitter kola (not related to the African masticatory *Cola nitida*) and *Uapacà guinense* were well tolerated by laboratory animals even at megadoses more than 50 g/kg. The three other plants have LD-50 within the range of the 10-30 g/kg reported for ginseng and eleuterococcus. Garcinia is used widely in western Africa as an occasional food plant, without any toxicity; similarly Uapaca and Sphenocentrum have enjoyed a long reputation as aphrodisiacs. Prolonged administration of Garcinia and Uapaca for more than 6 months increased the mean life span of rats, and stabilized anti-Salmonella flagellin and the antinuclear factor in aged humans of both sexes. All the plants showed no embryotoxic or teratogenic effects.

Further experiments should establish the effect of the drugs on human endurance, exhaustion, stress, stamina and age, and identify the active compounds involved.

The Palm Tree Climber – a new, portable device that may eliminate much of the hard work and pure danger involved in 'harvesting palms'.

A Climbing Device to Reach the Palmyra's Valuable "Toddy"

Trupapur Antony Davis

JBS Haldane Research Center, Nagercoil 2, Tamilnadu, India

Indian, born February 9, 1923. Director, JBS Haldane Research Centre; UNDP/FAO Coconut Agronomist. Educated in India; Ph.D from Indian Statistical Institute in Calcutta; Fulbright-Hayes Fellow in U.S.A.

It is my goal to devise a simple, relatively light-weight, manually operated gadget to enable workers to climb palm trees with minimum physical strain. The availability of such a device in palm-growing areas would greatly facilitate the economically important collection of Palmyra palm products.

Of the world's 2,500-plus species of palm trees, the Palmyra palm (*Borassus* sp.) is most important to man (next to the coconut palm) because it yields food and provides over 100 different utilitarian end-products. To obtain the majority of its benefits, the Palmyra needs to be climbed twice daily to extract the nutritious juice ("toddy") from its flower-bunches. It is this toddy, converted in several different methods, that is the basis for a wide variety of other products. Collecting the toddy is arduous, and often dangerous, work; the trees can top 30 meters in height.

The Palmyra grows in an area of about one-third of the circumference of the globe (Madagascar, Ethiopia, India, Sri Lanka, Bangladesh, Burma, Thailand, Malaysia, Indonesia and Northern Australia) offering sustenance and occupation to several million people of the semi-arid regions where other food crops are difficult to cultivate. The palmyra palm, known under various names in different countries, is classified into 7 species, the most popular among them being *Borassus flabellifer*.

Whatever may be the scientific name of palmyra, this non-clustering giant has a stout, erect, dark stem, and possesses several valuable products, most of which can be attained only by climbing the full height of the stem. Different cultures have devised a variety of ways of climbing the palmyra (and *Corypha*). Because of the great physical strain required for climbing, and the associated hazards, fewer and fewer young people enter this traditional profession, and so, only a small percentage of the potentially large source of food is tapped. Lack of men to climb the coconut palm in Indonesia has even led to the training of monkeys to harvest the coconuts, which they pluck and drop to the earth.

Hailing from the community of traditional palmyra climbers, I have been exposed to the perils of scaling the palmyra from my childhood. This has created a passion within me to improve the lot of palm climbers and their families. Starting by making a pair of oval rings to aid climbing, I have since assembled a number of various crude gadgets, including an elevator. The cash awards won by these inventions have kept my interest growing, as well as the challenges surrounding the task. The Khadi and Village Industries Commission of India announced a cash prize of Rs. one lakh (about Sfr 20,000) for a device to help climbing the palmyra. Unfortunately, no entry was awarded a prize, but worse, an announcement from a group of engineers at the Indian Institute of Technology in Bombay declared that it was not possible to devise a gadget suitable for aiding the climbing of palmyra palms. This verdict only increased the urge in me to intensify attempts to assemble such a device. Thus, at the Haldane Research Centre, we have a program devoted to improving climbing gadgets. In our work, however, we are aware of a number of other attempts in the area.

On June 2, 1981, the picture of a pole-climbing device developed in Lithuania appeared in The Times of London. Prof. N. W. Pirie, FRS, who is familiar with my gadget, commented that my palm-climbing machine is simpler than the Lithuanian model. However, since I am interested in introducing possible innovations in my device, I got in touch with the Soviet team. They invited me to visit Russia, apparently hoping I would enter into a trade agreement, apart from observing the working of my device. Funds were not available for me to make the trip, however.

There have been many attempts made in the area of developing these devices, so it will be helpful to describe the process of this project as follows:

— Making on-the-spot, critical field examinations of the components used, the principles involved, and any special features employed in the workings of three known, specific devices; the Swiss 'bicycle', the German 'forest tree-pruner', and the Soviet 'pole-climbing device'.

— Carrying out feasibility studies concerning how the best features of each of these devices may be combined in a superior new device,

— Locating optimum sources of purchase for suitable movable or immovable components required for making the final device,

— Working on final designing and assembling of the machine at our testing centre here in Nagercoil, South India, where inexpensive testing facilities are available.

The objectives for the device include the following characteristics:

— It must be simple, so that its use can be learned easily and quickly by unskilled people and so it can be maintained by them.

— It must be capable of manual operation, as power supplies are limited and expensive.

— It should weigh under 25 pounds, preferably, to be transportable from one tree to the next by one person, who should be able to assemble it on a palm trunk by himself.

As the palmyra stems are very strong, it is likely that the tool will be built to employ the cantilever principal. A key concern now is our interest in locating materials (metal or plastic) combining the necessary degrees of both light weight and strength.

In my position of Director, JBS Haldane Research Centre, my work involves carrying out and guiding research projects having to do with agricultural, biological, natural and social science subjects. As a temporary UNDP/FAO Coconut Agronomist in Indonesia (until the end of 1983), I have been carrying out agronomic trials with coconut palms, with the aim of improving coconut production, and training counterpart Indonesian scientists in designing and executing such trials as well as in processing the data obtained from them. Given this emphasis in my professional work on the development of the highly productive palms of the world, and seeing the potential lying ahead in developing better trees and end-products, it is natural that I am keenly interested in improving the means of harvesting the greater production we can expect to see in the future. The need for a better method of tapping the riches and benefits of the world of palm trees is clear; this device may be the answer we've long needed.

Producing Safe, Economical Private Planes for Flying by Disabled People

Yves Croses

22 Cours Franklin-Roosevelt, 69006 Lyon, France

French, born June 18, 1945. Maker of plane prototypes and machinery needed for their production. Educated in France; Baccalauréat Sciences Expérimentales; holder of 2nd degree Pilot's license.

My father, Emilien Croses, has for the last 20 years conceived and built different prototypes of very safe and simple planes whose drawings are sold to homebuilders. Many of these are now finished and fly in different countries.

This type of plane, which can be piloted with only one's hands, is perfectly suited to use by disabled people who cannot use their legs, thus opening a very wide new world of personal freedom for these people. As these people normally cannot build their own aircrafts from drawings, we intend to set up a small plant to produce both kits and finished planes.

In 1945, my father discovered the "HM14", a tandem wing biplane with a fixed rear wing. He learned how to pilot it, and then tried to improve upon it.

After having conceived, built and flown several prototypes, he then built the "LC6 Criquet". A 90 HP biplane for travelling use, it takes off at 60 km/h and cruises economically at 180 km/h on 75% of its power. The French government became interested in this plane, and financed all of its trials in windtunnel tests as well as in tests of strength. These tests are required to obtain the "CDN", a certificate of airworthiness necessary to commercialize aircraft.

Although the programme of tests and flying trials was successful, at this point we unfortunately came up against a difficulty; a safety Norm stated that a plane must be able to get out of a spin dive. Now, the Criquet cannot get into a spin, so you cannot get out of it...and we come up against an absurdity: The Criquet could not get the "Norm" because it was safer than the "Norm" requirements.

This ridiculous problem postponed the commercialization of the Criquet. (When you know that one fatal accident out of five is due to spin diving, you very much appreciate the fact that this plane cannot get into a spin.)

During the time of the trials of the Criquet, however, I looked forward to the eventual commercialization of the plane. After much study, I made a copy of the prototype wood Criquet out of glass reinforced plastics. This plane has been flying for 8 years, and is now based at Roanne.

Because of the problem concerning the Norm, we have not been able to commercialize the Criquet, although we have sold 105 drawings to homebuilders (30 of these craft already fly).

Among these amateur flyers, there are crippled people who have succeeded in having their planes built. Miss Clerc, for instance, who is deprived of the use of her legs, had her father build her plane, and, alone on board, completed the Paris-Dakar air rally. For this, she was made "Chevalier de la Legion d'Honneur".

Last summer, about 20 disabled people came to Mâcon Airport to fly on a dual-control machine built by my father. Their joy was absolutely incredible, of course; for the first time, they could freely move in an element they had mastered. They were so very happy that they asked us to arrange a piloting period of instruction, which will take place in July 1983. About 12 of them will be able to fly 15 hours each, thanks to an association in which, with the help of an approved instructor, a pilot can be trained and the members of the association can fly very cheaply (150 FF/hour), due to the low manufacturing cost of the aircraft.

While developing the Criquet, my father built another very light plane, the "Pouplume". Entering it at the Bourges Competition in 1961 and 1962, with a 175 cc, 8 HP engine, he became world champion of "Moto Aviette": taking off at 18 km/h, and flying at 80 km/h. Here was the first Ultra-Light Aircraft (ULAC). Seventy-six homebuilders have bought these drawings and are now building; 11 of the planes already fly.

Now, a new development has occurred. The ULAC movement has come from the U.S.A., with a quick and anarchical development. This has led the authorities to develop new rules for this type of aircraft. These rules allow the commercialization of kits and crafts without having the CDN, and we think this gives us the opportunity to begin developing the "Pouplume" of the Bourges competition, as well as two new 2-seater, dual control planes, one of which is an amphibian. According to ULAC norms, these three planes can be commercialized without any problem, and we begin next month with trial flights. After these flights, we will use these wood prototypes to form the moulds needed to make the planes in glass reinforced plastics.

As a specialist in the use of these materials for 15 years, I think they are the most suitable for the building of planes in series. This technique will allow us to offer, for the same price as the existing ULAC kit type machine, real planes with rigid wings, streamlined fuselages and effective controls allowing real and safe piloting. The labour saving is valuable, and the molds ensure exact reproductions of the prototypes.

Our planes are designed at the same safety factors as the planes which have got the CDN. They cannot get into a spin, so they are very safe, and their piloting naturally suits disabled people as well as normal pilots.

We work in a well-equipped workshop on our own grounds. Our two objectives are; 1) to create a company able to produce completely built planes, kits and spare parts, and 2) to enable disabled people, deprived of the use of their legs, to enjoy safe, easy piloting in planes that do not require the use of the legs, and that provide easy access to the controls thanks to wide door openings.

An Insulating, Rigid Foam — From Seaweeds

Arturo M. Kunstmann

B. Dibasson 885, Punta Arenas, Chile

Chilean, born September 28, 1945. Dean of Faculty of Engineering, and Lecturer in Chemical Engineering Department, University of Magallanes. Educated in Chile; Chemical Engineering at Universidad de Santiago de Chile, 1970.

This project's objective is to develop a product from seaweeds for use as a thermal insulating material that is safe and low cost.

The calcium alginate manufactured from brown seaweeds has special properties; namely, a fibrous structure and incombustibility. Subjecting this compound to certain laboratory procedures changes it into a rigid foam possessing desirable insulating qualities. This process has to be improved to accommodate an industrial procedure.

The raw material is extensively abundant and renewable. The process of preparing the calcium alginate is not complicated, and the product, if it is developed as expected, would surely contribute usefully to the economy of many people.

The ready availability of the raw material is a key reason for pursuing this development. There are many miles of coastline in the numerous fjords of the Southern Pacific ocean and in the zone of the Straits of Magellan. Very abundantly found in these waters is the brown seaweed, family *Macrocystis Piryfera*, commonly called kelp, which contains alginic acid as its major chemical component of interest (about 15-20% on a dry basis, depending upon season, depth and salinity of the surrounding sea water). The alginates are of value in various food and industrial applications (due to their ability to hold water, to gel, to emulsify and to stabilize), for which they are used in small quantities.

The purpose of this project is to develop a process for the industrial manufacturing of a rigid foam of calcium alginate (or other salt of alginic acid) with controllable properties of conductivity (thermal, noise, etc.), density and ease of handling, that will allow the use of the product as an insulating material. The process would be developed from the preliminary test techniques conducted at University of Magallanes (Punta Arenas, Chile) by the applicant, where it was proved that it is possible to form a cellular structure of solid calcium alginate with valuable properties.

The results of laboratory work with several water insoluble metallic salts of alginic acid (copper, iron, lead) resulted in the focusing of attention on calcium alginate. This compound exhibits a structure more suitable to the forming of a rigid sponge, as the final product should appear.

The tests were carried out with a solution of sodium alginate of 0.5 and 1.0 per cent weight (dry basis) in water. The sodium alginate is the chemical form of alginic acid directly attainable in the treatment of kelp with soda.

From kelp of 1.2 meters maximum height, a harvest can be made twice a year. The kelp is chopped, and then digested with soda, producing an algin liquor. This liquor, decanted for clarification and with the appropriate concentration, is mixed with a solution of calcium ions (chloride, for example) to obtain a fibrous, insoluble gel of calcium alginate. The way one mixes both solutions determines the characteristics of the end product. The main variables in the process are: concentration, pH, order of addition of reactives, and the use of a foaming agent and aeration.

In laboratory testing, the solution of sodium alginate was pressurized with air; the solution contained a foaming agent dissolved in it. Then, by means of a valve and through a special glass tube, the solution was bubbled under a coagulating bath of calcium chloride solution. The resulting foam that was formed was removed from the surface of the bath, drained off, and dried in an oven at 40°C

The different experimental conditions produced end products with densities varying from 7 to 10 lb per cubic foot, and thermal conductivity of 0.027 Btu/hr/ft^3/°F. This figure is in the range of magnitude of efficient commercial insulating materials.

The project now seeks to obtain an improved end-product through a technique that will be suitable to industrial implementation. The tests now planned are designed to assess the effect of modifying the conditions and active agents needed to produce the foam, as well as testing the use of new equipment to accomplish the process. It is thought that, with the availability of the elements required for a complete study, the goal of the project will be reached.

The replacement of materials commonly used in building insulation (such as polystyrene and polyurethane) with an incombustible, low-cost, alginate-based product would represent significant benefits, not only in the energy savings involved in the production of the standard plastic materials, but in the promise of the development of a new industry based on an easily harvested, renewable resource; the common, familiar seaweed known as kelp.

(Top) Four-month-old asparagus clump with enormous root system holding the soil in position on a slope. *(Bottom)* Special packaging mad for traditional festive season to boost the sale of asparagus as a means of encouraging local production.

Rescuing Mt. Kinabalu and its People – With Asparagus

Thean Soo Tee

Rolex Laureate – The Rolex Awards for Enterprise — 1984
Rural Development Corporation, Locked Bag 86, Kota Kinabalu, Sabah,
Malaysia

*Malaysian, born January 16, 1936. Managing Director, Rural Development
Corporation, and Technical Advisor/Horticulture. Educated in Malaysia,
Japan and U.S.A.; M.Sc. (Agronomy) in 1968, and Ph.D. (Genetics) in 1971
from University of California at Davis, California.*

Asparagus grown in the tropical highlands can be stimulated to produce
marketable spears within nine months from seed and continue to do so daily all
year round. Taking advantage of this technique and the high market value for
green asparagus, the project proposes a coordinated programme of short- and
long-term action to increase farm income of the rural poor living in the
highlands of Mt. Kinabalu, Sabah, Malaysia.

Short-term activities focus on extending this technology to the farming
community, involving a radical shift from cultivation of traditional annual
vegetable crops to perennial asparagus. The long-term programme is designed to reinforce and strengthen the short-term activities and to develop a
framework for a sound asparagus industry to benefit the farmers. Exotic
varieties would be introduced, evaluated and recommended. Agronomic
practices would be improved and asparagus growth patterns studied.

Equally, if not more important, the permanency of asparagus with its
massive root system will serve to alleviate problems of soil erosion caused by
continuous cropping on the precarious slopes of Mt. Kinabalu.

Background. Mount Kinabalu is the highest mountain in Southeast Asia,
rising to 13,455 feet (4,100 m) above sea level. Time, weather and geology have
conspired to produce this enormous landscape which seldom fails to create an
aura of greatness, mystery and inspiration for Sabahans and visitors alike.
Scattered on the high country between 1,200-1,500 meters are numerous
villages connected by dirt roads that converge on Kundasang, a small central
market place where temperate vegetables are collected and transported by
road to urban centres in Sabah, Brunei and even Sarawak. The population of
this region is estimated at 5,500.

Climatic conditions favour the growth of many temperate vegetables, and
because of the uniform temperature all year round, temperate vegetables are
cultivated throughout the year. However, the rugged terrain limits land
available for cultivation. Farmers plant vegetables on the slopes or on terraces

precariously cut into the mountain slopes. Vegetable production in this region has persisted as the traditional farming activity and is inherently labour intensive.

A wide range of vegetables is cultivated in the region. Cabbage and other cruciferous leafy crop form the major group. Tomatoes, onions, peas, carrots, celery, lettuce and capsicum are common. The quality and quantity produced are variable, and this, coupled with an inefficient marketing system, results in a market value for these vegetables that is erratic. Wholesale price is miserably low compared to retail price in the urban centres.

Almost all the vegetables grown are annuals, and farmers cultivate the land after each crop, exposing the loose soil on the slopes to rain and wind. Soil erosion is acute. It is a matter of time before erosion will carve the hillside bare, exposing rocks and boulders. This has already begun to happen.

The introduction of asparagus was intended to help change the annual cropping pattern of vegetable cultivation, in order to stabilize horticultural practices. The high market value of asparagus plus comparatively superior production will offer a better opportunity for farmers to earn more from the same unit of land. In addition, the enormous root system and permanency of asparagus would serve to protect Mt. Kinabalu's slopes from further erosion.

Asparagus on Mt. Kinabalu. Asparagus was first introduced in 1981 in Kauluan, some 4,000 feet (1,200 m) above sea level at the slope of Mt. Kinabalu. About 3 hectares of a 10-hectare Government farm were progressively planted with asparagus. The variety Mary Washington was first seeded in July 1981, and then transplanted in August 1981. Growth was vigourous and by May of 1982, commercial grade spears were available for harvest from the first planting. Now, after ten months of continuous harvesting, production is still in progress. Special techniques of pruning, fertilization and management are employed to sustain production. Asparagus plants that have attained sufficient foliar growth (at 3 months) are pruned.

Pruning is an operation that essentially removes apical dominance. Under ideal conditions of temperature, irrigation and maintenance of a proper fertilizer programme, the emergence of spears is both rapid and dramatic. Almost simultaneously, dormant buds burst forth. Two or three of the spears are allowed to develop into normal bushy plants, to continue the photosynthetic activity that is vital to sustain healthy growth and to maintain productivity. With this methodology, asparagus has suddenly become a potentially important vegetable for the highlands.

Our recently begun marketing strategy of developing oriental cuisine for asparagus has created demand for the product. The original price of M$10.00 per kilo wholesale has been raised to M$15.00 per kilo, for product packed in specially designed packages. Demand now exceeds the limited supply from the only asparagus farm in the country. It is postulated that rising affluence and promotional activities will create a tremendous impact on asparagus demand. The market potential for asparagus is good.

The Project. Our immediate activity in the project is to demonstrate the potential of asparagus to the local farmers. This is a formidable task as it is necessary to persuade the farmers to shift from their traditional system of annual cropping to permanent agriculture. The technology is new to them, and the concept of permanent horticulture is at a distinct tangent to their traditional system of annual crop culture. Pilot extension trials will be the most appropriate means of initiating an understanding of the practical aspects of asparagus cultivation, and of providing models enabling farmers to see the potential and benefits of asparagus in the highlands.

Five model asparagus farms are proposed in the region, each strategically located to give the greatest impact to the farming community in its area. All inputs to bring the model farms into production should be contributed by the Government. Local farmers will work these pilot farms to earn and learn. Most important, the farmers must experience the actual benefits, if they are to be persuaded to make the radical change to asparagus cultivation.

Over the long term, we wish to improve current practices, and develop locally adapted technology to serve as a guideline for the farming community's ability to produce high quality asparagus on a consistent basis. This programme would include:

Introduction and Evaluation of Exotic Varieties. We now grow only one variety, the Mary Washington from the U.S.A. Introduction and evaluation of commercial varieties from Europe and the U.S.A. will provide us the best opportunity to make rapid improvement in local asparagus production.

Development of Agronomic Practices. Accelerating the growth of asparagus requires a more exacting system of farming management than that now found locally. Current practice is based heavily on intuition, tradition and trial and error. More precise information is essential as technological support for the farming community. Crop husbandry studies would include cultural practices, crop protection (weed control), water management and harvesting techniques.

Studies of Growth and Development Patterns. The rapid, continuous growth of asparagus offers unique physiological studies opportunities. Current data on asparagus growth is based on temperate conditions where asparagus undergoes stages of activity and dormancy. No studies have been conducted under continuous growth and harvesting conditions before now.

Benefits. The primary aim of the project is to improve the earning capacity of the local farming community through the cultivation, production and marketing of asparagus, thereby ameliorating the standard of living of farmers and helping to prevent further soil erosion on the rugged slopes of Mt. Kinabalu. Additional benefits will include the stabilization of local horticultural practices, more efficient land utilization, creation of new jobs in packing and related services, and an opportunity for scientists to research and develop new technology for transfer to other similar areas of the world. This last could have far reaching implications if the noble asparagus could be made available to the world all year round, and probably at less cost. High tropical mountain regions of Asia, Africa and South American could benefit from this technology, which could give new impetus to their horticultural crops programmes.

Artificial Cell-Fusion to Create Drought Resistant Plants

Donald Frederick Gaff

Botany Department, Monash University, Clayton, Victoria 3168, Australia

Australian, born March 27, 1936. Senior Lecturer, Monash University. Educated in Australia and U.S.A.; Ph.D (Botany) in 1964 from University of Melbourne.

This program aims to produce plants with new combinations of characteristics, drawing on the resurrection grasses for their ability to survive complete dehydration, and on related non-resurrection grasses for high productivity. To avoid the delays and difficulties of conventional plant pollination procedures, new electrical procedures will be employed to fuse normal non-reproductive cells of the two types of plant.

Determinations of the desiccation tolerance of the 'hybrid' fused cells will provide information on the genetics of desiccation tolerance, the dominance of the characteristic, and the complexity of the genetic system.

Hybrid cells will be proliferated by tissue culture, and new hybrid plants generated by suitable hormone additives to the culture media. Hybrid grasses combining desiccating tolerance and high productivity should have direct application in grazing, reclamation of desertified land and eventually in crop-plant breeding.

Over the past 12 years, my research has concentrated on unusual higher plants which have an extraordinary capacity for drought tolerance. These 'resurrection' plants are able to survive complete dehydration, becoming so hard and brittle that they appear dead. They can remain in this condition for up to 2 years, or in some cases, longer, in air of 0% relative humidity. Yet their roots and foliage revive without injury within 24 hours of receiving as little as 10mm of rain. The ten resurrection species known among the higher plants before 1970 were regarded as rare curiosities. By undertaking extensive exploratory trips in Australia, South America, and above all Africa (including the source of the Niger River), I discovered a further 80 resurrection plants. As about 30 species are in the economically important grass family (Poaceae), direct practical application of the phenomenon now becomes feasible.

Since such grasses would appear to have enormous potential for grazing in arid regions and for land reclamation in desertified areas, a living collection of them has been established at Monash University to serve as a resource for future development.

Field trials are currently underway in Central Australia (in collaboration with Mr. D. Nelson and the CSIRO) in order to establish the relative dry-matter productivity of the various resurrection grass species and their suitability for grazing.

Current work also aims at elucidating the mechanism of desiccation tolerance. Reviewing this work and the literature on desiccation tolerant seed and pollen, it has been possible to draw up an initial hypotheses of the mechanism of tolerance. This places emphasis on a burst of protein synthesis induced by moderate water stress, producing protein stable to desiccation, and on a hormonal system for triggering such positive adaptive response to stress.

A first approach to the genetics of desiccation tolerance in *Sporobulus* has been already been commenced at a whole-plant and population level.

The information obtained on the genetics of desiccation tolerance is an essential prerequisite for any future attempt at genetic engineering to incorporate complete desiccation tolerance into the genetic constitution of crop plants. The success of this attempt is quite probable as the electrical fusion technique is sufficiently flexible to allow production of a wide range of polyploidy (a doubling or tripling, etc., of genetic information, often accompanied by increased vigour and extended climatic range; the development of modern "hexaploid" wheats is an excellent example) in the final hybrid cells, as well as different ratios of genetic material from either of the 'parent' species.

Seed are currently being harvested from resurrection grasses under cultivation at Monash, in order to commence this work in 1984. I propose to travel to West Germany where electrical fusion techniques have recently been developed, master these techniques and apply them to resurrection plant cells. In this method, cells are brought into tight membrane-to-membrane contact between two electrodes by dielectrophoresis, and then subjected to an additional electrical field pulse which produces an electrical breakdown of contiguous membranes. Cell fusion takes place within minutes, and the composite cells show no sign of injury. Sea urchin eggs, for example, can be fused repeatedly without losing their ability for fertilization by sperm.

The technique will be applied to cells from desiccation tolerant plants and related desiccation sensitive species in the same genus. Drought tolerance limits of the fused cells will be estimated in the first instance in solutions of different osmotic stresses; cell survival will be determined from cell membrane integrity, which will be tested by the standard neutral red method, followed by vapour-exchange techniques.

The data obtained should indicate the complexity and dominance of the inheritance pattern for tolerance, as it is expressed in single-cell systems. In order to check these patterns at the 'whole-plant' level, hybrid cells will be allowed to multiply in tissue culture. Hybrid plants generated in this way from fused cells will be tested for desiccation tolerance as intact plants by vapour exchange techniques.

Tolerant hybrid plants will be further tested for growth vigour, and productivity in field trials.

Solving Art Historical Problems of Mediaeval Manuscript Painting by Chemical Analysis

Mary Virginia Orna
39 Willow Drive, New Rochelle, New York 10801, U.S.A.

American, born July 4, 1934. Professor of Chemistry, College of New Rochelle, N.Y., U.S.A. Ph.D in Analytical Chemistry (1962) from Fordham University, Bronx, New York, U.S.A.

Althoung mediæval manuscript illuminations have been the object of intensive research in recent decades, very little research has been done on the pigments of the painters. This project's purpose is to initiate the first systematic application of the small-particle analysis techniques to the study of pigments in mediæval Armenian, Byzantine, Islamic and Crusader manuscripts. It is a pilot project to explore the problems of pigment sampling and analysis in a limited area and to establish a methodology that may be applicable to the study of mediæval manuscripts on a wider basis. The Middle-Eastern materials make an ideal starting point since many of the manuscripts are dated and located by colophons and inscriptions. This work may be able to shed more light on several art historical problems, including tracing lines of influence and/or interconnection between mediæval centers of manuscript production.

The project will involve the close collaboration of an expert in small-particle analysis, the Principal Investigator, with an expert in mansucript illumination, Dr. Thomas F. Mathews, working at the Conservation Center of the Institute of Fine Arts, New York University.

The pigments of mediæval painting have often been discussed in the context of mediæval artists' manuals, but the discussion is rarely founded on hard data, and the manuals themselves are often vague. In recent years, two techniques have been explored for providing better identification of actual pigments: in Munich, the technique of photomicrography has been applied to early mediæval manuscripts from Western Europe, and at State University of New York (Stony Brook) laboratory, neutron activation analysis has been applied to several nineteenth century forged manuscripts. However, neither of these techniques yields results as precise as the method proposed in this project.

For four years, the above-named investigators have been involved in a study of the Glajor Gospel Book, of the University of California at Los Angeles, an extensively illuminated manuscript of ca. 1300, hitherto unpublished. In the summer of 1979, they conducted a pigment analysis of the miniatures in this manuscript by X-ray diffraction and chemical microscopy techniques. The methods used permitted specifying each chemical compound employed, while pinpointing the exact spot where it was used and the artist who used it. The results demonstrated that different artists employ markedly different palettes. In addition, an art historical hypothesis, based upon the distinct styles of the painters, that the manuscript was composed in two different workshops was upheld by the establishment of two different palettes through chemical analysis. (Recent correspondence from S. der Nersessian, an expert in Armenian art history, suggests that the first of these workshops was located in Elegis, the center of Suinik' province, where Glajor was located.)

The presently proposed project will concentrate on several collections with significant holdings in Byzantine, Crusader and Islamic manuscripts, in collaboration with expert art historians in these respective areas. Plans are also being made to contact other significant collections in Europe (Paris and Venice) in order to include them in the project as well.

The choice of manuscripts and the location of pigment samples within each will be made by Mathews, and the taking of samples and their analysis will be carried out by Orna. Where possible, samples will be taken from smudges of paint transferred from facing pages; where such smudges do not exist, samples will come from margins or broken areas in the paintings; every effort will be made not to disturb intact paintings. Polarized light microscopy, microchemical tests, X-ray diffraction analysis, and other tests will be used to reach firm chemical identification of the pigments. At the same time, the colors as they appear in the manuscripts will be measured against the Munsell Color Charts to establish the visible palette achieved through the given pigments. Thus, it is hoped that the results will permit us to characterize regions, schools, workshops, and even individual artists by the profile of their palettes.

The observation, photographing and sampling of the first set of manuscripts (Chicago) will be scheduled for early May, 1983, with subsequent samplings on a half-yearly basis, and analyses in the intervening periods.

At the completion of the project, we expect to be able to describe with scientific accuracy the complete palette range of at least two separate regional schools of Armenian painting, one school of Byzantine painting, and the comparison of these three with several extant Crusader and Islamic manuscripts. We will try to correlate our results both with the techniques of painting known in Western Europe and with the techniques described in the Western European and Armenian painters' manuals. The data and the actual pigment samples will be available for consultation at the Conservation Center of the Institute of Fine Arts, New York University, as the start of a data bank of mediæval pigments. The program will have developed a methodology and an expertise in a new area of research which we hope will be useful to others in the discipline. Publication of the results is planned in *Studies in Conservation,* the *Journal of the Walters Art Gallery,* the *Revue des Etudes Arméniennes* and other appropriate journals.

Making optimal use of available local resources: a one-opening "Crescent" stove at work in a Kenyan village.

Better "Energy Tools" for Use in Third World Countries

♔ Waclaw Leonard Micuta

Honourable Mention — The Rolex Awards for Enterprise — 1984
5 Rue du Vidollet, 1202 Geneva, Switzerland

Polish, born December 6, 1915. Retired (1977) U.N. International Civil Servant. Currently specializing, in a private capacity, in the development, and utilization of primary sources of energy for developing countries. Educated in Poland; B.A. (Economics) in 1939 from Poznan University.

The project aims at the design, development and promotion of simple, yet efficient, implements which could be immediately manufactured in developing countries. Its emphasis is on providing rural populations with the means to increase their labour productivity and standards of living. It is hoped that they may thus be encouraged to remain on the land, where they are needed, rather than joining the migration to urban areas where there is usually inadequate food, lodging or employment.

An important part of the project is the creation of new rural training centres in Kenya to supplement the pilot project I established in Ruthigiti village, in 1981.

As indicated in *World Conservation Strategy*, "probably the most serious conservation problem faced by developing countries is the lack of rural development". This results in wasteful use of available resources, with often catastrophic social and ecological consequences.

Recognizing the need for grass-roots measures to complement the long-term programmes of inter-governmental agencies, I have concentrated on developing implements that will provide less privileged populations with the means to make optimal use of locally available resources to increase energy self-sufficiency, as well as to raise agricultural production and living standards.

Special attention has been paid to the more rational performance of the basic heavy agricultural and household duties that, irrespective of time and geography, must be carried out by all communities. All these chores require energy.

There are three major sources of energy used daily in the Third World: human muscle power, animal power and firewood. These energies are employed in a particularly inefficient manner. Rural populations are familiar with few, if any, of the simple tools and implements used for centuries by farmers in more developed countries. The introduction of these tools on a wide scale would result in a spectacular increase in the work accomplished, without additional fatigue.

Firewood, a vital source of energy for the poor, is now becoming so scarce that finding and gathering it presents a major problem in numerous countries where forest resources have been steadily depleted or destroyed. In many countries, the wood shortage has already attained dangerous levels. Wood is used largely for cooking, yet there are no simple, but efficient, wood stoves in general use. Nor have any substitute fuels or improved cooking pots been introduced to help diminish the present excessive consumption of firewood.

The efficient use of draught animals as part of an integrated farming system is little known in developing countries. Animals are badly harnessed and work and die in torture. Their draught power is usually half of what it is possible to obtain. Working life is drastically reduced and they yield little milk, meat or manure. If better harness were introduced, together with better animal driven equipment and animal husbandry methods, farms in many poorer countries could dispose of sufficient energy to cultivate fields, transport crops or supplies and provide families with a steady income throughout the year.

Since my retirement, I have gained particular experience in the design and promotion of fuel-efficient cookstoves. I have successfully demonstrated the results of this research at several international meetings, included many organized and sponsored by the UN.

In 1981, with the assistance of the Bellerive Foundation (Geneva), I published "Modern Stoves for All", a basic technical manual including descriptions and designs for a dozen fuel-efficient stoves selected on the basis of good service rendered during field tests in Europe, Africa and the Caribbean. The stoves form the focal point of a comprehensive cooking system involving the use of alternative fuels, together with more rational cooking methods and implements. All elements of the system are fully described in the text of this publication, which I have included with this application. The booklet also describes my philosophies in the domain of cookstove design and promotion.

In September 1981, in order to give practical application to my research, and again with the aid of the Bellerive Foundation, I established the first ever rural training centre for local stove makers in the Kenyan village of Ruthigiti, Karai location. The centre, which has been operating successfully for two years, provides basic instruction in the techniques for constructing fuel-efficient stoves, and ensures that village craftsmen are adequately supplied with the necessary tools and equipment. I have been particularly encouraged by the favourable reaction of the local population, which has readily accepted the technology introduced to date. Stoves are being manufactured by the villagers on a regular basis, and several requests have been received to instigate similar projects in neighbouring localities.

In January 1982, I established a second village stove making centre near Ruiru, Kenya. The scope of this project has been extended to include not only the introduction of fuel-saving stoves together with an efficient cooking system, but also the experimental plantation of fast-growing varieties of trees and shrubs.

For the production of biogas, I have developed a simple, inexpensive installation consisting essentially of two plastic film sheets. Prototype units were installed in Kenya in 1981, and have successfully completed two years of field trials under real conditions. In addition, units are currently being used daily to run the "Community Stoves" built during my last visit to Kenya at the Thika "Parking Boys" Centre, an institution caring for homeless youths in

Nairobi. As semi-dry manure is employed, the installations are particularly suited to use in regions where there is a shortage of water.

I have also devoted several years to research on efficient draught animal harnesses. With the assistance of the remaining harness craftsmen in Switzerland and France, as well as farmers in those countries still using draught animals, I have gradually evolved a model for application in developing countries. The new harness was successfully introduced in Kenya in 1982. In January 1983, I founded a special unit within the Agricultural Machinery Department of the University of Nairobi for the manufacture and use of the new harness. The unit is manned by a Kenyan craftsman, whom I trained and equipped with all the necessary tools and implements.

The new harnesses significantly increase the efficiency of draught animals and enables them to work in greater comfort. They may be adapted to any draught animal and have already been successfully used by me in Kenya on donkeys as well as on oxen. I am currently preparing the text of a book describing in detail the results of my research in this area.

I am also collecting, and conducting research into, hand tools and implements that increase the efficiency of human and animal muscles in the performance of basic agricultural duties. This study is being undertaken in close cooperation with European farmers who are still using equipment of proven efficacy that is virtually unknown in developing countries. I have also assembled in Geneva an extensive set of agricultural machines, including village power gears that transform animal energy into mechanical energy, which can then be used for such activities as threshing, grinding, chopping and other procedures for primary food processing. The machines are available for inspection and for possible future promotion.

All the projects mentioned in this submission have already been commenced on a small scale and are progressing well. After two years of successful operation of the pilot centre in Ruthigiti village, and in response to requests for assistance received from neighbouring villages, I am currently pursuing efforts to create similar centres in other Kenyan villages.

The basic philosophy underlying the establishment of the rural training centres is to build upon local initiative and to involve the communities is the solving of their own problems. There are several advantages in such an approach: 1) Communities are far more likely to adapt new ideas if they feel that they have participated in their development, 2) Specific local conditions are taken into account, and the designers are afforded the opportunity to work in close consultation with the intended beneficiaries, thus adapting to specific needs and preferences, and avoiding costly mistakes by use of small-scale beginnings, 3) Experience shows that effects spread rapidly to neighboring areas, 4) Use of local skills and materials allows significant contributions to be made by small private groups or individuals, 5) Aid funds are applied directly and efficiently, reducing bureaucratic or middlemen erosions, 6) As self-help, the funds are invested in future prosperity, rather than merely written off as relief aid, and 7) The scope of each project is readily adaptable to the amount of funds available.

A Solar Drier for Agricultural and Piscicultural Products

David Rechnitzer

P.O. Box 1005, Nahariya 22110, Israel

Israeli, born September 4, 1924. Technical manager in own workshop, designer of solar collectors. Educated in Yugoslavia and Israel.

I have designed a solar drier that radically improves on existing methods of drying produce by free exposure to the sun.

This drier enables a regulated and accelerated process of solar drying, hitherto not achieved, and offers improved hygienic conditions for such drying.

Drying food by means of solar energy is an ancient process applied wherever food (any agricultural/piscicultural product, and also tobacco) is found or produced, and climatic conditions make it possible. It is achieved by direct exposure of the food to the sun. This method, as presently applied, obviously suffers from many disadvantages:

— Food is exposed to the sun's U.V. range, affecting its quality.
— Food is contaminated by dust, insects, bird, etc.
— Food is also contaminated by human contact, as it usually has to be reversed several times during the drying process.

— the process is slow because the sunshine hours are limited, necessary temperatures are not reached even during the sunshine hours, the food cools at night, and eventual reabsorption of moisture from surrounding air may take place during darkness.

— the impossibility of creating drying conditions (air velocity, temperature) specific for the different sorts of food (e.g., the drying of whole prunes, or sliced apples, requires different temperature and air velocity).

A simple installation to eliminate the above disadvantages and other deficiencies, but to keep the benefits of solar drying, was envisaged and designed.

The project was conceived to achieve the following objectives:

— To eliminate the contamination of food during the drying process.
— To furnish microclimatic conditions to attain improved quality and hygiene of dried goods and diminish waste.
— To make an extremely simple, inexpensive unit, of readily available materials, with very little know-how required for assembly, so that the drier is attainable by poor producers in non-developed regions.

Considering these objectives, the drier has been designed as follows:
Description of the Drier. The drier consists of an insulated chamber which

contains several rows of rails at several levels. These support trays which hold the product to be dried. The roof of the chamber and the front (sun-facing) wall consist of solar collectors for air heating. From the exits of the collectors, short pieces of flexible tubes conduct the air to the inlets of the drying chamber. At the chamber inlets, simple vanes distribute the flow of the hot air evenly over the entire area and to all levels. Velocity of air through the system is controlled by a manifold pipe, connected to a vane-containing exhaust tube. Tray bottoms are of wire mesh to expose the lower part of the product to the hot airflow.

Functioning of the Drier. In the exhaust pipe of the drier, an upward movement of air is created by a wind-driven fan, a small electrical fan, or by the thermo-siphon effect owing to the height of the pipe. This air movement causes a suction effect at the collector entrances, pulling air from outside into the collector system. Passing through the collectors, the air is heated to a temperature of about 85°C, and conducted via the flexibe tubes to the inlets of the drying chamber. Here, the vanes distribute and direct the airflow so that it passes, in equal volumes and velocity, above and beneath all the product-bearing trays. The hot air heats the product, causes vaporisation of its moisture, absorbs the moisture and, having reached a high degree of saturation, leaves the chamber via the outlet manifold and exhaust.

The air temperature in this drier is a function of air velocity. Thus, regulating the air temperature is accomplished by controlling the air velocity through moving the angle of the vane situated in the exhaust pipe.

Structure of the Drier. The collectors are simple "U-shaped" shallow "channels" of wood or available materials, 1 m square, and about 15 cm high, with a layer of insulating material on the bottom, covered with a blackened aluminum sheet and topped with glass or polyurethane.

The drying chamber is a framework made of rectangular tubes to which double walled panels made of materials similar to those used for the collectors are attached. Hinging on the panels allows insertion and extraction of the trays, which sit on simple angle iron profiles.

The entire construction is modular, the basic chamber being 3m long (3 rows of collectors), 2m wide (2 collectors in each row) and about 1.2m high. The complete drier may be enlarged by adding panels to the chamber, increasing the number of collectors to a row and increasing the number of rows themselves.

Transportability. Since the capacity of the drier even in its smallest size — the basic modular unit has about 6 cubic meters of drying space, i.e., 30 trays of about one square meter each — could serve more than one farm, it is designed to be quickly dismantled to sub-assemblies, and easily transported and reassembled at a new site.

Uses. A vast amount of fruits and vegetables, fish, seeds, spices, tobacco, etc., is being dried by insolation wherever conditions permit.

Dried fish and dried vegetables constitute the main or secondary nourishment of millions of people around the globe.

Making this drier available to these people could provide them with significant benefits. Assistance, or partnership, in the project is welcome.

Nearing Perpetual Motion, with an All-Magnetic Power Generator?

Juan Manuel Gómez
Apartado Aéreo No. 54694, Medellin, Colombia

Colombian, born October 27, 1952. Chief of maintenance in textile factory. Educated in Colombia and via correspondence in U.S.A. (Applied Science).

This extremely efficient power generator is a torque-producing device that taps the energy of the magnetic field created by unpaired electron spins (atomic currents) in ferro-magnetic elements and alloys, using specially magnetized rods that mimic the circular magnetic fields generated by current carrying conductors. Several of these rods are mounted radially on a shaft and made to interact with dipole horseshoe magnets (or electromagnets) that displace the rods at right angles to the applied field, forcing the shaft to rotate on its own axis; thus, obeying the same laws that govern conventional electric motors.

The only difference consists in that, while a conventional motor requires an external power source, this generator creates its own by harmonizing electron spins and the intrinsic magnetic fields of its constituent elements.

I understand the purpose of my invention has been a long sought-after "impossible" dream, but my radical and rational departure from traditional approaches deserves another careful examination.

Within the realm of basic laws of magnetics, there is the phenomenon of north and south poles being 'born' in any magnet whenever the continuity of its atomic arrangement is broken, thus imposing a more difficult path for the lines of force to reach the other end.

An unbroken magnetic circuit is the key to success in obtaining energy from permanent magnets. If a piece of ferromagnetic material having the shape of a ring (toroid), is placed in a strong field like that of a coil wound around it, a permanent magnet is formed with a magnetic field having not only no definite poles, but also a magnetic circuit having most of its flux contained within the specimen and very little leakage.

By extending the 'ring' so that its length is at least ten times its diameter, we have a hollow center through which we can insert a rod made of diamagnetic material; this forces the inner lines of force outward, thus reinforcing the flux density within the rod.

When this circularly magnetized rod is placed in a dipole (horseshoe) magnetic field, it will not be attracted by the poles but will, instead, experience some force at right angles to the field (the 'right hand rule'). This means that if the rod is coupled to a shaft, and numerous, closely spaced, laminated dipole horseshoe magnets are evenly arranged around it, a working motor is obtained, the energy of which will not run down easily unless the applied load exceeds the coercive force of the rods, in which case the tiny magnetic domains would turn away from their perfect magnetic circle.

The two basic components of the invention are: A rotor, which holds severally specially magnetized rods (they are the heart of the invention), and a stator made up of laminated horseshoe magnets or electromagnets (in a more practical design).

To make the rods, we take a hollow tube, which has pressure-applying sealing pistons placed at either end, each with a hole allowing insertion of a brass rod and filing material (a mixture of magnetic powder and a quick-curing plastic matrix). A strong direct current is fed into the brass rod, whose circular magnetic field aligns the plastic suspended domains in the same direction as the rod's field (just as would a compass needle). Meanwhile, high pressure is applied on each piston at the ends of the tube, compressing the particles and thus increasing the final field strength.

Once the plastic matrix is set, electricity is stopped in the brass conductor, leaving a tube filled with compressed magnetic power whose internal magnetic field is similar (and behaves accordingly) to a current carrying conductor's field. The outer tube is machined away (it was only a mould), and the brass conductor threaded on one end so that it can be screwed onto a hub that is mounted on a shaft.

The rod's magnetic field is substantially closer to its surface than a conductor's, because it is contained mostly within itself (it is basically a long toroidal magnet).

When the rod is placed in a dipole magnetic field, it is permeated (soaked) in that flux, and since there are no north-south poles in the rod to align with the applied field, the whole rod is displaced at right angles to the field, leading to continuous motion around the shaft.

My invention should not be contemptuously regarded as "another perpetual motion machine trying to violate the second law of thermodynamics". It should be taken more as a 'magnetic battery', essentially differing from electrochemical ones in the following respects:

— It stores energy in the form of a magnetic field supplied to the rods by an initial surge of high current at the moment of manufacture.

— Its materials are much lighter than those used in equivalent powered batteries.

— Quick recharging of the magnetic rods can be accomplished by a strong, momentary input of current.

— Most important is that, unlike electrochemical batteries whose charge duration is determined by their amount of mass, this invention's working life is determined by the coercive force of the magnetic rods (capacity to retain magnetism under reverse fields) and the amount of load applied.

If high-standard modern materials are used, and the applied load doesn't exceed certain 'turning points' of the magnetic domains in the rods, this generator could outperform all of the clean sources of energy so far known to mankind.

One elegant, compact example of a new line of devices that could replace the traditional spirit levels of engineering and architecture.

Elegant and Precise New Levelling and Protracting Instruments

Yuk Wah Lee

12A Whampoa Street, 5/F, Hung Hom, Kowloon, Hong Kong

Chinese, born October 24, 1945. Senior Technical Officer, Sites and Buildings Department of The Hong Kong Telephone Co., Ltd. Educated in Hong Kong; The Hong Kong Technical College (Building Technology).

My invention involves a new approach to devices used in measuring, levelling, protracting and transiting requirements. It utilizes an enclosed, non-mixing fluid/liquid combination to provide very high degrees of accuracy.

Conventional spirit levels comprise a slightly bowed length of transparent tubing filled with liquid, apart from a small air bubble. A pair of space mark lines are provided to judge whether the bubble is in the centre. This tubing is positioned within a levelling ruler, on a tripod, or the like. The user then has to judge levelness by observing whether the bubble is between the two marks. This cannot be very accurately achieved, and the uses for such a device are therefore limited; it cannot satisfy the requirements for high standard measurements and surveys.

It was the objective of this invention to provide a measuring and levelling device which provide higher accuracy.

My invention provides a closed vessel having at least one transparent region, half of the internal volume of the vessel being occupied by a liquid and the other half by a fluid that does not mix with the liquid and visibly contrasts with it. The interface of these two is visible through the transparent region, graduations are provided adjacent to the interface, and the vessel is symmetrical about the interface at least in planes parallel to the plane of the graduations and coinciding with an axis for the vessel.

The vessel can be used on its own, or combined into an elongated block, or the like, for use as a level measuring device. It is particularly useful as part of those surveying instruments that must be set up strictly level; e.g., theodolites and flat surface levels which are mounted on tripod legs, as well as other scientific levels.

The liquid used will normally be water, or coloured water, but could be other liquids such as oil or spirit, and the other fluid can be air or another gas, although any visibly contrasting fluid of a different density than the liquid is possible; e.g., mercury. The interface will therefore stay strictly horizontal irrespective of the orientation of the vessel, and so, by means of the graduations, one can easily read off the inclination of the vessel relative to the horizontal.

The graduations can be 360° markings formed in a circular scale and/or cross as horizontal and vertical axes. Other markings, such as radial lines, can be provided. There can be a single set of graduations in one plane, or these can be duplicated on, or adjacent to, more than one face of the vessel. Also, a circular scale of graduations can be extended to the side faces of the vessel; the cylindrical surface in the case of a cylindrical vessel.

Preferably, the vessel is made entirely of a transparent material. If it is, the vessel can also be symmetrical in one or both other axes, and thus give readings of levelness in more than one orientation. In such a case, two or three sets of graduations are required in planes oriented in differing directions.

The interior of the vessel can be of a wide range of shapes, provided its section is regular about the centre point in a plane parallel to the plane of the graduations. Thus, the vessel could be of casket shape, contain internal hollow bodies (e.g., of cube or prism shape), or could be spherical or globular. The thickness of the casket shapes and depth of any prism must be equal, whilst spheres and globes are calibrated by diameters. The cut sections of a ring shape must be similar and equal, but the thickness may be irregular and need not be constant. Simultaneously, these thicknesses and depths are calculated by the thickest and deepest dimensions of the vessel, but those dimensions must be regular.

Because the interface divides the volume of the vessel, the surface area of the interface and the lengths of the edges for the surfaces of the liquid and fluid are identical each with one another at all times, even though the actual shape of the interface may vary depending upon the inclination of the vessel.

This invention consists of applying the principle of cross lines and graduations on the dial of the vessel to the internal area and volume of the vessel which is divided precisely equally between liquid and fluid. By rotating the graduated vessel, the diameters, circumferences, diagonals, and projected lines allow precise calculation of required degrees, etc.

The characteristics of these instruments can be designed to suit many kinds of devices, and the liquid level can be turned to any degree of the graduation. Such instruments can be formed in any shapes desired, but the cut sections must be similar and equal. The size of such instruments can be as small as a ring, or as large as a pool. Usual forms include:
— Spherical, globular or polyhedra shapes.
— Caskets and prism shapes.
— Annular ring shapes.
— Combined cubical shapes.

In application purposes, the surfaces of these vessels are graduated with radials, degrees, symbols, etc., and indicate horizontal and vertical axes. By the simple expedient of placing the vessel's marked centre point at the centre point of any concerned matter, all angles, degrees, horizontals and verticals may be clearly discerned and marked.

The instrument has multiple applications in many fields, including building and civil construction work, decoration and engineering, earthquake and landslide forecasting, toy manufacturing, and others.

It can be used independently as a leveller protractor, combined with or fixed to a flat surface levelling tripod as a levelling instrument, combined with or fixed to globes and astronomical instruments, distance surveying instruments or levelling instruments as a protractor, etc.

The conventional bubble tubing device appears to have been invented by the

Greeks in roughly 500 B.C. Little has changed in its application since then. I believe that the use of my principle can sharply improve the performance of even the most elegantly constructed measuring devices in use today, including the theodolites of the professional surveyor.

Initially, I should like to continue with the extension of the patent coverage I seek, and to commence production with levelling rods (such as the mason's level) and my "thread levelling coupling device" which is of great practical value in the field of interior decoration. Secondly, I would welcome aid in the marketing of the products. Thirdly, I should like to publish works on the principle of my invention.

A New Compliance Aid for Patients Using Medication

Bart Joseph Zoltan

152 De Wolf Road, Old Tappan, New Jersey 07675, U.S.A.

American, born December 26, 1946. Research Engineer/Project Leader with Lederle Labs. Educated in U. S. A.; M.Sc. in Engineering from Fairleigh Dickinson University.

This project will develop and bring to market an aid to patients who are taking medication. The Compliance Aid for Pharmaceuticals (CAP) is a modified digital timpiece, built into the top of a standard pharmaceutical vial for the purpose of indicating to the patient the last time at which he/she had taken his/her medication. For many individuals, the failure to remember whether the last dosage of medicine had indeed been taken is a major problem. The CAP will display for the patient the day of the week, the hour, and the AM or PM at the last time the medication was taken. The device requires no action on the part of the patient, and will be of help to hundreds of millions of patients world-wide.

Thus, this project's aim is to develop a low cost device which will help patients who must adhere to a regimen of medication.

The problem of lack of patient compliance with regimens of medication is a well documented one. Many, especially the infirm, the elderly, and the confused have great difficulty in taking medication on the prescribed schedule.

The sequella of non-compliance includes continued illness, or the use of more potent medication by the unwary physician who assumes that the original dosage was insufficient to be efficacious. The side effects to be expected from the more potent medication are usually more severe.

Aids to compliance have included counseling of the patient, dispensers of the type used with contraceptives, and calendar packs in which each day's medication is affixed to that day on the calendar. These methods are expensive, and rarely used.

With the advent of micro-electronics, it is possible to build a cheap and very useful compliance aid. A standard pharmaceutical vial can be modified in such a way as to always display the time of its last usage. A modified digital timepiece can be built into the cap, which also includes a sensor to indicate whether the cap is secured to, or is off, the vial. The timepiece is modified so as not to advance to display of the time while the cap is on the vial. Thus, the patient can look at the vial and see displayed on it the time of his last dosage, the day of the week, and AM or PM. At the next time of removal of the cap, the time displayed catches up to the correct (current) time. This time is once again held fixed after the cap is re-secured.

The device described can be manufactured for under US $4.00, and would be reusable.

The time-keeping device is incorporated into a typical container for medicinal products without the need for complicated container construction or complex mechanical parts or expensive electronic circuitry. The device displays the time and day of the week, when the container was last opened by the patient-user, and continues to display the same, even after closing of the container, to serve as a reminder. The device may also be provided with settable alarms to visually or audibly alert the patient as to when the next dose is to be taken. It can be provided as a convenient separate element, or as part of the cap or cover of a container. It therefore may be adapted for use with standard containers and need not be integral with, or part of, a medication container as such, but rather can be utilized as a reusable item with fresh containers.

Although I am employed by a pharmaceutical company, my employers are not interested in developing a product which is not drug based. The initial investment in an integrated circuit to perform the desired function is high, and I lack sufficient funds to underwrite it myself. In the last year, I have spent my own time and money pursuing the Compliance Aid for Pharmaceuticals.

That the idea is new, and original should be clear. It is inventive in that it applies a new technology, that of digital electronic timekeeping, to a hitherto identified but unaddressed problem. Interest in the finished product will be great. In the United States alone, 1.4 billion prescriptions are filled annually, and in developed nations the average elderly patient takes six different medications, often on varied and complex schedules.

With minimal support, the likelihood of successfully completing development, and bringing this item to millions of potential users is, in my opinion, certain. The procurement of working prototypes will take about three months. Clinical evaluation will take about one year. The best and most detailed description of the Compliance Aid for Pharmaceuticals is in my patent application, copy of which is included with this application to the Rolex Awards for Enterprise.

A New System for Man-Machine Synergy in Highly Automated Manufacturing Systems

Laszlo Nemes

154/B Brasso ut, 1118 Budapest, Hungary

Hungarian, born December 28, 1937. Head of Mechanical Engineering Automation Divsion, Computer and Automation Institute, Hungarian Academy of Sciences. Educated in Hungary and Japan; Candidate of Technical Sciences/Dr. Eng., Hungarian Academy of Sciences.

In order to stem and reverse the dehumanization of labour in an environment of increasing automation, greater attention is being paid to the systematic design of attractive man-machine communications in a computer controlled workshop, where man and machine work together to achieve an effect of which each is separately incapable. In various studies, I have examined the criteria for designing such systems. New procedures and equipment have been developed on the basis of a number of novel ideas. The first phase of the research project was begun in 1970. Recently, I have initiated a new phase, due to the ever increasing relevance to present day problems of the social effects of automation. New, improved criteria and evaluation methods have been established, congenial interactive graphic communication methods and shop-floor work-styles are being investigated, and alternative solutions evaluated through intensive user participation.

At the beginning of the seventies, numerical control techniques became well established in advanced manufacturing. Due to the increasing number of control functions and the heavy demand for editing control information and correcting numerical values on the machines, the control cabinets themselves were very soon equipped with sophisticated operator panels having a large number of switches, buttons, lamps and displays.

There was no question that the machine operator could master the thick volumes of manuals, with all their restrictions and conditions. Highly qualified personnel had to be trained for long periods, they had to memorize very many codes in order to understand system messages and to intervene in the machining process. Machine operators and supervisory staff were alienated from the creative activities of material forming and pressure accumulated on them since they controlled machines, computers, and processes that were beyond their understanding. They even entertained fears of becoming button-pushing subordinate components of a system that dominated them.

I envisaged the creation of workplaces where humans and machines work together in complete synergy; where they live together in total interdependency, they help each other, they act together and they tolerate and compensate each other's deficiencies.

The project was started more than ten years ago, and several pioneering systems were developed, ranging from large manufacturing complexes to single CNC machines. All of them incorporated the basic philosophy of dependency, where machines depend on human decisions and humans entrust the checking, calculations, executing and administrative tasks to computers. The key factor in this approach is the development and design of an appropriate communication interface between man and machine.

The requirements were set: The information flow between man and machine was carefully studied, and its contents were minimized, in order to make it easy to follow changes in time. New display facilities were built and various comprehensive tests carried out. To avoid special training, I first used a simplified dialogue system for communication. The computer prompted the possible technological questions in the form of suggested commands and the operator simply accepted or rejected them. In the latter case, the computer advanced to the next possible question and communicated it to the display with the vocabulary of any of the predefined languages (e.g., Hungarian, English, French, German, etc.). Many references in the periodical literature praised this method, and new developments even copied it. People who work with it in the workshops appreciate its conveniences, since it is easy to learn how to use it.

There are several built-in guarantees against misusage and these free the operator from stress, providing pleasant job satisfaction instead. Nobody really notices that there is a computer behind the system, since it is carefully hidden behind this unique method. Although this was the first dialogue-oriented machine control system in the world, I never thought to stick to this solution forever. Now I have taken a fresh look at the basic requirements, and we are developing special graphical, interactive, input/output devices. These incorporate much of the workshop-level technological know-how, which makes their use highly convenient.

The development of this new system began with setting up new design criteria and evaluation methods. The next step was the animated simulation of various proposed communication methods, using an extended minicomputer configuration in order to allow designers, the ultimate users and shop-floor managers to compare the different ideas. Taking the best solution to emerge from this participative phase, a new prototype will now be built and tested in a real industrial environment.

The previous phase of the project was partly financed by the State Office for Technical Development, Hungary. Currently the Hungarian Academy of Sciences is financing the project but the yearly budget (about 1 million Forint) cannot be converted into hard currency to buy the necessary equipment and to maintain the international contacts needed for rapid and efficient work that will be useful to the international manufacturing community.

Fifteen days after this 'mother sponge' has been transferred to its culturing position, it is brought up, still healthy, to be chopped up on board for further regeneration.

Testing Commercial Sponge Culturing in the Sea and Aquariums

♛ Jacques Le Mire

Honourable Mention — The Rolex Awards for Enterprise — 1984
Hôpital de Djerba, Tunisia

*French, born July 9, 1946. General physician/pediatrician; in charge of
Djerba's Hyperbare boxes and sponge divers' health. Educated in France;
Medical School of Marseilles University (1975).*

This project has a triple objective: protection of the environment, protection of
a means of living and protection of the health of sponge divers.

Tunisia is the world's leading supplier of sponges, producing about 100 tons
a year (about 50% of world production), for a turnover of three million dinards
(U.S. $4.8 million). In spite of the inroads of synthetic sponges in the market,
natural sponges are considered irreplaccable for the cleaning of certain
scientific instruments, for the polishing of optical glasses and china, etc.
Demand is normally higher than production, which has led to certain sponging
grounds becoming poorer in quality due to intensive diving, and the taking of
younger and younger sponges. Tunisia is presently facing this deterioration of
sponging grounds, which represents a triple problem; dangerous depletion of
a valuable natural resource, potential loss of a major source of income, and,
increasingly, danger to the health and lives of the divers involved, as they are
obliged to make their dives deeper and longer. Added to the overall problem is
the discovery of offshore oil in the sponge-diving areas, with the associated
anticipated leaks and damages to the natural environment of the sponges.

Living on Djerba Island, in the Small Syrte Bay, which is the main field of
sponge production, being medically responsible for the divers, and having a
natural passion for the sea, I have inevitably taken interest in the problem of the
spongi-culture. The story has ancient beginnings.

In 350 B.C., Aristotle noted that if a sponge is cut off, but not uprooted from
its support, it regenerates on what is still rooted. Sponges also have a sexed
reproduction, and many of them are hermaphroditic.

Little, however, is known concerning the sponge reproduction system, and
many doubts remain. This is due mainly to the fact that commercial sponges,
belonging to the Demosponges family, are extremely sensitive to various
ecological factors, which makes it difficult to keep them in aquariums for
experimental studies. Though a number of attempts have been made to
culture sponges, all with promising starts, none was carried through, even
excluding all serious conclusions concerning their profitability. Men or
elements have each time stopped the experiments. The hostility of fishermen,

the taking of experimental sponges, destruction by storms (due to lack of depth), epidemics among the sponges (still unidentified), and other problems have occurred.

Reading of the various attempts, I got the impression that there is really no serious hindrance to developing spongi-culture, and that it would be necessary only to avoid the errors of the past.

With this in mind, I looked for, and found, a site fulfilling the requirements for developing a culture technique that: would not be too expensive, would have a certain profitability, and would be easy to learn. Already filled with commercially valuable sponges, the site was obviously favourable to their growth. Easily located via landmarks, it does not need beaconage to relocate, thus avoiding curiosity and malevolence. Reachable by boat in 15 minutes, it can be visited frequently. It is naturally sheltered from storms, and not affected by river or wadi run-off (softening of the water is known to be unfavorable); water density has been measured repeatedly at 1.034 without any variation.

The culture has been installed at a depth of 40 meters (120 feet), the depth of the site offering several advantages:

— In the same variety, sponges found deeper down are always of a better quality; the stream is less strong, presenting less risk to the installation.

— The depth provides additional protection against storms and visitors.

— Local fishing boats do not anchor at such depths, thereby not risking damage from lines, anchors, or nets.

There are also disadvantages to the site:

— The depth limits the time for diving.

— The tide streams, while favorable for the growth of sponges, require precise working hours, corresponding with high or low tides, which is difficult from a commercial exploitation viewpoint.

These difficulties have been freely chosen, however, as they offer a maximum chance for the success of the undertaking, permitting total devotion, first of all, to the variants of the techniques necessary for ultimate profitability.

The Technique. The basic principle consists in taking off a sponge, chopping it into pieces, fixing these pieces on a support, controlling their growth and picking them up when they reach a commercial size.

The first stage is installing the support system on which the cuttings will be fixed. There are different solutions. I chose one using nylon lines and bricks, which are cheaply and easily found in Tunisia.

Thirty meter (90 foot) nylon lines are held to the bottom with bricks tied on the ends, and located every ten meters.

Then a large 'mother sponge' is taken from the surrounding area and made fast to the bottom on a brick, via a thread passed through it by a needle. It is then left for 15 days, to observe any possible defect due to its transfer. When its vitality proves to be undamaged, propagation by cutting is carried out. A diver descends, releases the nylon line from the brick, fixes the mother sponge on a carrier line, and brings everything up to the boat. The mother sponge is transferred on board in a bowl full of water, and chopped up with a sharpened knife. The cuttings are stitched with fishing thread to the carrier line, and when the entire carrier line is equipped with cuttings at about every meter, the diver returns the line to its bottom-based brick, and the operation recommences.

There are still questions outstanding about propagation by cuttings. The first has to do with the age, and therefore size, of the mother sponge on the growth rate of the cuttings. I have found no written documents on this, so the problem is still to be resolved. The second has to do with the period of the year when propagation cutting should best be done. Personally, having started my experiment in September 1982, and the water temperature having fallen since then, I have found that the healing of the cuttings is slower at 10°C than at 13°C, proving a reduction of metabolism by hypothermy. This might be considered favorable, reducing the trauma of the operation, but still needs to be resolved. The third question concerns the ideal size of a cutting. It is known that growth rapidity is exponential. Thus, a cutting of 1 cm in diameter multiplies its volume by 75 in four years, to 314 cm³. On the other hand, a cutting of 7 cm in diameter, by only doubling its volume in the same time, gives a result of 2,872 cm³. We must, therefore, find the best ratio between the cutting size and growth rate in order to obtain a commercializable size in the shortest possible time.

Experiments. After having found a favorable site, and developed a technique of propagation by cutting, I found that the first 25 cuttings had healed in one month's time, all showing signs of good health. I therefore enlarged the installation, to receive 510 cuttings. These will be divided in 17 lots of 30 each; every lot being a mixture of different parameters which are still unknown and to be studied. When this site is equipped, others are planned in places already located which are more suitable for commercial exploitation. In parallel to this field work, I have started the construction of aquariums, to attempt to preserve these sponges long enough to see them reproducing themselves in a sexed way, with the hope of setting the grubs (which appear to grow faster than cuttings) in place at culture locations.

As this experiment is designed to show that spongiculture exploitation is absolutely possible, its profitability remains to be proved, after the long delay of perfecting the process. The data I give now are an approach to the financial and material aspects of the problem, which will need to be verified.

Based on present calculations, a sponge-diver team of three could cultivate sponges at a desirable income level (U.S.$800/month/man) from an installation of 85,000 square meters. Allowing for full growth over seven years, a new installation should be emplaced each year for seven years, calling for a total 'sponge farm' surface area of 59.5 hectares (147 acres). This allows 5 square meters per cutting, and is based on readily achievable (mixture of time for planting and time for harvesting, etc.) work load producing about 15,000 marketable sponges (20 cm diameter, 10 cm high, weighing 50 grams, with a value of U.S.$40/kilogram) per year. (A detailed description of costs, equipment, time factors, etc., was submitted with this application — Ed.)

An Economical, Quiet Electric Generator Powered by Freon

Kwang Jin Kim

AID. Apt. 24-407, Joong 2 Dong, Haeundae-Gu, Busan, South Korea

South Korean, born January 1, 1946. Senior mechanic at Kori Jobsite of Bechtel Int. Inc., for construction of Korea Nuclear Units 5 & 6. Educated in Korea; Government Certificate in Auto Maintenance, 1975.

I have invented the "Freon Generator", which is a generator that applies Freon-22, a refrigerant in the fluorine family, to spin a turbine.

Freon gas easily turns into freon liquid at a sudden drop of its surrounding temperature and pressure when it is in a state of very high temperature and pressure. It turns into liquid at as low as −41°F at standard atmospheric pressure. This characteristic can be used to advantage by putting freon into a cyclic system that initially starts in a state of vacuum, as explained here.

This type of generator has many advantages over the other conventional types such as the hydraulic or thermal generators. Hydraulic generators require very large areas of land and water, very long construction times, sophisticated techniques, high degrees of precision, very large amounts of manpower and are very costly. Thermal generators use great amounts of fuel, pollute the air with poisonous gas and fumes, and have the great disadvantage of wasting time and failing to produce electricity energy for the hours when they are required to be shut down for frequent repairs due to their many parts and components.

My Freon Generator — by applying gas pressure as the means of turning a turbine — becomes almost semi-permanent once the system is charged. It saves on expensive conventional fuel, has no harmful smoke, fumes or noise, and will not suffer shutdown due to failure of sophisticated technology. Thus, it is possible to construct such a power plant in an urban area, with resulting savings on the transportation of electricity to users.

The structure and function of this generating system is based on four major elements: an evaporator system, a turbo-generator system, a condenser system and an oil heating system.

The Evaporator System. This includes a bottom tank, a top tank, connecting pipes, heat exchanger, safety valve, high pressure gauge, thermometer, pressure switch, gas control valve, high pressure pipe, transmitter and 3-way valve. The system maintains high pressure by converting liquid freon into gas freon by means of hot oil.

The bottom tank contains liquid freon which is evaporated in the heat exchanger by means of hot oil heated in the oil heater. This evaporated freon is sent to the top tank, where, as a high pressure gas, it is released by the control valve to the turbine.

The Turbine and Generator System. This is composed of dual low and high pressure turbines, spun by the jetting freon gas, and providing mechanical power to the generator, which converts it to electrical power. The expended freon gas, after passing the turbine, goes into the condenser, to commence its cycle once again.

The Condenser System. This has a secondary high pressure pipe, condenser, cooling tank, low pressure gauge, thermometer, water pump, cool water thermometer, gas charging valve, feed pump, pressure gauge for the feed pump and the third high pressure pipe. The system condenses gas freon into liquid freon.

The secondary high pressure pipe extends from the exit of the turbine to the intake of the condenser. It has a larger internal diameter than that of the primary pressure pipe so that more freon gas can pass more rapidly through the pipe to the condenser to keep the turbine from being overloaded. It also has a cooling pin to give off as much heat as possible to the atmosphere.

The condenser turns gas freon from the turbine to liquid freon, by passing it past many bars of cooling pipe in the cooling tank to its condensed form at the bottom.

The capacity of the condenser and the cooling tank must be in proportion to the capacity of the evaporator, in order to fully condense the freon gas coming out of the turbine and to keep the turbine from becoming overloaded.

If the temperature and the pressure of the freon gas at the time of entering the condenser were 150°F (65.6°C) and 384.5 lb/in^2 (27 kg/cm^2), and the temperature of cooling water inside the cooling pipes were 50°F (10°C), the temperature and pressure of the liquid freon at the outlet of the condenser would be 60°F (15.6°C) and 102.44 lb/in^2 (7.1 kg/cm^2). Thus, the difference would be 90°F (32.2°C) in temperature and 282 lb/in^2 (19.8 kg/cm^2) in pressure.

The Oil Heating System. This has an oil heater, burner, pump, reservoir, temperature switch, heater control panel and solar heat collector. The system heats the oil in the oil heater and sends it to the heat exchanger to turn the liquid freon in the bottom tank into high pressure freon gas.

The heater is similar in construction to those of conventional type steam boilers. Oil is heated by the burner or solar collector, and sent to the heat exchanger by the oil pump. Generally, the oil is SAE #10, for this oil does not erode metal, and keeps its heat well once it is heated, while hot water or steam have adverse effects on the metal and heat. Use of the solar collector, or industrial waste heat, is envisaged whenever possible.

The Freon Gas Generator can be made to any generating capacity ranging from thousands of kilowatts to hundreds of thousands of kilowatts for industrial use. It is highly attractive in its ability to save costly power generating fuel owing to the semi-permanency of the freon gas. Once it is charged, as long as there is no leakage, it makes no fumes and no noise in operation.

Low Cost Wheels for Use in Developing Countries

Henry James Fenroy Gerrand.

25 Haldane Street, Beaumaris, Victoria 3193, Australia

Australian, born October 11, 1923. Mechanical Engineer at Aeronautical Research Laboratories, Melbourne. Educated in Australia; Diploma of Mechanical Engineering, Gordon Institute of Technology, in 1944.

The need for large numbers of low cost wheels is obvious to anyone with an engineering background who has seen the heavy loads carried on the backs of workers in developing countries like India. If low cost wheels were available in hundreds of thousands, then it might be possible to significantly reduce the number of workers who suffer from back injury as well as to improve the flow of goods from producer to consumer.

I believe that I have found a means, which can be easily learned, whereby useful load carrying wheels can be made in large numbers at low cost, in cottage industries set up for this purpose in developing countries. The wheels would be used for barrows, wheelchairs, hand carts and bullock carts.

Wheeled transport in India is a pressing need. Although in some crowded places in India, the only space for carrying a load is in a basket on the head, there are many places where a single-wheeled barrow would allow the wheel to carry the majority of the load. For this to be practical in India, the barrow must be cheap. Bicycle wheels can be, and are, used to carry heavy loads, but they are too expensive and the tires damage easily.

Of perhaps more pressing importance, it is estimated that 14 million more bullock carts will be needed in India over the next decade. The traditional 4 ft diameter wheel costs approximately 500 Rupees, or about 100 day's pay for a labourer. I estimate that my design can be made for 1/3 of that amount, so the savings are significant. The major unknown factor is the price that a worn-out truck tyre is sold for to people who cut out the sides and then sell the annular rings to other people who convert them to sandals.

Traditional wheel making requires a craftsman, sometimes with skills passed from father to son. With my design, anyone having the skill needed to be able to mend a puncture in a bicycle tire could make wheels, if given minimum instruction, the required components and a simply assembly jig.

I estimate that my design uses less than 1/8 of the timber used in a traditional bullock cart wheel. If multiplied by even 1/10 of the numbers of bullock cart wheels anticipated to be necessary in the next ten years, the savings of wood and forests would be large. With my design, only the ends of the spokes have a function, so the sides can be rough sawn. As most timbers are strong in compression along the grain, most timbers can be used for my design of spokes.

While travelling overland from London to Calcutta in 1973, I became aware of the need for low cost wheels in developing countries. Since then, I have spent a large part of my spare time making various types of wheels for slow speed use. My objective was to design wheels suitable for manufacture in cottage industries in these countries, using the cheapest materials and requiring only easily learned skills to manufacture them.

The novel feature of my wheels is that the 'wire', which is in the bead of all automotive tyres, is put into tension and formed into a polygon by using radial spokes of the correct length. The spokes are held in compression against the hub at one end and the tyre bead at the other end. This is the inverse of the means by which a bicycle wheel takes its load, where the rim is in compression and the spokes are in tension.

My original bullock cart wheel used a 7.50 x 20 tyre, and required dismantling the wheel to mend a puncture. The wheel now being endurance tested at Melbourne University uses a 9.00 x 20 truck tyre, and allows the interior tube to be removed for repair without dismantling the wheel. (This is achieved by putting only one of the beads into tension to hold the spokes in compression; when the six retaining plates are removed, the 'free' bead can be pulled axially to remove the tube for repair.)

This later design has been carrying a load of 770 kg for a distance (on rollers) of approximately 15,000 km without any major component failures. (Some of the 5/16 inch diameter retaining threaded rods failed, and have since been replaced with 3/8 inch diameter rods.) The average bullock cart travels about 3,200 km per year.

Similar considerations apply to barrows and hand carts, with the concept of used tyres forming the basis of lower cost wheels, and their commensurate benefits to people and transport.

The wheel itself utilizes an arrangement needing only one 'needle bearing' (60 pence in U.K., 1980) for each wheel, simple pipe for axle and hub, wood spokes and the tyre beads (one of which provides the necessary tension to the structure). In India alone, the savings in destruction of forests, the widened access to wheeled transportation on the part of people who cannot now afford the cost of standard devices, and the benefits to be gained by widely dispersed use of these low cost wheels, would be considerable.

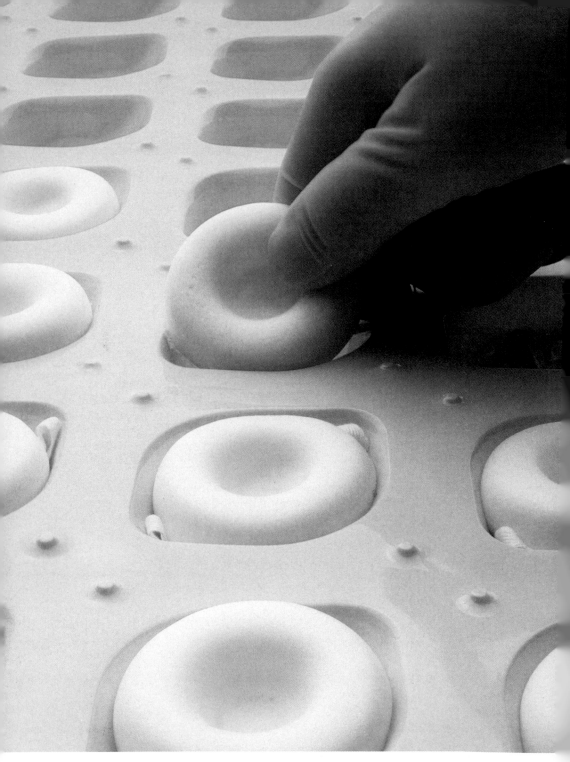

The new, and highly successful, VLI Vaginal Contraceptive Sponge; a potential break-through for Third World family planning.

Third World Family Planning — A New Alternative Method

Bruce Ward Vorhauer

17902 Butler, Irvine, California 92715, U.S.A.

American, born November 22, 1941. President and Chairman of the Board of VLI Corporation, Cosa Mesa, California. Educated in U.S.A.; Ph.D. (Biomedical Engineering) from West Virginia University (1968), M.B.A. from Northeastern University, Boston (1973).

This project will develop the technology and systems necessary to transfer a new contraceptive technology, the VLI™ Vaginal Contraceptive Sponge, to lesser developed countries (LDC's). This method is being received with wide acceptance as an alternative to systemic (pills and injections), implantable (IUD's) and surgical (sterilizations) methods of family planning in the developed countries, but it is still too expensive for most LDC's. This project will concentrate on simplifying and economizing the manufacturing process and packaging to appropriately transfer the technology to the 800 million LDC women of childbearing age in dire need of such an alternative.

Since 1976, and funded entirely with private capital, I have been developing a new type of vaginal contraceptive, the "Today Contraceptive Sponge". This new contraceptive is most analogous to a disposable diaphragm; the major distinctions are that only one size is needed, it has the spermicidal compound built into the sponge during manufacturing, and it can be worn continuously for two days, then discarded.

The Today Contraceptive Sponge works in three ways; 1) It inactivates the sperm by continuous release and availability of spermicide; 2) It acts as a barrier by blocking the cervix so the sperm cannot enter, and 3) The absorbent sponge soaks up semen, reducing the number of sperm available to enter the cervical canal.

Unlike oral contraceptives, the Today Sponge is non-hormonal. The active ingredient, nonoxynol-9, is a safe and highly effective spermicide which has been in widespread and continuous use for more than twenty years in other forms of intravaginal drug products.

Interestingly, the vaginal sponge method is not a new idea — the first mention of such a technique was of a sea sponge and/or tampon soaked in citric juice (a mild spermicide) in the Egyptian Ebers papyrus of 1550 B.C.! Even today, crude tampons of sea sponges, plant substances, and often rags are used in lesser developed countries for contraception as well as during menstruation. With the availability of new biocompatible polymers such as that used in the Today contraceptive, the irritation and toxicity problems of these crude materials can now be eliminated.

The Today Contraceptive Sponge was approved by the United Kingdom health authorities for marketing as a new drug in mid-1982, and by the United States Food and Drug Administration in March, 1983. It offers distinct benefits over contraceptive methods currently available:

— Unlike the diaphragm, the one-size-fits-all 2-inch-diameter disposable sponge does not require fitting by a physician, and can therefore be purchased without a prescription.

— The spermicide is self-contained in the hydrophilic sponge and released during the wear time of up to two days — even with multiple acts of sexual intercourse. Thus, there is never a need to apply messy creams, jellies or foams, which, when used with a diaphragm or alone, are only effective for one act of intercourse.

— This unique method of dispensing spermicide means there is no need to interrupt sexual activities or destroy spontaneity.

— The sponge is completely reversible — no after effects with use.

— In laboratory tests and more than five years of human clinical trials worldwide (over 2,000 women), there have been no reports of significant side effects such as those associated with the pill or IUD (intra-uterine device).

— Clinical trials have proven that most couples are not aware of the sponge's presence as they are with diaphragms, condoms and foam. This is because the sponge's soft texture feels like vaginal tissue. The sponge also be used for contraception during menses.

— High effectiveness (Life-table at 90%), which is essentially equivalent to the diaphragm and other vaginal contraceptives, has been scientifically and statistically proven.

The Today sponge is simple to use and is easily inserted/removed by the woman or her partner.

Since it can be inserted up to two days in advance of intercourse, it allows sexual spontaneity — lack of such spontaneity is the major drawback of other vaginal or 'barrier' methods such as the foaming suppositories or condoms.

All of the clinical testing has been conducted independently by Family Health International, a non-profit family planning research organization formerly affiliated with the University of North Carolina, in Research Triangle Park, North Carolina. These trials have been funded entirely by the United States Government, Agency for International Development and National Institutes of Health.

The U.S. FDA and the U.K. DHSS have scrutinized and approved the extensive New Drug Approval applications we have submitted to them. Additional approvals are now issuing in major countries worldwide (e.g., Switzerland also approved, in February 1983).

While the product is receiving enthusiastic acceptance in the developed countries, it is still too expensive (about U.S. $1.00 at retail) to achieve significant use in the Third World. VLI has contractually agreed to sell the Today Sponge at low cost (essentially non-profit) to the U.S. State Department for free distribution to the lesser developed countries. But, with the sophisticated and redundant manufacturing and packaging systems which are required and economically possible for developed country markets, the price, even at cost, is still significant.

In order to develop this unique product, it was necessary to be innovative not only in clinical development, but also in manufacturing techniques. In its final form, our manufacturing process is a compact and modular microprocessor-controlled system that can be established in repeatable units anywhere in the world.

This project would use the Rolex Award funds toward the design of both manufacturing and packaging systems more appropriate for the LDC markets, thereby achieving lower costs and making the method more feasible for the approximately 800 million women of childbearing age (14-45) in the Third World.

Specifically, a containerized manufacturing line in which the container itself would serve as the production "clean-room" could be developed, substantially improving the portability of the system and making local manufacture feasible. Further, simplified packaging could be designed, tailored for specific markets. (The FDA-approved packaging is currently the most expensive component of the product.)

Please understand that these proposed developments are complex and would involve considerable exploration and innovation. It is true that the major hurdles have been passed in developing and proving the Today Contraceptive Sponge; funds expended to date exceed U.S. $6 million, not including the clinical testing conducted independent of VLI. However, the recognition and support of the Rolex Award would substantially encourage the move to make this contraceptive alternative more available to those who need it most.

The human race — now at 4.6 billion people — is expanding at the alarming rate of 146 people per minute, 8,790 per hour, 210,959 per day..., 77 million each year. India alone increases by over one million people every month. This rapid expansion continues despite wars, starvation, overcrowding and birth control advances. And it is seemingly unaffected by an increasing worldwide concern. Social and political pressures are likely to increase dramatically as migrations to urban areas continue. By the close of this century, more than half of the world's people will live in large cities.

The growing tensions caused by overcrowding and the increasing pressure on available space and resources could increasingly pit poor nations against rich, region against region and race against race, adding significantly to global instability.

Understandably, the potentially devastating effect of increasing population growth has become a matter of grave concern to political leaders worldwide. Family planning programs, while certainly not new to international politics, are gradually becoming a central issue in many parts of the world. In spite of progress in the area, there is today a growing mistrust of birth control methods that are taken internally, injected, implanted, or that involve systemic changes in the human body. In many parts of the world, there is resistance to any birth control methods which threaten the user's privacy or the personal control of the user over his or her own body. For these reasons, women all over the world have been turning to non-systemic methods of contraception.

The Today Sponge can be an effective answer in the search for a solution to the population growth problem.

Does Motion of a Light Source Affect Velocity of Emitted Light in a Vacuum?

Marija Sesic
Zmaja od Nocaja 13-A, 11000 Beograd, Yugoslavia

Yugoslavian, born March 10, 1952. Graduate student in Ph.D. programme. Educated in Yugoslavia; M.Sci. (Materials Science) from University of Belgrade (1983).

According to extant scientific thought and Einstein's Special Theory of Relativity, the answer to the above question is simple — no! The motion of the light source, as it is considered, does not affect the velocity of the emitted light. The First Postulate of Einstein's Theory of Relativity foresees the validity of the so-called additional theorum:

$$C + V = C \qquad (1)$$
$$C - V = C \qquad (1a)$$

V = velocity of the light source
C = velocity of the light in a vacuum

These relations are valid for the motion of light in a vacuum, and do not refer to the propagation of light in the optical mediums. The so-called Doppler-Fizeau shift, the lengthening or shortening of the wave length of light (the red/blue shift) emitted from the movable source, effects only a change in the wave length, though not in the velocity of the light. In the case of the Doppler shift, we can apply the traditional relations which are found in any elementary physics textbook. Depending upon whether the light source is moving toward or away from the observer, there will be an observed change in the wave length. We also know that the velocity of light in a vacuum is the function of the wave length, and that the frequency of light is an important characteristic since it represents the measure of energy of the photon of a given wave length.

Logically, it follows that the photon (or light wave) undergoing a Doppler shift should have changed its frequency (and energy) in order to satisfy the above relationships. On the other hand, it is considered that the frequency of the light which passes through optical mediums does not change, though both its velocity and wave length change, as presented by Snell's rule on refraction of light in the case of transitions from a vacuum into an optical medium, or through any two mediums. Moreover, it has been experimentally ascertained (as early as 1890, Lecher) that while passing through different mediums, the frequency of electromagnetic waves does not change, though both their velocity and wave lengths change.

The possibility that electromagnetic waves, i.e., photons, i.e., light, always maintain their frequency while passing through different mediums provided us with the idea of attempting to test, through specially designed equipment, whether or not light that has undergone Doppler shifting maintains its frequency. If this were shown to be the case, it would reopen the discussion on the velocity of light and the First Postulate of the Special Theory of Relativity.

The equipment to be be used in this experiment is a modification of that used in the 1900 experiment by Byelopolsky in which he achieved the Doppler shift on earth. A light source, based on excitation of a metal vapour under low pressure, emits a monochromatic light with a given wave length. This light beam is directed in between a pair of very rapidly rotating wheels, whose prismatic side surfaces are highly polished mirrors. As the mirrors move toward one another with the relative velocity of 2v, the reflected light beam will undergo the Doppler shift for every reflection of each particular mirrored facing. The total relative velocity of any beam leaving the the system of rotating prisms will be 2vN (N = the total number of mirrors). The wave length of this light will then change. Depending upon the direction of the rotation of the wheels, the light will undergo either the blue Doppler shift or the red Doppler shift.

The emitted light beam with the altered wave length, after leaving the system of rotating prisms, heads for an ampule filled with vapour of the same metal used in the light emitting source. Interaction of the emitted light with the altered wave length, and the vapour of the metal in the ampule could be:

1. Photons of the emitted light, with altered wave length, are absorbed in the ampule's system (exciting atoms of the vapour in the ampule), or.

2. Photons of the emitted light, with altered wave length, are not absorbed in the ampule's system (could not excite the atoms of vapour in the ampule).

We will analyze each of these two cases separately:

Case 1)   If the Doppler shift *did not* cause a change in the frequency of light, the the absorption of photons (excitation of atoms in the ampule) with the altered wave length will occur in the same way as if these were photons with the unaltered wave length associated with the metal vapour. This can be registered by an absorptive spectrometer (or, in the case of excitation, by an ordinary spectrometer). Such result would mean that the Doppler shift affects *only* the change of wave length, and not its frequency, thus raising the question of the constancy of the velocity of light.

Case 2)   If the atoms of the metal vapour in the ampule do not absorb the photons with the altered wave length, this would mean that the Doppler shift affects the change in both the wave length and the frequency, and the velocity of light is indeed constant.

Such a result from our experiment would be the *first* confirmation of the additional theorum, i.e., of the First Postulate of Einstein's Special Theory of Relativity in terms of *both classical and quantum mechanics*.

"Energy Carpets": A Source of Renewable Energy

Hemant Madhukar Ranadive
Hetkari Mahajan Wadi, Ranade Road, Dadar, Bombay 400 028, India

Indian, born January 13, 1950. Vacuum Pumps Sales and Service Engineer. Educated in India; Diploma in Mechanical Engineering (1970).

When an ordinary human being walks on a road, he or she exerts a certain amount of force on the ground. The amount of that force depends upon:

1) The weight of the person,
2) The style of walking of that person,
3) The speed of walking.

If this force were to be converted into useful work, there would be a potentially valuable new source of energy available to the world. In my opinion, it is a valuable source of gravitational energy untapped until now. There are two immediately apparent means of converting these forces into useful energy; 1) Pneumatic or hydraulic systems, or 2) Piezo-electric systems. In this project, I am concerned only with a system for converting these readily available forces through pneumatic means.

Consider a normal man weighing 60 kilograms, walking 100 steps on a flat road. In this process, he is transferring his weight of 60 kilograms from one leg to the other 100 times. That means he has 'compressed' the road 100 times with a force of 60 kilograms. In fact, the 'compression force' is slightly greater than this, due to the added impact of the foot coming in contact with the road.

Now, if this force of walking man can be converted into useful work, there is an enormous source of energy available for mankind.

This wasted human energy can be converted into useful work in the following manner.

Consider a small air compressor of the following specifications: 1) A piston having a cross sectional area of 10 square centimeters, 2) A piston stroke of one centimeter, and 3) A clearance volume of zero. This gives a swept volume of the compressor of ten cubic centimeters per stroke, or, in the case of your walking human, for each step placed on the piston. The maximum pressure of the compressed air can be around six kilograms per square centimeter. This results in the swept volume of 100 steps equalling 1,000 cubic centimeters.

Calculations of work done via the compression of this air result in the figure of 22.23 kilograms/meter/100 steps. If we now consider a crowded place, such as Church Gate, a western railway train terminus in Bombay, we can anticipate some startling numbers.

Almost one million people move in and out of this station in the peak hours, i.e., from 9 A.M. to 7 P.M.. Let us assume that they walk at least one hundred steps before coming out of the station. The "work" done by these people is 22.23 x 1,000,000 kilograms per meter in ten hours!

Now, the above is only theoretical. In fact, when a person walks, he lands on his heel first. Then the weight is transferred from heel to toe. The actual work done would be more than that calculated above.

As it is very difficult to calculate the exact amount of work to be expected on a theoretical basis, this project is concerned with an experimental means of determining what useful energy could be extracted from actual conditions. I wish to develop a 'piece of land' of known area, containing closely spaced cylinders. This construction can then be placed in a heavily crowded area, where tens of thousands of people will traverse it on a regular basis. With the help of proper metering devices, actual values of pressure and delivery of compressed air can be obtained, in terms of 'per unit area'. Then, the actual energy available for any given area of interest can be found out.

Hereinafter, let us call this device an 'energy carpet'. These energy carpets could be placed in crowed areas such as railway stations, airports, bus stands, theatres, big shopping areas, entrances to office buildings, etc. (There would even be an extra advantage in that heat generated from the air compression could be used to warm buildings.)

In my design for the energy carpet, the piston to be used is almost six centimeters in diameter, or about the size of a normal heel. An ejector spring for returning the piston to surface level after compression maintains the piston in a constant state of readiness for further compression. Arrangements for dust-proofing and water-proofing the cylinders will need to be tested under field conditions.

Given the prospect of the useful amount of work that could be freely and regularly achieved through 'energy carpets', I wish to test an initial installation in order to gain practical knowledge regarding costs, maintenance and feasibility. In theory, such energy carpets, once perfected, could continue to provide pollution-free, inexpensive energy for very long periods of time.

fontain ✳ herzog

Stunning examples of a uniquely complex and sophisticated range of beautiful fountains that change colors, shapes, and will 'dance' to electronically programmed music.

Beautifully Illuminated "Dancing Fountains"

Johann Herzog
Hof 10, A-5310 Mondsee, Austria

Austrian, born May 12, 1924. Constructor of fountains. Educated in Austria; with three years apprenticeship as an automotive mechanic.

My invention is a fully automatic, electronically controlled, high jet fountain. The control mechanisms for my devices make it possible to create beautiful and intricate 'fountain programs' lasting up to one hour in length.

Description
The fountain head contains 20 different nozzles which move synchronically in all directions. As the fountain head itself is capable of rotation, rotating figures are able to be constructed in the air. Specially devised means of controlling water pressure through the nozzles allows for the height of the fountain figures to be between one and ten meters high.

A further program provision illuminates the figures with colored lights in a virtually endless variety of color combinations. From time to time, the fountain head itself submerges, which produces the effect of bubbling jet figures.

Construction
The construction consists of the fountain head, surrounded by the colored lights, all mounted on a stand which is securely placed within a water basin. The installed unit is connected to a master operating console, in which the control mechanisms are contained. These include electronic steering equipment, a magnetic tape, a colored light mixer, as well as various switches, pumps, a time switch and fuses. The overall construction is supplied with pressure from a five horsepower pump. The material used is Hydronalium and bronze with a sintered plastic covering to protect against corrosion. The necessary steel parts are hot-dipped galvanized, and hammer-effect enamelled. All of the color filters are unbreakable, covered with hard glass and sealed against water pressure.

Background
I have been working on my invention for more than thirty years, and am still in the process of improving and perfecting in.

I started the invention as a result of a tragic accident in the summer of 1951.

The "SIGN WRITER™" NEWSPAPER

Valerie J. Sutton

The Movement Shorthand Society, Inc., Center for Sutton Movement Writing, P.O. Box 7344, Newport Beach, California 92660, U.S.A.

American, born February 22, 1951. Publisher/Editor of SIGN WRITER™ NEWSPAPER, Inventor, Society founder/president. Trained as a ballet dancer in U.S.A. and Denmark.

The SIGN WRITER™ NEWSPAPER writes the news in different sign languages, using Sutton Sign Writing®, and in different spoken languages using the Roman alphabet. The first newspaper of its kind, it employs deaf reporters who interview and write in sign language. Sign language is the fourth most used language in the United States, and the statistics must be similar in other countries. There are many sign languages in the world, and Sutton Sign Writing® is an international way to write them all.

20,000 copies of the SIGN WRITER™ NEWSPAPER are distributed free to 41 countries, the majority sent to sign language users in the United States and Denmark. At present, it is a quarterly, with hopes of more frequent publication with increased funding.

I came to Sign Writing® and the problems of deafness (I am not deaf myself) through my invention of Dance Writing™. Trained as a ballet dancer, I went to Denmark in 1970 to dance and research the Bournonville ballet technique unique to Denmark with Edel Pedersen, oldest exponent of this 19th century training. Inspired to notate what was being lost, I invented Sutton Movement Writing and Shorthand, and wrote its first textbook in 1973. In 1974, I returned to teach the Royal Danish Ballet Company to read and write dance.

Since that time, I have dreamed of a newspaper written in sign language that deaf people could read with minimum training. I knew the detailed form for research I developed at the University of Copenhagen would have to be modified for lay people to enjoy. And I understood that a self-teaching text was mandatory before the dream could happen. In 1981, first the book and then the newspaper came out. In 1983, response is enthusiastic and deaf people cut off from words now have new stimulation and opportunity.

Why Write a Newspaper in Sign Language?

People who are born deaf, or are pre-lingually deaf (become deaf at a very young age, before developing spoken language), often have trouble learning to read and write spoken language. Learning to read is not just dependent on eyesight. It is necessary to *hear* the sounds each letter in the alphabet represents. Children who hear learn to read by sounding out each letter and piecing the words together. A born-deaf child is deprived of that and must memorize the letters without sound recognition.

It is estimated that only 5% of the born-deaf population go to college. Those 5% are the exceptionally gifted individuals with whom hearing people come into contact. But what about the other 95%? Some of them lead isolated lives away from hearing people. They may have trouble reading a newspaper, because their reading level may be quite low, and they are cut off from television and radio because of their deafness. Some deaf people do not know about world events and crave and need contact with the world.

The SIGN WRITER™ NEWSPAPER gives them that contact with the world. It provides information written in both the sign language and spoken language used in their country (currently English and Danish). The world is opened by the knowledge that their own native language can now be put onto paper. For many, this alone has given them a feeling of added self worth — to know that their language can now be side by side with spoken language in print. Many deaf people felt ashamed of sign language in the past, and that attitude is changing. A written form has given their language added respect.

Specific Goals for the Newspaper

The goals of the newspaper are: 1) to teach reading to deaf people with minimal language skills, 2) to teach sign language to hearing people, 3) to teach foreign sign languages to deaf people, 4) to educate hearing and deaf people about each other, 5) to inform deaf people, cut off by illiteracy, about the news in the deaf community, 6) to give deaf people training, jobs and careers in sign language journalism, 7) to preserve sign languages and the unwritten literature of the deaf community, and 8) to provide a tool for sign language researchers in the study of natural languages of deaf people and the grammar and syntax of languages heretofore never written.

The SIGN WRITER™ NEWSPAPER is Providing Jobs and Careers for Deaf People

We have created a new profession called Sign Language Journalism. Never before could a deaf person get a job writing articles in his or her own sign language.

The newspaper provides free training for deaf people, from learning Sign Writing® through to learning about layout and design, interviewing people, getting advertisers for the paper, and the production of a paper. There are now seven deaf reporters on the newspaper staff, five from the U.S.A. and two from Denmark. The reporters interview people using sign language in person, by mail, or over a tele-typewriter. These interviews are first written in Sign Writing®, and then the staff translator puts them into spoken language.

At present, the Sign Writing® symbols in the entire newspaper are written by hand. The deaf Assistant Editor of the newspaper copies the Sign Writing® in each article with a fine ink pen for publication. Computerization of Sign Writing® is now developing. Soon the newspaper will be typed by computer. Until that time, our deaf staff members devote many hours to making the newspaper look beautiful. It is a labour of love.

We hope that with further funding more deaf people can benefit from this work.

Fantasyscape with acrylic *(Top)*
Photomicrograph of shattered acrylic under polarized light, and telephoto
picture of sun with 200 mm lens through red filter; double exposure.

Moonrise, Panjab University *(Bottom)*
Photomicrograph of crystals of sodium thiosulphate under polarized light,
telephoto picture of moon with 200 mm lens; double exposure.

A Search for Art with Microscope and Camera

Ashok Sen-Gupta

Gastroenterology Division, Beth Israel Hospital, Harvard
Medical School, 300 Brookline Avenue, Boston, Mass. 02215, U.S.A.

*Indian, born July 1, 1948. Assistant Professor, Department of Biophysics,
Panjab University, and Chief of Optical Microscopy and Photography for
the Department. Educated in India, Ph.D. in Zoology from Panjab
University, Chandigarh in 1976.*

This project has four objectives. (1) It envisages a detailed, full-time exploration of the possibilities of using the microscope+camera combination to generate art images. The methodology involves "fusion" of two or more images, at least one of them photomicrographic, *or* the use of the microscope's controls alone to alter light so that graphic forms are generated. Several thousand such pictures will be made, and the best will then (2) form the material for a book entitled "Art Through Camera and Microscope", and (3) a travelling exhibition of the same name. (4) Lastly, a device is under development that allows the photographer to fuse two or more images on the microscope in any fashion he desires. It is my aim to perfect the device so that it can be brought within the reach of any practising microscopist for a sum not to exceed U.S.$50.

The project is already under way, and examples of work are enclosed.

Like many microscopists before me, I realised early as a biological researcher that the microscopical image can be both informational and beautiful. Many years elapsed, however, before I realised that a new and ethereal dimension could be imparted to microscopical objects by superimposing them against images of the outer world. My initial success rate was very low, mainly because of problems of standardization and lack of a good microscope and camera.

These barriers have been surmounted now that I am in charge of a sophisticated Carl Zeiss Jena large research microscope, which allows me to work in brightfield, darkfield, phase contrast, polarization, fluorescence and interference modes, with choice of epi- or trans-illumination, or both. By this time I have also purchased a Nikon F2 camera with microscopy viewing screen type M, and been able to calculate the precise exposure correction needed when the Nikon camera is coupled to the CA microscope via an improvised adapter. It is essential to use the Nikon because the microscope's own camera does not take normal, wide angle or telephoto lenses and does not measure exposure or permit multiple exposures: these are all indispensible to the idiom of the present work.

Early work consisted mainly of marking a spot on the camera's viewscreen and placing the first object (such as the sun or moon shot through a tele lens) there. Then the film was rewound and the camera placed on the microscope. The microscope image was then located in such a way that the marked area was left in darkness, and the double exposure was made.

However, it will be apparent that all microscopical objects do not leave dark areas (most leave *bright* backgrounds) and certainly not many leave dark central areas even under polarized light. Thus the necessity arose for a device that would allow blocking of the microscopical image centrally or peripherally, regardless of bright or dark backgrounds.

While it is a simple matter to block out part of the scene in front of a camera lens with a hand or cardboard, it is very difficult to block out any part of the microscope's image field because nothing can be placed on the eyepiece without ruining the image. With time, though, I have been able to work out a method for suspending a spot between eyepiece and film, and can now block out any area, regardless of size, from the microscope's image field; this area can then accommodate a second image, microscopic or otherwise. The method at this time is extremely difficult and awkward, but it works.

Furthermore, it was realised during the course of this work that the microscope itself offers fascinating possibilities of creative control through its own devices, regardless of the principal object. Thus, severe decentration of condenser and lamp, normally a serious fault, could create a novel "sun" over a "landscape" of broken acrylic.

Ultimately, perhaps even the broken acrylic is not needed. Ernst Haas stated many years ago about his abstractions that "...it is not necessary to have a subject, because light itself is the subject..." and that applies fully to the microscope as well. Thus I have recently begun to work on the "subjectless" pictures, where the intention is purely to photograph light, be it focused or defocused, altered or unaltered by filters, and demarcated only by the simplest of devices, such as pinholes or a scratch in a piece of tinfoil.

For such work, the microscope is replete with creative possibilities. Its mechanical stage can offer precise, graduated movements for multiple exposures, while its pancratic magnification changer (zoom eyepiece) can "explode" objects in graduated steps or steplessly. Its rotating stage can spin objects around concentrically or eccentrically; its condenser can be raised or lowered and filters can be interposed in its light path at will. Besides, the same object can be seen in a multiplicity of modes, and be photographed on the same piece of film.

I believe it is possible that the microscope is the most powerful tool for creative photography, and one whose potentials we have barely begun to explore in this field.

The project is already five years old and crippled for want of time (owing to heavy teaching and research commitments) and funds. The situation is made harrowing in light of our limited earnings and the Government's policy of treating photographic equipment as luxury goods dutiable at 320%. Thus a Durst 'Laborator' enlarger costs over 70,000 rupees after duty and agents' commision. A simple color analyzer, which sells in the U.S.A. for $130 (about Rs 1,300) is selling here now for Rs 6,000. For a man earning Rs 2,000 per month, completion of this project within a lifetime is impossible without aid. Unless help arrives from somewhere, this project will surely die.

My University permits me to take 26 months' leave without pay, which I will avail myself of, if aid is available, to devote my full time to this project. I plan to make several thousand art images and select the best for publication in book form. Each illustration will bear accompanying text. Large versions of these illustrations will make up the exhibition. Lastly, the microscope "blocking and superimposition" device will be perfected and offered to independent lens makers for manufacture. In order to make it accessible to my fellow microscopists everywhere at low cost, it will *NOT* be patented.

Using Bacteria to Restore Ancient Monuments

♛ K. Lal Gauri

Honourable Mention – The Rolex Awards for Enterprise – 1984
Department of Geology, Natural Science Building, Belknap Campus,
University of Louisville, Louisville, Kentucky 40292, U.S.A.

*American, born October 16, 1933. Professor and Chairman, Department of
Geology, and Director of Stone Conservation Laboratory, Unversity of
Louisville. Educated in India (B.Sc. and M.A.) and Germany (Dr. rer. Nat.
from Bonn University in 1964).*

In recent times, the intensified attack of sulfur dioxide on marble has disfigured many ancient buildings and monuments. If means are not found to stop the destruction brought on by the affluence of the industrial society, the residual stock of human heritage in architectural art will soon fade away. It is incumbent upon modern science to discover some means to arrest the ruthless rotting of precious structures.

The effects of sulfur dioxide attack on marble are expressed as black gypsum crusts which eventually exfoliate, causing massive destruction of sculptural forms. They are also expressed in the less obvious scouring by acid rain, which reduces reliefs to unrecognizable hunks of weathered rock. Between the black crusts and the acid scoured surfaces, the brown surfaces are future martyrs in the making. In the less industrialized countries of the world, the marble monuments are chemically in an excellent state of preservation.

Present technology lacks the methods to treat the gypsum to recovery. Gypsum may be removed by treatment with hydrofluoric acid and ammonium bifluoride; chelating agents in surface-active poultices; air, water, steam, and dry- and wet-grit blasting; laser beams and gamma rays. However, some measure of caution is in order. These methods clean the marble, but in the process they cause massive surface reduction. The only hope of treating sulfate-crusted objects lies in the *in situ* conversion of gypsum into its parent calcite. Nature has already revealed the mechanism for this in the epigenetic conversion of hundreds of feet of gypsum and anhydrite into calcite ($CaCo_3$) in the cap rock of the salt domes of Louisiana and Texas.

Salt diapirs are masses of deformed halite that have intruded into the country rock. The cap rock of the ore halite mass consists of, in ever-younger sequence, a granular anhydrite rock followed by a sulphur-bearing limestone transition zone, followed by barren cavernous limestone. This salt plug, i.e., the halite-anhydrited-calcite sequence, is enveloped by the rock. It is a matter of common geologic knowledge that the halite and the anhydrite are coprecipitates from a saline basin, and that the presence of calcite is, from the point of view of sediment genesis, a rather anomalous phenomenon.

Most of the salt domes occur at depths of 100-1000 meters from the earth's surface. In Culberson County of West Texas, however, more than one hundred hills of limestone rise 3 to 30 meters above the Gypsum Plain. The limestone possesses such features as lamination, microfolds, brecciation and crystalloblastic texture. These are the characteristic structures of the underlying anhydrites. One wonders whether the calcite has chemically replaced the anhydrite while preserving the structures of the parent rock.

The major petroleum deposits of the Gulf coast are associated with the salt diapirs. It has been suggested by many authors that the hydrocarbons and other organic materials contained in the crude oils are capable of reacting with anhydrite, though the process has been calculated to take more than 150 million years to cause a direct reduction of sulfate. The geological evidence — salt domes penetrate Pleistocene Strata — suggest that the formation of sulfur must have taken place in the last million years.

The sulphur-reducing bacteria *(Desulphovibrio desulphuricans)*, by providing enzyme catalysts, have the capability to enhance the sulfate reduction at the more rapid rate of 1000 mg per liter in 24 hours. Growing in an anaerobic environment, they utilize sulfate ions, rather than free oxygen, to oxidize carbon compounds, which are their energy source. When introduced into the sulfate rock in the anaerobic environment of petroleum, they are able to dissociate sulfate from anhydrite and provide the organic carbon for conversion of anhydrite into calcite. That this really is the case is supported by studies on the fractionation of sulphur and carbon isotopes. The conclusion is that limestone deposits are bioepigenetic, in having originated by alteration of lithified gypsum and anhydrite. As seen in the buttes of Culberson County, this limestone preserves the textures and features of the parent rock. This leads us to the proposal to convert gypsum crusts on the surfaces of monuments into the parent marble by the agency of sulfate reducing bacteria in an anaerobic environment of petroleum.

We plan to carry out bacteriological investigations on the rate of growth of the genus *Desulphovibrio* and on the reduction of sulfate by these bacteria. Petroleum-sulfate reaction in the presence of *Desulphovibrio* will be examined for rate of reaction in several ways. Petroleum poultices containing the microbe will be applied to the gypsum crust of weathered marbles that have been prepared for later subjection to various instrumental techniques designed to provide measures of reduction of gypsum and increment of calcite counts. In addition, the application of the poultices to accessible marble monuments is planned, and the conversion, if any, of gypsum into calcite studied.

Extending a Wave-Powered "Seawater-to-Freshwater" Conversion System to Arid Coastal Zone Agriculture

Charles Michael Pleass

University of Delaware, College of Marine Studies, Newark, Delaware 19711, U.S.A.

British, born March 23, 1932. Lecturer at University of Delaware. Educated in U.K.; Ph.D. (Chemistry) from University of Southampton in 1955.

I have recently demonstrated a simple, wave-powered system for producing fresh water from seawater. It is designed to be used in developing countries and coastal zones. One major advantage of the system is that water cost is unaffected by changes in price or availability of fossil fuel or its derivate, electric power. Salt is removed from the seawater by a process known as reverse osmosis. Seafloor pre-filtering is used to gain long unit lifetimes, and sea trial data suggest very low product water costs, which will justify the use of the system as a source of drinking water and irrigation water. The sea trials are in progress now; the project described is the first step toward experimental coastal zone agriculture using water from this wave-powered system.

One of the major problems facing society is the uneven distribution of resources and population in the world, which causes extreme variations in the supply of food and water. As global population increases, desertification of the planet becomes a major problem, compounded by changes in rainfall patterns.

It is well known that sandy desert soils, such as those in the Sahel, can be very productive if they are irrigated. Since 1976, I have been working toward the long-term goal of providing inexpensive freshwater on desert coastlines using desalination systems powered by seawave energy. I believe that the forced redistribution of population by drought is fraught with social and economic problems. The tremendous movement to cities in developing countries has created false hopes, urban poverty and social unrest. If copious supplies of freshwater could be produced along the eastern edge of the Sahel, for example, new communities could be nucleated around agricultural enterprises that begin in the desert coastal zones, distant from the existing towns and cities. Given our present awareness of techniques for the use of wastes, such communities could become virtually self-sufficient. They would offer adequate food and clean water to people chosing to work in them, and I believe that they would begin to attract people naturally. Simple new community infrastructures could be developed around the agriculture and the water supply, reversing the present-day trend to the crowded, unsanitary conditions and blight of existing cities.

Seawater desalination by the process known as reverse osmosis is essentially high pressure filtration at a molecular level. I have shown that the energy for reverse osmotic seawater desalting can be most appropriately supplied by seawaves, since the energy and raw material for the process coexist, and the development of hydrostatic pressure is a natural method of seawave attenuation. A typical seawave formed in the tradewind regions of the northern and southern hemispheres has a height of 1.3 meters, a period of 5 seconds and a wave length of 45 meters. Each meter width of wave crest will carry circa 7 kw in the direction of movement of the crest. Waves travel for great distances with little loss, and their arrival at a coastline therefore represents the accumulation of energy. In the example given, if the wave power were absorbed with 15% efficiency, each meter width of this wave front would could yield circa one cubic meter of freshwater every six hours.

Many devices have been described for the purpose of harvesting seawave energy. However, until I began my work on wave powered reverse osmosis in 1976, all these devices addressed the western need for electric power. Most attempted to convert wave power into the rotary motion of a turbine and generator combination. The wave powered desalting system that I call DELBUOY combines a very simple seafloor filtration unit, a linear wave powered seawater pump and a reverse osmosis unit. It differs from almost all wave powered systems in that it does not generate electricity, and it differs from all previous reverse osmosis desalination systems because it does not utilize electric or diesel power.

Small prototypes of DELBUOY are now under sea trial in the Caribbean, sponsored by USAID, with encouraging results. More than 1 cubic meter of water containing less that 500 ppm of salt can be obtained each day from a 2-meter diameter DELBUOY in seas with an average height of 0.8 meter and period of 4.5 seconds. Data show that an array of scaled up DELBUOYs should produce water at about U.S.$0.30 per cubic meter. For comparison, local water costs on arid Caribbean islands range from about $1/m³ (Key West) to $20/m³ (Curacao).

Present sea trials involve single small DELBUOYS. The natural progression of this project is to assemble arrays of DELBUOYS that feed larger, more cost effective reverse osmosis modules, that in turn would be linked to experiments in shoreside agriculture on sandy desert soils. At present, I seek to pursue this experiment on the arid coast on the SW corner of Puerto Rico, and to design and install a drip, trickle or hydroponic irrigation system specifically suited to use in conjunction with DELBUOY.

GUESSES	SCORES
B E F G K	2
A B C G M	3
B E F G M	2
C E H K M	2
A C D G M	4

An example of the simplified mastermind-tableau. This one has two solutions; could you find them both?

A Culture-Fair Measurement of Logical Mental Ability by Computer

Francis Martin Mitchell
10 Meadow Close, Stoney Stanton, Leicester LE9 6BX, England

British, born January 20, 1949. Nuclear engineer, using FORTRAN computer programs in simulating nuclear reactor plant. Honors Degree in Engineering Science from Southampton University.

The parlour-game "Mastermind©" has already been used for psychological and intelligence testing. It is 'culture-fair', and ideal for setting up on a micro-computer. A simplified form of this game may be treated mathematically using Matrix Algebra. Thus it may be possible to develop a mental test in which the difficulty of the problem and the effectiveness of the player's response can be measured and correlated in strictly quantitative terms. This could avoid many disadvantages inherent in present-day I.Q. tests.

The existing game is played between a Code-Maker who selects a hidden permutation of colored pegs, and a Code-Breaker who attempts to guess the permutation in successive moves. Each guess is rewarded with score pegs which indicate the numbers of correct colors and placements.

The game may be simplified, however, by ignoring the positioning of the pegs and allowing only one of each color in a combination. This retains the essential character of the game as a test of logic, while enabling it to be treated mathematically in terms of Linear Matrix Algebra.

I have written a monograph showing how a simplified "Mastermind©"-type game might be set up on a microcomputer, using letters of the alphabet in place of colored pegs. In this illustration, the 'hidden code' consists of 5 letters selected at random from a set of 10, and each 'score' corresponds to the overlap between the 'guess' and the 'hidden code'.

The computer would act as Code-Maker, and the human player's logical ability as Code-Breaker would be tested. Within the computer, the successive guesses and scores would be handled mathematically in Matrix form. The problem confronting the human player can be translated into a set of simultaneous equations, and the constraints upon the solution of these equations would enable the complexity of the problem and the effectiveness of the player's move to be evaluated at each stage.

I have already developed the first stage of the process, which would probably make use of an algorithm known as Gauss-Jordan elimination. My monograph gives an example of this, and a program in BASIC for the Sharp MZ80K computer which demonstrates the process. Although the BASIC language is suitable for this kind of experiment, it may be too slow and unsophisticated for further stages in the finished program, which may need to be written in the Assembly Language, or Machine Code, of the Z80 microprocessor. I have already written an Assembly Language program which analyses positions in the standard "Mastermind©" game, and I am familiar with the additional flexibility offered by this form of programming, which utilises the fundamental operations of the microprocessor itself.

In the immediate future, work on this project will be primarily mathematical, and centered on the parametric solution described in the monograph. This summarises the logical possibilities for the hidden code, and ways must be found to quantify the information contained in it. The main quantities to be studied will be: the structure of the set of possible solutions; the number of degrees of freedom involved; the effectiveness of the player's guess, which will depend on the variety of possible scores from the set of solutions; and the time taken to decide upon the next guess.

Assuming the applied maths and computer programming are successfully concluded, I plan to test the system on myself and close acquaintances. This pilot test may well reveal fresh aspects of the problem in practice, and perhaps a need for some additional conceptualisation. It will probably yield a great deal of quantitative data in itself; the difficulty of the game may be varied by increasing the number of symbols in the hidden code or decreasing the time allowed for a player to decide on his move.

The "Mastermind©"-type of game is particularly interesting because it involves not only logic, but some degree of imagination to visualise the possible hidden code, and even creativeness to invent suitable guesses for gaining information about it. Such sophisticated mental processes have been studied more adequately by philosophers than by psychologists in the past, but I believe that the quantitative study of simple games could enable experimental science to draw conclusions about them. If large-scale trials revealed a correlation between logical skills and random skills, this might have profound implications.

If I do not succeed in obtaining a neat mathematical solution to the Matrix model, there is still a backup option. I can take advantage of the speed of Assembly Language to use the 'brute force' technique, with a program which simply examines all the alternative "Mastermind©" combinations. In fact, I had already done much work with a program of this kind when the more elegant Matrix method occurred to me. Alternatively, in view of the large number of combinations involved, it might be necessary to arrange for a representative random sample to be taken. Such a statistical method would avoid excessive computing time in the running of the program, while remaining equally reliable in the long run.

I already have all the equipment which I am likely to need in the foreseeable future, in the Sharp MZ80K which I own. The Gauss-Jordan BASIC program I am using is my own creation, and should demonstrate my familiarity with Matrix algorithms and computers.

Because of the technical nature of this project, I decided to write up the mathematical ideas in a separate monograph. I am hoping that this will enable a reader with some knowledge of Linear Algebra and Matrices to see that I have evolved a workable approach, without going into lengthy detail.

I became interested in the varieties of mental ability while I was an Engineering student at university, and later when I became Local Secretary of

Leicester Mensa, and visited various Mensa groups composed of people with I.Q.s in the top 2% of the population. In both cases, it was clear that academic qualifications and intellectual talent are not inevitably connected. I became convinced that many people, particularly Asians and other immigrants in Britain, would benefit if intelligence tests were regarded as more significant. Now, as the need for microprocessor-related skills increases, I expect that it will become essential to identify people with high logical abilities whatever their level of education.

However, standard I.Q. tests have been denounced by educationalists and others because they have been known to give bizarre results. A person's score can be dramatically improved by familiarity and practise with an I.Q. test; conversely, an inexperienced university student can be scored as sub-average. Some tests depend heavily on the English language, and are thus misapplied when give to immigrants. It has been alleged that I.Q. tests are unfair to people of different races and cultures in any case, leading to controversy about possible racism and sexism. Also, adult I.Q. scales can only be validated statistically; thus they become less accurate and more expensive at the higher and rarer levels of ability.

This simplified "Mastermind©" test itself is limited in scope, but could lead to the development of other similar tests to cover a wide range of abilities. Because the test can be generated endlessly by computer, the testee would have an unlimited opportunity for familiarisation and practice. Depending on the quantitative nature of the results, there might be little need for statistical validation compared with present I.Q. scales. The game of "Mastermind©" is as independent of race, sex, class and culture as can be imagined. Its worldwide popularity indicates an inherent appeal which may counteract the suggested 'poor motivation' of some test subjects; this appeal is shared by computerized games in general, a fact which might lead to easy public acceptance of such tests of ability.

It seems to me that a test of abstract logic, such as this one, relates directly to the kind of skill involved in practical thinking.

Note that the object of my mathematical study is the puzzle itself, and not the human being who attempts to solve it. The Matrix Method is not intended to be a model of human thought-processes in any sense. It is rather an instrument to provide quantitative data about the efficiency of a person's mental perceptiveness. The advantage of the game is that it requires a high degree of conceptual abstraction to solve it, while being strictly quantifiable in this way. The underlying rational of the experiment is that while mental abilities cannot be measured directly like physical abilities, one may set them to work on a problem which can be measured. The result should be a far more immediate measure of mental ability than that provided by an I.Q. test.

Self-Initiated Rescue from Space

Peter A. Rundquist

7235 Lake Drive, Orlando, Florida 32809, U.S.A.

American, born August 2, 1930. Sales Representative, Sun Electric Corporation. Educated in U.S.A.; in Business Administration.

This project seeks to engineer and test the equipment and procedures to enable astronauts to safely return from earth orbit to earth's surface should a spacecraft malfunction preclude normal de-orbiting.

As a Senior Navigator in the U.S. Air Force from 1948-1970, from which I retired as a Major, I held the position of Master Parachutist. From 1965-1968, I was Chief of Pararescue Operations, Headquarters Aerospace Rescue and Recovery Service. From 1968-1970, I was Pararescue Officer, 41st Aerospace Rescue and Recovery Wing. I developed, tested and trained APOLLO Spacecraft Recovery techniques for the U.S.A.F., and presently hold United States Parachute Association expert rating D-7458, with over 370 jumps to date.

My project separates into two phases, the first of which involves development and testing up to the point of high altitude performance, and the second of which involves live testing from a space shuttle orbit.

The details of Phase I include the following:

1)   To engineer and high-altitude test the Orbital Escape Vehicle (OEV) patented by Mr. Caldwell C. Johnson. This ingenious craft is to be made in the form of a flexible material casing, capable of being folded in order to reduce its dimensions for stowage aboard an orbiting space craft. The outer surface of the casing is covered with a heat ablative material, while the inner casing is lined with an insulative material.

A key factor in the design of the vehicle is an inflatable interior 'bladder' that serves two critical functions; A) It contains the 'support' for the passenger, and B) It forms the flexible vehicle into a stable (round) aerodynamic shape for its initial descent from orbital altitude. Additionally, the craft will contain a small, solid-propellant retro-rocket system, a gaseous oxygen supply for breathing and cooling purposes, and be fitted with a small double-panelled window for reference point sighting. The nozzles of the retro system will expel carbon dioxide and used oxygen, allowing the passenger to orient the vehicle for a proper re-entry attitude.

Use of the vehicle will be fairly straightforward. Wearing his extra-vehicular pressure suit and backpack, the crewman unstows the escape vehicle from an external compartment of his disabled spacecraft. He then unzips the 'entrance' port, switches his oxygen supply from his backpack to the internal oxygen system in the OEV, dons the parachute and related survival gear and enters the OEV. He closes the zippered opening, positions himself to work the retro-rocket assembly, and sights pre-selected reference points on earth through the window. Firing the retro-rocket inflates the internal bladder, and forces the vehicle into an areodynamic shape that will provide stability during re-entry

The heat ablating material of the covering will dissipate dangerous re-entry heat until the escape vehicle reaches the lower regions of the atmosphere. At this point, atmospheric pressure will cause the bladder to automatically deflate, signalling to the crewman that he may 'unzip' and exit the vehicle, making a standard jump with his parachute. Survival gear and an oxygen bottle are attached to the parachute harness.

2)	Modifying an existing spacesuit to contain an adjustable harness.

3)	Building a parachute package for attachment to this harness. The package should contain a parafoil type main parachute plus a reserve parachute, as well as necessary survival equipment.

4)	Live test the entire package from high altitude to obtain performance data and operating procedures.

5)	Develop nominal re-entry footpaths to establish visual reference points for the manually operated de-orbit retro-rocket firing.

Phase II Details

After the above testing, a minimum of two tests from the space shuttle orbit are envisaged. I would ride a space shuttle craft into orbit to conduct this Phase, following the steps outlined above. Following exit from the OEV, I would use freefall techniques to track to the best apparent landing area, and then deploy the parachute for a safe landing.

As it was with the Wright brothers, early space ventures required a judicious balance of safety and practicality. Weight, space availability and system redundancy overshadowed a requirement for a rescue recovery capability. As normal flight capabilities increased, the evolvement of parachutes gave pilots the safety margin they needed. So, too, must the space program advance.

Spacecraft weight and space availability have improved to the point that crew safety must again become paramount. While system redundancy is still there, accidents can occur. Past space missions have not been without mishaps. Someday, the degree of seriousness of an accident may negate these redundant systems, leaving spacecraft crews unable to de-orbit. The cost of maintaining a rescue shuttle on standby is prohibitive. The space shuttle program is requiring more crew members and more involved misssions. The day may come when these astronauts could be trapped in orbit by an unforeseen accident. A self-initiated rescue system has been theoretically available for many years. Let's provide it before it is too late.

Deepwell Jet Pumps for Rural Area Drinking Water

Mohan Dhondiram Patil

Anusaya Equipment Engineers, Plot No. 15/196, Industrial Estate, Sangli, Maharashtra 416416, India

Indian, born October 19, 1950. Industrialist, owner of engineering company. Educated in India; B.Sc (Hons.) in Chemistry from Shivaji University Kolhapur in 1972.

In many rural areas, the critical shortage of adequate water supplies for domestic use (drinking, cooking, etc.) is not due to the lack of available water but rather to its inaccessibility. To solve this problem, communities have used centrifugal and submersible, or turbine, pumps, both of which have advantages and disadvantages. This project concerns a new pump invention that has successfully overcome the problems of previous pumping systems, and which is being accepted for use in rural communities in India.

Centrifugal pumps are inherently limited to a maximum lift of about 25 feet, and so are unsuited for deepwell service on their own. In many Indian villages, suitable water lies over 100 feet beneath the earth. Access to this water, until recently, has been only through the use of very costly turbine pumps.

To overcome this problem, we have developed the "VARSHA" Packer Type Deepwell Borewell Jet Pumps. Depending upon the engine power used, these pumps are now providing Indian communities with appropriately scaled capabilities. For example, a one-horsepower engine has a suction lift capacity of 120 feet, a three-horsepower engine draws from 200 feet, and a five-horsepower engine will give water from a depth of 250 feet. The pumps are very much cheaper than submersible or turbine pumps.

The working of a 'jet pump' is initiated by ejecting a stream of high pressure water through a suitable nozzle so that it enters into a large volume of low pressure water and forces it to a higher level in the system. When combined with a centrifugal pump, part of the output from the centrifugal pump is directed back to the bottom of the well through an injector, to boost the flow up the delivery pipe. A limitation of this method of boost is that higher lifts are only obtained at the expense of lowered capacity and overall efficiency. Thus, jet pumps are usually of two basic types: 1) Low-lift, high capacity (most efficient), and 2) High-lift, low capacity (least efficient).

Our "VARSHA" Jet pump is of 'vertical monoblock type' construction. Its electric induction motor is mounted vertically on the pump bearing housing. The two opposed impellers (i.e., the suction and pressure impellers) are fitted to the same shaft of the electric motor within two adjacent but separate casings.

On the delivery side, there is a pressure meter as well as the non-return adjustable valve. The Packer head is a foundation on which the pump is vertically installed; from this same packer the suction and pressure pipes extend into the bore along with the Jet assembly. The Jet assembly consists of the jet body, a footvalve body, a needle, and a Venturi and nozzle.

In this connection, we would like to inform you that we have recently made an invention regarding the testing of the pump portion on our existing test bed, from which we are getting readings in the approximate range of from 30 meters to 70 meters. We have a test report and performance curves of the three-phase, 3 Horsepower Bore Well Pump manufactured by us.

Until now, there is still no one in India who can specify the Testing Procedure for the pump we have developed, and we have thus put the matter before the Indian Standard Institution for resolution. To date, they have also been unable to test the pump and develop appropriate specifications for the testing of jet pumps.

There are other jet pumps on the market, but they are of the 'twin type', which is to say that the suction and pressure pipes are different and are used separately. Our "Varsha" Jet Pump jet is connected by two pipes, one inside the other. This results in our pumps requiring very little space, providing trouble-free operations, and making them very easy to maintain.

We are currently executing a Government order for the supply of Borewell Pumps to 400 villages in Maharashtra, as part of the Drinking Water Supply Scheme in these villages. Having once upon a time been unemployed myself, it is a pleasure to report that, through the development of this pump, I am now providing employment for 80 to 120 workers directly.

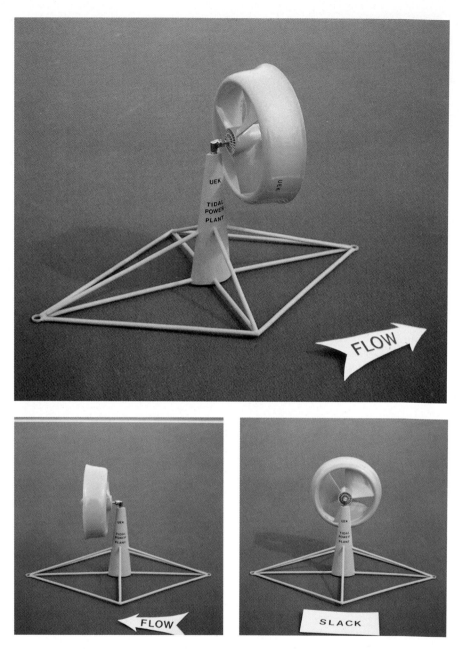

For tapping the energy of tidal waters, the underwater "Electric Kite" (as modelled here) may be a long-sought answer

Innovative Control for an Underwater Electric Kite in a Tidal Environment

Philippe Vauthier

P.O. Box 3124, Annapolis, Maryland 21403, U.S.A.

Swiss, born April 7, 1936. President and Treasurer of UEK Corporation, an energy R&D firm. Educated in Switzerland, Canada and U.S.A.

This project will investigate the control of an ocean-current electric generator to be used in a tidal current. Generation of electrical energy in remote areas near tidal estuaries and river is promising, if control of the machine during slack-tide periods can be assured. Three options for such control have been derived, and will be tested. The best will be selected and designed into a pilot operational system. From this experiment, design information will be developed for use in areas where low-cost, low-technology, reliable, renewable-energy sources are required with minimum environmental impact, using water current.

What is an Underwater Electric Kite?

Attempts to extract energy from nature's flowing waters have resulted principally in the use of turbines for mills and hydroelectric dams where adequate water heads are available. The economical use of ocean-current energy has been elusive, however, because the low current velocity demands energy aggregation on a vast scale. Transformation of the slow stream into a flow with sufficient head to drive a conventional, moderate-speed turbo-electric generator has awaited an innovation in hydraulic systems. Such an innovation has appeared with my invention of the Underwater Electric Kite, or UEK.

Acceleration of a slow stream by the centrifugal force of a rotating structure is a concept which promises simplicity, efficiency and reasonable device size. A large, current-turned structure would accelerate water entering at a hub inlet, flowing outward through radial tubes, and exhausting at low-pressure rim outlets. This raised-head flow would efficiently drive low-cost, conventional generators through turbines mounted in the stream. The electrical power would be routed through slip rings and the hub's tethering cable to the anchor system and then through cables to the shore. This concept and associated systems considerations are now under development.

Results of calculations and experiments in towing a 1.2-meter diameter model have show that a full-scale device anchored in the Gulfstream is feasible to generate tens of megawatts using today's technology. Experts have agreed that building a prototype of this "Underwater Electric Kite" is the next practical step toward realizing the energy potential of stream currents. Progress is now being made toward an analytical model and engineering solutions to the many challenges of sizing, construction, emplacement, control and maintenance.

Control System:

In a conventional UEK operating on a tether cable, its position in a steady current is determined by current velocity sensors at the ends of the four control arms, and the structure's depth is sensed by an acoustic sounder system. These data are input constantly to a microcomputer to find the difference between the present position and the desired position. The computer sends error signals to controllers of two cable winches that slightly shorten or lengthen the control cables, in turn causing the structure to dive, rise or move laterally, "flying" it into the highest stream-velocity position.

The systems to cause lateral and vertical movement to search for favorable current, and to maneuver for deployment, retrieval and maintenance have been tested satisfactorily. Design refinements are now progressing.

Description of the Proposed Project:

The proposed project is an investigation of a control system for an Underwater Electric Kite when used not in an ocean, but in a tidal current. Generation of electrical energy from rivers and estuaries would make the UEK attractive for many remote areas, but such an application requires a different control system because the current varies in direction and velocity with the tidal flow. This project would blend innovation with the characteristics of a conventional UEK system, leading to its beneficial use in a tidal environment as well as in large rivers and ocean currents.

Statement of the Problem:

To operate stably, a UEK depends upon a continuous, unidirectional current of water with only small variations of velocity and spatial position. However, if a UEK were to be deployed in a tidal current with tether cables, at slack tide the slightly buoyant structure could surface and wander out of control, possibly obstructing surface navigation or fouling its own anchorage. Because the UEK was not originally intended for tidal currents, a new design solution is necessary if the highly desirable use of tidal energy sources is to be considered.

Clearly, for the optimal solution, the device must stay submerged, in position, and totally under control during slack tides, which in most locations occur four times daily for about an hour.

Selected Approach, Option A:

The control of a UEK can operate at extremely low velocity (20 cm/sec. or less) because of its relatively large active surfaces. However, in this operating state, energy extraction is practically nil. The slack-tide solution proposed in this option is to reverse the attitude of the machine. That is, operate the generator as an electric motor, activating the second stage turbine and causing water to flow through the machine. Then, a low-pressure suction is caused at the central frontal inlet, and a low-velocity jet flow results at the peripheral deceleration outlet.

But, since the objective for control is to create a load on the anchorage controlling system, we must also reverse the natural sense of rotation of the first stage turbine. It appears that the only accessory necessary to accomplish this is a flow deflector on the peripheral deceleration outlets, which will reverse the structure's rotational direction. A similar technique is successfully used on jet aircraft, operating in parallel with the aerodynamic brake and flaps to reduce speed when landing.

The re-established operative load on the mooring because of this small alteration will permit the UEK to execute, under complete control, a 180-degree turn around its anchorage, to be ready to generate electric energy again with the next active tide cycle. This control mode will need only a fraction of shore power for a short time, compared to the long time period in full-power generation mode.

This control mode can be compared to a ship at anchor that puts its propulsion slow astern and executes a slow 180-degree arc around the anchorage with the rudder a few degrees off dead-ahead position. The UEK is similar to the propeller, and moves in response to its own second stage turbine in a motor mode. The flow in the turbine penstock is reactivated and the direction of the structure's rotation for "slow astern" is accomplished by the peripheral deflector.

Selected Approach, Option B:

In estuaries, confined tidal flows, or race currents found in locations usually closer to shore, the flow-seeking characteristics of a modified UEK system as in Option A may not be needed. However, the centrifugal and hydrodynamic concentration which permits the utilization of efficient standard turbo-electric generating equipment remains preferable.

For implementation, we would implant three short piles in the river bottom at the operation site. At the equicenter of this pile pattern, we would mount a pedestal around which the UEK machine can rotate in a controlled, 180-degree arc. This simplified design fixes the position of the UEK, eliminating the need for spatial and depth control.

Selected Approach, Option C:

For tidal-current operation in a very close, restricted area (like New York City's East River at Hell Gate), the same approach as Option B is envisioned, but the UEK would be suspended in the current on the end of a boom structure attached to the shore at the other end. The end which supports the machine will repose on a pontoon maintained in a perpendicular position, related to the shore, by two guy cables.

In Options B and C, the 180-degree travel is restricted. A cam action (loaded by the 3-5% buoyancy, or by ballast) would bring the machine to rest the in 90-degree station (slack period). The resuming tide flow would turn the structure the remaining 90 degrees to its new operational position. The cam system will be limited to prevent a full 360-degree rotation, thereby avoiding the need for slip rings to deliver the power to shore.

The project will provide answers to a wide variety of fundamental questions regarding these innovative control techniques, and should lead to determination of how best to proceed in the building of an effective, efficient "tidal" Underwater Electric Kite.

Developing a New Aid to Reproductive Management of Endangered Species.

Nancy Marie Czekala-Gruber

Route 1, Box 74A, Del Mar, California 92014, U.S.A.

American, born April 10, 1948. Research Associate and Laboratory Supervisor in Endocrinology Laboratory, Zoological Society of San Diego. Educated in U.S.A.; B.A. in Biochemistry from University of California — Berkeley.

This project is directed toward the development of simple urinary hormone assays which can be used to enhance the captive propagation of endangered species. Based on recent developments from the author's laboratory, these assays will be in the form of kits that can be used by the non-scientist outside the laboratory to obtain rapid, objective appraisals of reproductive function and status. Such information will subsequently provide more thoughtful management techniques for rare and endangered species. Ultimately, well-managed captive colonies will safeguard many animals from extinction.

For species to survive they must reproduce, and to obtain optimal reproduction, animals must be properly managed.

The techniques for monitoring ovulation and determining and detecting pregnancy presently exist for many species. These procedures, however, are performed by a few select laboratories and the real need is to have a simple technique that can be used by animal handlers and veterinary staff in the field. These rapid evaluations should be made available without costly time and transportation factors. Historically, the evaluation of an animal's reproductive state (i.e., fertile, non-fertile, pregnant, non-pregnant) has been performed only by measurement of hormones from daily blood samples. In exotic species, however, the danger and stress factors of venipuncture prohibit this approach on a daily basis. In the last decade, there have been advances in the ability to monitor hormones from urine, which is a more realistic and practical means of sample collection in exotic species.

Reproductive processes are reflected by specific changes in hormone production. All reproductive events can be evaluated or monitored by observing the changes that take place in hormone production over a given time. From radioimmunoassays (RIA) of total estrogens, we moved to the ability to measure steroid conjugates (such as pregnanedil glucuronide, or PDG). With this more rapid method, we have been able to detect pregnancy as early as eight days post-ovulation in the gorilla, 12 days in the lion-tailed macaque, and 15 days for the tamarin. In species with low estrogen levels, PDG monitoring has proven effective. The okapi is an example, since normal okapis always exhibit estrus following the fall of PDG. Our laboratory currently monitors all captive female okapis in the U.S. for ovulation and pregnancy detection. We currently know the reproductive status of each of these individuals (fertile, non-fertile, pregnant, approximate parturition date).

Steroid conjugate measurements are both evaluative and diagnostic. This procedure, however, is restricted to laboratory use only because of the elaborate equipment needed and the use of radioactive material. Only a few zoological laboratories are equipped for these expensive procedures and much valuable time is lost in shipment of samples to available laboratories.

Fairly recently, radioimmunoassays (RIA) have started to be converted to enzyme immunoassays (EIA). This conversion eliminates the use of radioactivity by having steroids labled with enzyme rather than radioactive tracers. The endpoint of the assay is now visual (colorimetric) rather than requiring the use of radioactivity detectors. Enzyme reacts with an added substrate to produce varying color changes, according to the quantity of hormone. Potentially, this new assay method could be put in the hands of non-laboratory personnel without the need for expensive equipment.

Working with Dr. Sabine Gallusser, a Swiss veterinarian, we propose to take our valuable estrone conjugate RIA and PDG assays and modify them to EIA so non-laboratory personnel anywhere can use them, as well as to distribute them to various zoos, preserves, etc. The basic qualities our kit needs are:

— Simple (anyone should be able to use it).
— Transportable (reasonable shelf life).
— Rapid (faster reaction time than the laboratory procedure).
— All or none endpoint reaction (whether this be a color reaction, lumination, or pH change).
— Visual detection (no machinery necessary to measure results).

We propose to set a positive endpoint for ovulation. For example, a first positive reaction in the gorilla, would indicate ovulation. Another positive reaction 10 days later with continuous positive reactions after that would indicate pregnancy.

The benefits of kits like these would be great. With our present knowledge, exotic species of animals could be monitored outside the laboratory. This number can only grow. And, thus, with proper management, individual populations will also grow.

Although this technique is not presently functioning, it is within our capabilities. It is something I have dreamed about for many years, but had no way of realizing. The chance is here now; we can make it a reality, and put proper reproductive management in the hands of the animal handlers. With a little human help, we can stop the ever rising numbers of vanishing exotic species caused by human expansion.

Seaweed Aquaculture in Argentina

Alejandro Miguel Santiago Mayer

Sanchez de Bustamante 1159, 7° A, Buenos Aires 1173, Argentina

Argentine, born April 13, 1950. Research Scientist for National Council for Scientific and Technical Investigations. Educated in Argentina; Ph.D. in Biological Sciences from University of Buenos Aires in 1978.

The objective of this project is to achieve the massive aquaculture of *Gracilaria sp.*, an agar producing seaweed, along the extensive coastline of Argentina. *Gracilaria sp.* is presently the principal object of the seaweed industry in Argentina, approximately 3,000 to 3,500 dry tons being produced annually from the naturally existing beds which are restricted to the Province of Chubut, Southern Argentina. For various reasons, production has not increased in the last 20 years. I propose a 3-stage program to increase *Gracilaria sp.* production sharply by mass cultivation. In the first stage, laboratory scale experiments shall be performed to study the effect of the different environmental factors affecting *Gracilaria sp.* growth. In the second stage, the study will investigate the growth and yield of *Gracilaria sp.* in a series of outdoor continuous flow seawater cultures. Different methods for massive cultivation shall be analyzed. Our expectancy of yield is between 50-60 dry tons of *Gracilaria sp.* per hectare, which is 50% of the maximum theoretically possible yield. In the third and last stage, I plan to develop a theoretical production scale seaweed model farm. It is expected that if all 3 stages are completed and the necessary venture capital has been secured, massive cultivation of *Gracilaria sp.* in Argentina will become a reality in 1986.

Though Argentina is a food producing country, it has vast expanses of land that are extremely poor for agricultural or livestock rearing alternatives. Along the 2,500 km coastline, however, much land seems very suitable for intensive use as 'aquatic farmland'. Current seaweed production derives from 3 private-ly owned enterprises, and there is also an agar producing factory which industrializes the available seaweed. The industry is stagnated, however, as it is presently impossible to increase the natural production of the seaweed beds. Lack of scientific knowledge, lack of trained scientists and pertinent equip-ment in Patagonia, lack of either governmental or private funding to do the needed research, and lack of vision concerning the potential of massively increased production have all contributed to a lack of interest in developing this science.

But why seaweeds? Seaweeds are plants that represent a very important commercial commodity. Their major value resides in their high content of easily accessible colloidal polyschaccharides, such as agar and carrageenan, polysaccharides of the red seaweeds (Rhodophyceae), and algin from the brown algae (Phaeophyceae). These colloids find widespread use in the food, cosmetic and pharmaceutical industries as thickening and gelling agents. Another and increasingly important use of seaweeds includes human food, animal fodder and fertilizers.

Much research has been done with respect to the cultivation of the carrageenan and algin producing seaweeds, with a resulting increase in production of cultivated varieties. China leads the world, with production of approximately 100,000 dry metric tons of brown seaweed annually. With the agar producing seaweeds, however, the situation is different. Only in China and Taiwan has research on, and cultivation of, *Gracilaria sp.* taken place.

In Argentina, there exists no attempt whatever to cultivate the only important species present, the agar-producing *Gracilaria sp.* The object of my project, therefore, is to develop all the necessary techniques to mass culture *Gracilaria sp.* on coast-line land in Southern Patagonia.

In the first stage of the project, a laboratory shall be established in Chubut Province, near the *Gracilaria sp.* beds. Studies of the effects of environmental factors such as temperature, light intensity, light period (hours of illumination), salinity, and pH on the growth of *Gracilaria sp.* under optimum, though artificial, laboratory conditions will yield fundamental information on the biology of this marine algae. This phase should run rather fast, as other investigators have studied other species of *Gracilaria sp.* in other parts of the world, and their studies will be used as guidelines.

The second stage will investigate the growth and yield of *Gracilaria sp.* in a series of outdoor, continuous-flow seawater cultures. These studies will attempt to compare different methods of artificial cultivation in 1) a floating algal culture system, and 2) onshore tanks or raceways in which dense cultures of *Gracilaria sp.* are maintained in continuous suspension. Of key importance will be the amount of biomass produced per unit area per unit of time. Successful completion of stage 2 of the project should demonstrate that Argentine *Gracilaria sp.* is capable of producing yields similar to those achieved for other *Gracilaria* species in other countries, notably the 40 dry tons per hectare produced in special *Gracilaria sp.* ponds in Taiwan.

The third phase of the project will be the construction of a model production scale seaweed farm. I am presently confident that my proposed mariculture farm of 50-100 hectares will be obtainable, and will serve as an example for the development of numerous such small farms along the Argentine coastline.

A Back-Packable "Ski Tow" for Uphill Skiing

Giovanni Allisio

6 Vicolo Lucinicco, Rivoli 10098 (TO), Italy

Italian, born March 24, 1948. Technician, Automobile body designer. Educated in Italy.

This project relates to a self-propelled unit (SPU) able to propel a person wearing snow skis along a rising snow path. It comprises a frame, rolling means (treads) for the movement of the SPU along the snow, an engine for driving the treads, a transmission connecting the engine to the treads, and an engine control unit.

Self-propelled units of the aforesaid type to be used by a skier in overcoming a rising path are already commercially available. The varieties offered up to the present time, however, are in the form of actual vehicles of considerable weight and overall size, and are high in cost.

The main characteristics of this SPU are that it is of such dimensions and weight as to be portable, and that it is also provided with means for its releasable connection to the skis and poles of the skier. It is configured in such a way that, when in use, a large part of the skier's weight is supported by the SPU while in the ascent mode.

By virtue of these characteristics, this device can be used advantageously by a skier in order to overcome an upward snow path, following which it can be carried on the shoulders and conveyed downhill during the next descent on the skis.

Further characteristics and advantages of the present invention will be apparent from the following description.

In the ascending mode, the skier's weight is predominantly resting on the SPU itself, and control is obtained by means of a handle connected to the ski poles, and flexible cables connected to the unit. At the end of an ascent, the skier separates the SPU from the skis and poles, and straps it on his back, much as a knapsack would be carried, so that he can carry it on his shoulders during his subsequent descent on the skis.

The SPU itself comprises a frame on which there are front and rear wheels with peripheral recesses into which the ends of transverse bars (forming part of the crawler track) engage. Two lateral protection walls are screwed to the frame, from which project two lateral supports. On these supports, there are mounted two rotatable brackets that project frontwards, and the arms of a fork which are joined to a plate designed for rapid releasable fixing. A bayonet connector takes the lower parts of the ski poles, so that the poles are able to rotate.

The free ends of the brackets each comprise an articulated connection unit for releasable fixing to a corresponding connection plate situated on the rear end of each ski. These articulated connection units enable a support to rotate around both horizontal and longitudinal axes.

Because of this arrangement, when the articulated connection units are connected to the skis and the plate is connected to the poles, a large part of the skier's weight acts on the SPU because of the rigid connection constituted by the connect brackets and support. At the same time, the skis and poles are given the necessary freedom to enable the direction of movement to be controlled and any necessary edging to be undertaken.

An internal combustion engine of small piston displacement is mounted on the frame of the SPU in a forward position between the two wheels. The drive shaft of the engine connects to the rear (driving) wheels with the kind of automatic clutch used in mopeds. A clutch and drum brake complete the motor mechanics.

Additional devices on the SPU include a crawler track tension adjuster, the fuel tank, a starter pulley, and an exhaust silencer for the engine.

When the skier wishes to ascend, he connects the bracket ends to the skis, and the central part of the support (with seat and control handle) to the poles, and starts the engine. The automatic clutch gradually transmits the torque from the engine to the wheels. The automatic speed variator sets the optimum transmission ratio at all times in relation to the load conditions and the slope of the path.

An SPU of this type enables a skier to overcome ascents up to a maximum slope of 30%, at a maximum speed of about 15 kilometers per hour. Braking is accomplished by means of the handle, and direction changes are made by the usual method in skiing; namely, the shifting of the skier's weight from one ski to the other. The handle also includes a control for stopping the engine. Within the principles of this description, the constructional details and shape embodiments of such a self-propelled unit can be widely modified.

At the moment, there is only a prototype of this SPU, made by myself and named "Cucciolo" ('pup'), with which we have made practical tests in ascents. It has been patented in Italy, and other patents are pending in many countries. To date, the SPU has roused much interest, and I am continuing to seek ways to reduce its weight, to further improve its portability.

Using Mining Methods to Produce Petroleum

Ahmet Pekkan

Yesilyurt Sokak No. 7/3, Cankaya, Ankara, Turkey

Turkish, born May 15, 1922. Mining and Metallurgical Engineer. Educated in Turkey and U.S.A..

Billions of tons of petroleum remain in abandoned oil fields, even after use of secondary and tertiary petroleum production methods. This project seeks to use a new, patented means of extracting the remaining oil through the use of deep-boring mining techniques. After boring a 1,500 meter deep mine shaft, to reach below the oil-bearing strata, holes will be drilled *upward* into the oil pool. Gravity alone, providing drip-drainage, will allow collection of the oil at the bottom of the shaft, for subsequent pumping to the surface.

A High-Technology Attempt to Control Snoring

Anthony Russell Dowling

7 Wentworth Road, Vaucluse, Sydney, New South Wales 2030, Australia

Australian, born January 20, 1935. Container Manager for shipping lines agent. Educated in Australia.

Snoring is a social problem of vast dimensions. Previous devices have failed to solve the problem because they were either ineffective or impractical or both. This project's invention is a device that uses advanced micro-electronic technology, and accepted principles of conditioning. Based on a miniature ear-plug transducer, acting as either microphone or loudspeaker, it reacts to beginning snores with a pulsed aversive noise, switching back and forth as required. It is expected to reduce snoring to negligible levels.

Development of Unique Sintered Thermoluminescent Dosimeters

Mirjana Prokic

"Boris Kidric" Institue of Nuclear Sciences, Vinca, Institute of Radiation and Environmental Protection (OOUR-100), P.O. Box 522, 11001 Beograd, Yugoslavia

Yugoslavian, born February 7, 1941. Scientist and principal investigator. Educated in Yugoslavia; Ph.D. in 1977.

This project has developed unique thermoluminescent dosimeters (TLDs) using MgB_4O_7:RE and $CaSO_4$:RE phosphors. Inventing a technique for producing TLDs in solid form, in which dosimetric characteristics stay unchanged and are equivalent to those of TLDs in powdered form, it is now developing specific sintered TLDs which are suitable for separate measurement of gamma and neutron dose equivalents, as well as special TLDs for beta-ray dosimetry.

A New Entry Port for Medications into human Blood System

Joseph John Makovich

38 Weatherbell Drive, Norwalk, Connecticut 06851, U.S.A.

American, born March 12, 1927. Dentist in General Practice. Educated in U.S.A.; D.M.D. in 1954.

Many patients, e.g., diabetics, require frequent or daily intermuscular injections of medications through a hypodermic needle, resulting in complications through scarring of tissue. This project is developing a system of utilizing the teeth as an entry port for continuous delivery of medication. In an extension of root canal therapy, a ball valve is to implanted in a tooth, allowing medicine from a fixed or removable reservoir to be transmitted directly into the pulp canal, and thence into the blood stream.

The Tacloban Bio-Gas Automatic Bio-Digester

Lollie Villegas Nebasa

93 Real Street, Tacloban City 7101, Philippines

Filipino, born December 28, 1938. Contractor/Company President. Educated in the Philippines.

Mrs. Villegas Nebasa has invented and constructed a new device that improves the performance of bio-gas operations. It is a single, compact unit that contains an automatic bio-digester, septic tank and leaching chamber, each housed in separate compartments. It is automatic, in that decomposed waste is run directly to the septic tank, on through the leaching chamber and ultimately out to sewage. The units is self-cleaning, easy to maintain, and built for very long life. Installed in Tacloban, the unit processes pig wastes in seven days to usable gas that is odorless and pollution-free.

Electron-Beam-Induced Recoil Implantation in Semiconductors

Takao Wada

Faculty of Engineering, Mie University, Kamihama, Tsu 514, Japan

Japanese, born July 23, 1930. Professor (Research and Training), Faculty of Engineering, Mie University. Educated in Japan; Ph.D. in Electrical Engineering in 1965 from Nagoya University.

EBIRI in solids was originally discovered by Dr. Wada in 1980. The technique employs an impurity sheet in contact with the semiconductor surface, which is bombarded with high energy electrons. By using EBIRI in powdered materials of Zinc-doped ZnO phosphors, which are largely used in low voltage fluorescent applications, an increase of the cathodoluminescence intensity is obtained. To interpret the mechanism of EBIRI, a new theory for anomalous sputtering and anomalous room-temperature diffusion is being developed.

Investigating a Plant with Possible Anti-Carcinogenic Properties

Gerardo Antonio Quijada Diaz

5a. Calle Poniente No. 21, Santa Ana, El Salvador

Salvadoran, born December 8, 1965. Third year high school student.

In the coastal area of El Salvador, a tropical plant called the "Mangle" (Rhizophora mangle) is used by local people to make a preparation that appears to provide effective treatment of skin ulcerations. As some ulcerations have cancerous origins, this project is seeking to identify the active substance that cures or alleviates the lesions. To date, animal tests indicate that the preparation is virtually non-toxic, even in high doses. Clinical studies are being attempted, in collaboration with a doctor, to determine the effectiveness of the preparation.

Phytochemical Screening of Indonesian Traditional Medications

Priambudi Kosim-Satyaputra

Department of Chemistry, Faculty of Science, Padjadjaran University, Jalan Singaperbangsa, Bandung 40132 Indonesia
Indonesian, born September 19, 1956. Fifth year student in Department of Chemistry, Faculty of Science. Padjadjaran University, Bandung, Indonesia.

Traditional herb medications are widely sold throughout Indonesia, some having clear therapeutic values and some of dubious efficacy. Most of the so-called medicinal plants or herbs are combined into formulas, often held as family secrets and subject to uncertain modifications. This project is undertaking the phytochemical screening of 50 or more such medications, to determine which may contain harmful compounds. Results will be communicated in all possible forums, in order to warn the public of potential dangers.

Design and Construction of a Wind-Turbine-Powered Vessel

James Gordon Bates

P.O. Box 1043, Whangarei, New Zealand

New Zealander, born March 30, 1924. Development engineer and commercial fisherman. Educated in New Zealand.

Following earlier experience with a yacht modified with a wind turbine, this project is now well along on the design for a 45-foot, 16-ton displacement, long-line fishing vessel to be powered by a large wind-turbine. Using turbine blades of 11.69 meters in length, the craft will survive violent changes in wind velocity, have reduced roll, possess excellent maneuverability for fishing in confined areas, and offer major savings on fuel. Though the initial craft is intended as a model for fishing vessels, it should prove the feasibility of wind turbines as being suitable for ocean-going craft up to 4,000 tons.

Developing Apiculture to Aid Subsistance Tropical Economies

Jean Fauchon

Ornex, 01210 Ferney-Voltaire, France

French, born January 10, 1925. Retired International Civil Servant. Educated in France.

Though bee-keeping has become a sophisticated operation in temperate countries, it is virtually not practiced in tropical countries. This project seeks to improve the nutritional and economical situation of farmers and other rural people through the organization of cooperatives that can create small apiculture centers for the production of honey. Following the initial research on local requirements, the program is designed to create units possessing approximately 1,000 beehives belonging to 200 farmers, producing about 20,000 kilograms of honey annually.

Disposable, Continuous Wear Contact Lenses from New Protein Polymer Hydrogels

Orlando Aloysius Battista

5268 Trail Lake Drive, Fort Worth, Texas 76133, U.S.A.

American, born June 20, 1917. Company President, Research Institute Director. Educated in Canada; Honorary Sc. D. from Saint Vincent College (U.S.A.) in 1955.

Until now, contact lenses (both soft and hard) have been made of non-protein, man-made synthetic polymer raw materials. Polymer chemist Battista's project was to develop disposable natural protein contact lenses that would be put in the eye sterile, worn continuously until they become cloudy, then discarded and replaced with a very low-cost replacement. He has patented a new protein polymer hydrogel, which now has been produced in lens form acceptable for human clinical testing. A major optical company has signed for worldwide rights to the product.

Developing a Unique, Energy Efficient, Low Cost Personal Hovercraft

Christopher John Fitzgerald

Neoteric-USA-Incorporated, Fort Harrison Industrial Park, Terre Haute, Indiana 47804, U.S.A.

Australian, born May 24, 1944. President of Neoteric-USA-Incorporated. Educated in Australia and U.K.; diploma in Mechanical Engineering.

Since 1959, this engineer-businessman has pursued the goal of making hovercraft transportation available at a reasonable price for private users. He founded the Australian Air Cushion Vehicle Development Group, and has researched hovercraft technology on a worldwide basis. A prototype vehicle was built, and the sale of 'do-it-yourself' plans and kits have succeeded in bringing the operation to a point where large-scale production can now be envisioned, utilizing new technologies and world-wide sourcing of materials.

Non-Polluting, Biological Control of Ship Fouling

Uriel N. Safriel

Department of Zoology, The Hebrew University of Jerusalem, Jerusalem, Israel

Israeli, born December 24, 1936, Chairman of Department of Zoology, Associate Professor of Zoology. Educated in Israel, France, U.K. and U.S.A.; D. Phil. in Ornithology from Oxford University in 1967.

Barnacles and algae, attached to ship hulls, create excess friction that impedes ships' speed and requires increased oil consumption. Routine cleaning with toxichemicals pollutes the marine environment. Limpets are crawling snails that crush delicate organisms such as settling barnacles' larvae and algal spores, and have been found to control barnacle growth on experimental panels. This project is researching means of using limpets to control ship fouling by 'culturing' limpets on ship hulls in ports.

A High Resolution X-Ray Instrument with Depth Perception Quality

Russell Maurice Tripp

15231 Quito Road, Saratoga, California 95070, U.S.A.

American, born July 12, 1916. President, Skia Corporation. Educated in U.S.A.; D.Sc. in Geology from M.I.T. in 1948.

This project involves developments in several disciplines which are being combined to provide an instrument for producing X-rays (without exposing the patient to as much ionizing radiation as now required) that have improved optical resolutions for fine detail, and that present the results in such a way that the observer perceives the true three dimensional or spatial nature of the subject. These characteristics are being achieved with state of the art developments in fiber optics, electron optics, microprocessors, crystal growing and three dimensional display techniques.

New Anti-Seismic System for Foundations and Houses

René Gaston Marcel Barnier

42 La Beaucaire, 83200 Toulon, France

French, born April 23, 1902. Independent research worker, retired from building materials trade. Educated in France.

In an ingenious combination of simple design elements, this project seeks to provide protection against earthquake for buildings through the use of easily made and assembled components. Foundation columns with downward-turned, concave, steel "feet" rest on large steel balls, which in turn rest on upward-turned concave surfaces. Horizontal tremors are 'damped' by the rolling ability of the balls, coupled with the inertia of the structure above. Vertical shocks are damped by heavy duty shock absorbers contained within the foundation columns themselves, thus affording high levels of seismic protection at low cost.

Utilizing Magnetic Liquid to Construct a Propelling Motor

Bachir Georges Kassir
Mirna Chalouhi Est, Blvd Sin-el-Fil, P.O. Box 55108, Beirut, Lebanon.

Lebanese, born May 31, 1963. First-year physics student at Lebanese University.

This project seeks to develop a new type of 'motor', based on the the principal of maintaining a magnetic liquid between the poles of properly oriented magnetic fields in such a way that it will support a fluid (not mixed with the magnetic liquid) on one side of a connected chamber at a different height and pressure from the other side. Once the system is fixed in equilibrium, a closed chain of metallic balls placed within the fluid should theoretically be subjected to different lifting forces, and thus drive the chain. Perpetual motion?

Instrument for Continous, Automatic Measurement of Refractive Index

Arvind P. Kudchadker
Department of Chemical Engineering, Indian Institute of Technology Bombay, Powai, Bombay 400 076, India

Indian, born February 16, 1934. Professor of Chemical Engineering. Educated in India, U.K. and U.S.A.; Ph.D. (Chem. Eng.) from Texas A&M (U.S.A.) in 1967.

This new instrument is designed, fabricated and tested for the continuous and automatic measurement of the refractive index and for recording and control of Iodine Value, or any other physico-chemical process dependent on the refractive index, without the need for manual intervention. Extensively tested up to 15 bar pressure at a plant site, it can cover any range of Iodine Value from 200 to 20 units with suitable zero set. Fabricated from readily available parts, its unit cost is estimated at about U.S.$1,000.

Balance Wheel Power Source Generator for Use in Electronic Watches

Stephen Marwa
International Computers (E.A.) Limited, P.O. Box 16176, Nairobi, Kenya

Kenyan, born January 18, 1946. Computer Engineer. Educated in Kenya, with advanced education via University of London correspondence.

The sudden loss of battery power in electronic watches is annoying, and replacement of batteries is difficult (and expensive) in some areas. The problem could be overcome with a new approach to the powering of these low current devices. This project has developed the design of a system that would use the potential energy stored in a normal watch spring (either manually or automatically wound) to power an electronic watch, thus providing electronic accuracy, powering the additional features increasingly found in such watches, and rendering battery replacement unnecessary.

Technology Adaptation in Tinnitus and Non-Invasive Diagnostic Investigations

Dr. Sayed Tewfik

Post Office Box 158, Maadi, Cairo, Egypt

Egyptian, born August 15, 1939. Private practice M.D., Consulting M.D. Educated in Egypt; M.D. in Otolaryngology from Cairo University in 1969.

Tinnitus, the disturbing ringing or other noise in the ears, is often ignored by doctors, though it represents a troubling symptom for patients. Pulsating tinnitus can be even worse. This project has developed models for medical technology adaptation in the problem of tinnitus, non-invasive diagnostic researches and new otological instruments. These include 1) Phonocephalography, 2) Tympanic plethysmography, 3) Symptom feedback, and 4) New otological accessories. Their use promises easier, less expensive and more effective diagnosis and treatment of patients.

Micro-Surgical Correction of Lymphoedema: A Need for World-wide Expansion

Dr. Leo Nieuborg

4 Vijverlaan, 3062 H.K. Rotterdam, Netherlands

Dutch, born December 26, 1951. General Surgeon and Consultant in Lymphoedema treatment. Educated in Holland; M.D. from University of Amsterdam in 1982.

Swelling of the arm or leg after infection of the lymphatic system (filariasis), or after surgery or radiation therapy, means a significant handicap for the patient involved. After performing 150 operations on such patients with a microsurgical technique, the results obtained indicate the success of this new low morbidity technique. This project seeks to provide more training facilities and trained surgeons to make the treatment available to the many patients suffering from lymphoedema.

The "Flomor Thermodynamic Cylinder" — A Fuel Saver and Anti-Pollution Device

Florencio T. Morales

Flomor Industries, Crossing, Calamba Laguna 3717, Philippines

Filipino, born November 7, 1924. Owner-President of Flomor Industries. Educated in Philippines; trained in Mechanical Engineering.

This patented device is a novel, two-stage, pre-heater for fuels prior to their entry into the combustion chambers of internal combustion engines. Using hot water from the cooling system of the engine and the heat of exhaust gas, it raises the fuel's ignition rate to maximize combustion, minimize exhaust pollution, boost engine power and save fuel. It has been tested by government agencies, and certified to give fuel savings of from 10-27.5%.

Hormonal Control of Growth in Crustaceans with Aquaculture Potential

Ernest S. Chang

Bodega Marine Laboratory, University of California, P.O. Box 247, Bodega Bay, California 94923, U.S.A.

American, born December 7, 1950. Assistant Professor of Animal Science at University of California-Davis. Educated in U.S.A.; Ph. D. (Biology) from University of California-Los Angeles.

To meet future protein food demands, production of aquatic animals through controlled culturing techniques must be increased. Having shown that arthropods can be induced to 'molt' (and thus grow more rapidly) through the removal of the eyestalk glands, this project seeks to identify the hormonal factors involved, characterize them chemically, and suggest ways in which this information could be applied to aquaculture.

A New, More Efficient, Small Capacity Oil Burner

Dag Gotskalk Johnson

Ovre Ferstadveg 10, 7000 Trondheim, Norway

Norwegian, born August 12, 1909. Professor emeritus (Industrial Heat Technics) from the Institute of Technology of Norway. Educated in Norway, Holland and U.S.A.; Dr. techn. from Institute of Techology of Norway in 1949.

This patented device is a light oil burner for very small loads, down to 0.5-2.0 kgs of oil per hour. It will work modulating (not on/off), bringing the advantages of continuous, clean combustion, and a stable flame. The main feature is the use of a small fluidized bed for *vaporizing* the oil, instead of atomizing it just before combustion. Even small modulating burners can thus obtain the best combustion conditions, reduced oil consumption and improved environmental performance.

Mono-Lever-Controlled Automobile for Children and the Handicapped

Yoshikata Muguruma

15-2, 1 chome, Sojiji Ibaraki City, Osaka Prefecture, Japan

Japanese, born December 24, 1944. Head Chief of Research Institute. Educated in Japan; Applied Physical Science degree from Hokkaido University in 1969.

This gasoline-engine car was invented for use by children or the handicapped, for whom ease of driving is critical. Using a torochoid propelling system, it is a four-wheel-drive six wheeler (four small wheels in front, two in back), and requires no steering wheel, gearbox, accelerator pedal or brake. A single control stick provides the 30 km/h vehicle with all speed, direction (it can turn in place) and braking requirements, making it ideal for persons with limited physical abilities.

To Fight Against Hookworm, an Unusual Cell "Excretion" is Studied

Sydney Chukwunma Ugwunna

Department of Biological Sciences, Faculty of Science, University of Lagos, Akoka, Lagos, Nigeria

Nigerian, born October 24, 1942. Science Lecturer in Biology at University of Lagos. Educated in Nigeria and U.S.A.; Ph.D. from Wayne State University (U.S.A.) in 1979.

Hookworm is a dangerous parasite that causes death in both human and domestic animal populations around the world, particularly in tropical regions. To control the hookworm population, and thus infection, understanding the development of the hookworm gametes is a key need. Using advanced equipment, this project is researching the potential significance of a hitherto unknown excretion of the hookworm gamete that could provide a clue to the control of this widespread disease.

Determining the Impulse Response of the Ear Using an Adaptive Digital Filter

Ricardo Pradenas

1523 Sycamore Avenue, Fullerton, California 92631, U.S.A.

American, born August 25, 1951. Electrical Engineer, Technical Staff, Hughes Aircraft Corporation. Educated in U.S.A.; M.S.E.E. from Brigham Young University in 1980.

To date, mathematical models, all based on observed, measured and interpreted physiological data, have been developed that describe: middle ear functions, cochlear dynamics, basilar membrane displacement, and neural and hair cell action. Measurement accuracy and observation interpretations limit the validity of these models. This project realizes the impulse response of the human ear, without these limitations, using new technologies from the fields of electrical engineering and audiology, including adaptive digital filters and brainstem evoked responses.

New, Safer Ski Equipment: Non-attached to Foot and Leg

Kenneth Anthony Henson

1 Regency Court, Dundas, Sydney, New South Wales 2117, Australia

Australian, born September 6, 1943. Medical Practitioner. Educated in Australia; M.B., B.S. from University of New South Wales in 1966.

This project has developed the Sloping Terrain Vehicle (STV), consisting of a set of skis joined toward the front by a hinged crossbar system, a handle on each ski and a place for each foot of each ski with low side rims. Designed to avoid lower limb injury associated with ski boot and bind systems for downhill skiing, it allows for effective skiing, using normal techniques to control speed and direction. It allows cheaper footwear, rapid learning and the joys of skiing for virtually anyone, including many handicapped persons.

EXPLORATION
AND DISCOVERY

The projects appearing in this section were submitted in competition under the "Exploration and Discovery" category, which was defined in the Official Application Form as follows:

Projects in this category will be concerned primarily with venturesome undertakings or expeditions and should seek to inspire our imagination or expand our knowledge of the world in which we live.

An example of the ancient, endangered petroglyphs of Nevada; American Indian art in need of protection and preservation.

A Petroglyph Catalog: Ancient Indian Rock Art in Nevada

W George Thomas Appleton

Honourable Mention — The Rolex Awards for Enterprise — 1984
3400 Florrie Avenue, Las Vegas, Nevada 89121, U.S.A.

*American, born June 1, 1926. Retired high school teacher, photographer
and editor of Mensa sub-group newsletter. Educated in U.S.A.; with
Masters Degree from Michigan State University.*

My primary goal is to preserve, through photography, a form of rock art,
found in Nevada and done by Yuman and Pueblo Indians, that dates back to at
least 3,000 B.C. Five thousand years of people chipping away at large rocks with
small ones, to draw all sorts of figures; from simple straight lines to animals, to
humans, and to intricate and complicated designs. The least of these was
difficult to do; most took hours of very hard work (with how many smashed
fingers, accompanied by cries of pain and other appropriate comments?).
Some were done by an artist sitting comfortably on the ground, some are on
cliffsides so steep there is barely a place for a man to hang on, much less use his
hands to inscribe his ideas in stone.

They are scattered all over Nevada — major sites (usually in a canyon near
water) with hundreds of these drawings, smaller areas with only a few; most on
solid rock or huge boulders, but some on rocks small enough to be picked up
and carried away, which people have done. I know of one home in Las Vegas
with a "native stone" fireplace in which some of the stones have petroglyphs on
them.

No one has ever tried to preserve all of them on film. There is not even a
single index of Nevada petroglyph site locations. There has been no concerted
effort to do anything about these examples of pre-historic art, especially in
Nevada. A few 'masters theses' have been written about them (though more
often in California than here), concentrating on one particular area or another.
I am in possession of an unpublished study by Stan Rolf and others of a small
valley near here, apparently a camping site for Indians passing through, in
which — really as an aside — they located and made drawings of the few
petroglyphs nearby. In the late 1920's, a man named Hizer did a paper for the
University of California at Berkeley on a major Nevada site some 75 miles south
of Las Vegas, and a woman is currently doing a study of some glyphs just south
of that. But, no effort has been made to catalog the whole of them, and very little
effort has been (or, I suppose, can be — other than saying it is illegal to deface
one) made to preserve them.

One reason that preservation is necessary, if only on film, is the sheeer growth in numbers of people in this area. There are more people now (some 600,000) in the Las Vegas area than there were in the whole state of Nevada when we moved here in 1963. These people are far more mobile because of the availability of 4-wheel-drive and "all terrain" vehicles. Graffiti and guns are a way of life; both totally destructive of this really rather fragile rock art.

A second real value of the project is that it *is* difficult to get out into the more remote areas of this state. There are, for instance, only three main north-south highways in Nevada, and on one of them there is a stretch of 110 miles between gas stations — between much of anything for that matter. It is nearly 450 miles (713 km) between the two main cities in the state — Las Vegas and Reno — while the petroglyph sites in Clark County (southern Nevada) alone are scattered over a 7,874 square mile area. Just the element of time involved in trying to reach a number of sites is a limiting factor in what many students of pre-history can do.

So, a catalog of photographs to include such things as repatinization, size, compass orientation, etc., should be of great help to people involved in building up knowledge of Nevada's early inhabitants. And, because I am retired, there is no restriction on the amount of time I can spend on this project.

What I propose is basically a four-part task.

First, to find and photograph all of the major petroglyph sites in the state of Nevada. The climate here is such that I can use the winter months working within about a 150 miles radius of Las Vegas, and from April to October cover the rest of the state over the next 3-4 years.

During this time, I would expect that in my own travels and through people I already know and/or expect to meet along the way (BLM staff people, prospectors, miners, ranchers, sheep herders, the few folks who still live in some of the old ghost towns, etc.), I should be able to find at least most of the lesser known sites around the state, locating them on topographic maps as I go.

While it is not possible to photograph each glyph separately (there are simply too many of them), I plan to concentrate on small groupings and on specific types (round or linear, filled-in or 'empty space' figures, for instance). Each photograph would be marked with at least one measured petroglyph, and with the compass bearing from which the picture was taken. I already suspect that, where the artist had a choice, most glyphs were done on north or east canyon walls.

Second, to try to interpret as many glyphs as possible while on the sites. Only one writer, LeVan Martineau, himself an Indian, in his book *The Rocks Begin to Speak,* has made a strong attempt at the meaning of the rock symbols — and his efforts are not given much credence by some other 'experts'. Which is the problem. There *are* no experts, only people who have looked at one site, or a few, and who have greatly conflicting opinions. I've corresponded with Mr. Martineau and feel that his book is the best, most valid attempt at "reading" the glyphs so far, though he leaves many, many questions unanswered.

Petroglyphs *do* have meaning. After all, these were people living and travelling in small groups, in harsh country — hunter-gatherers who *had* to spend most of their time simply getting enough food to stay alive. To have found the time for scribing on rock would have been nearly impossible had it not been very important for them to do so.

Third, to assemble and correlate the photographs. First by area — *this* is what was *there*. By apparent age. By type — linear, circular, maps, direction pointers, etc. By style — which seems to have changed over the centuries. Is there a particular kind of glyph that shows up in various sites, such as the "man with headdress", that might have been done by one man, or men from one group? Is there a similarity between certain petroglyphs and "modern" Hopi Indian clan symbols? This stage, too, would be a time to study them for their meaning.

Also, I would inform the anthropology and archaeology departments of colleges and universities in the southwest that these photographs are available for them to use and study.

Fourth, the writing of an illustrated book to serve as a guide to petroglyph areas around the state that are fairly accessible, and to give the reader an understanding and appreciation of what a legacy we have from the Anasazi, or "Ancient Ones", in this rock artistry. Nevada's economy is heavily based on tourism, while its population is growing rapidly as people come from other states to live here. Both visitors and residents should be a ready market for the book. I would think the state magazine, published by the Nevada Highways and Parks Department, would use an article, or perhaps a series of articles, especially covering sites within state parks and recreation areas.

In the final analysis, I don't know that I can make an ultimate determination that "this is what petroglyphs are, this is why they were made, and this is what each one means".

I do, though, think I can add to ideas that others have had and pass those conclusions along for people who will be studying them in years to come.

And I *know* that I can make the first complete, comprehensive, photographic survey of pre-historic Indian rock drawings in Nevada — before any more of the sites are vandalized or further destroyed by the natural processes of weathering and erosion.

The painstaking, exacting process of copying Mayan murals; here in Chamber 4 of La Pasadita Structure I *(Top)*.
A Mayan mural, brought 'back to life' by a dedicated artist/archaeologist *(Bottom)*.

The First Complete Documentation of the Mayan Murals of Mexico and Guatemala

♛ Martine Maureen Fettweis-Viénot

Rolex Laureate — The Rolex Awards for Enterprise — 1984
9 Route de Saveuse, 80470 Ailly-sur-Somme, France

Belgian, born December 16, 1947. Archaeological Researcher. From the Institute d'Art et d'Architecture at Belgium's Louvain University, she received two B.A. degrees (one in History, one in Archaeology and Art History) in 1968, and her M.A. degree in Archaeology (with "Grande Distinction") in 1973. Mrs. Fettweis-Viénot then earned her Ph. D. (awarded "Très bien") in Mesoamerican Archaeology in 1981, from the Ecole des Hautes Etudes en Sciences Sociales in Paris.

My project is to prepare the first complete catalog of the Mayan murals of Mexico and Guatemala: to gather all the existing documentation on the question, to seek still unknown paintings, to photograph all the murals, to make very precise and intelligible copies of them, to prepare their iconographic analysis and to publish them in order to make this new material available to anyone.

Background of the Project. The Maya left us a quantity of direct information on their way of life, such as scenes painted on the walls of their palaces and temples, a kind of illustrated story books, ritual almanacs, religious manuscripts and mythological tales; all of them permanently open for the visitor.

Most of the remaining documents are still unknown, even to many archaeologists, as too many paintings have never been surveyed. The copies of some others, made a long time ago, are inaccurate or incomplete, such as those of Chacmultun and Dzula (Yucatan), and the well-known compositions of Bonampak, Chiapas, Mexico, which are lacking accuracy for details and glyphic inscriptions.

Another concern is that these documents are fading away slowly and will soon be lost. In order to save them, and to compose an exhaustive catalog, I began copying and studying every trace of mural coloring, figurative as well as non-figurative, however small or little representative.

The Area. I wish this corpus to cover the whole Mayan zone, i.e., the Yucatan Peninsula and the Peten area of Mexico and Guatemala. In fact, most of the paintings are to be found in three specific zones, which are different culturally and ecologically:

— The East Coast of Quintana Roo, Yucatan, Mexico: a flat, swampy, rainy, low and thick bush along the Caribbean sea, with high forest in the interior.

— The Puuc area, Yucatan: a hilly and relatively dry region of savannas and medium forest.

— The Peten area: a rainy lowland, heavily covered with tropical jungle.

I have also discovered that several mural fragments are scattered in private collections in America and Europe, and I wish also to gather that documentation.

The Paintings They are generally situated inside stone constructions built between the 7th and 15th centuries (those on the exterior walls have generally vanished). They are painted on plaster layers laid on walls, vaults and even ceilings (capstones), varying greatly according to local culture and buildings.

I find them usually in very bad condition, due to the awfully wet climate, the partial collapse of the buildings, alterations by calcite or green microscopic lichens, bats' excrement, insects' galleries and some cases of vandalism or looters' work.

This explains why so little attention has been paid to them, and why a special care is required to produce reliable copies.

How to Reach the Paintings Most of these documents are found in very remote sites, but can be reached using 'range-rovers', and then by walking, depending on the rainy season. Sometimes I have used a motorcycle because the tracks were too muddy; I have also used a motor canoe, and once a helicopter, for especially difficult cases. Usually, I carry my hammock so that I may camp anywhere on the spot until the work is achieved, in order to save transportation time.

To make a careful copy is physically very exhausting because of the unusual body positions I have to hold for hours and hours, in spite of the heat and nuisance of the mosquitos and flies. So, I try to make the work position as comfortable as possible by scaffolds and ladders made by my Indian guides from branches and creepers. To reach the ceilings, we build hanging platforms that allow me to work lying down.

The Field Work Census of the paintings. After my bibliographic investigation, I check the data on the field and complete the information with local inquiries of the Indian hunters, gum collectors or ruin guardians, and I systematically explore all the buildings likely to shelter murals. I also make inquiries to list private owners of mural fragments.

Recording the data. I think that the utility and value of a copy depends on the objective I've planned for it, the technique I use and the care I take to achieve it. After a few problems in the beginning of the work, I developed a technique which fulfills the need for precision, accuracy and completeness of my project, while also corresponding to field necessities. I must not forget time and money limitations, heavy gear which must be carried on long hikes and its resistance to heat and humidity, the bad state of conservation of the murals and the very poor light inside the buildings.

Technique. I copy full size, directly on the walls or ceilings. The support medium I use is a pryphane sheet, the waterproof cellophane used by florists. It is a perfect material; light to carry, perfectly transparent and waterproof, and does not crumple or react to local temperatures. For graphic utensils, I use very sharp permanent markers of five different colors.

Color notation. By patient and methodical observation, I "rediscover" the most faded motives and I identify the original tints. I trace with precision the exact edges of every line and every colored surface, as well as the contours of missing or erased parts. Then, in every area so delimited, I note the references of the color. For the best results, and to permit general comparisons of color uses throughout MesoAmerica, I wish to use the Munsell Book of Colors (a color code used by American scientists and industries). I've worked occasionally with a borrowed copy, but have not been able to buy one as it is fairly expensive.

Picture shooting. Most of the time, an ordinary photograph does not provide a comprehensive view of the murals — except for details. The surface accidents come out stronger and clearer than the faded pigments, and obliterate the paintings. Nevertheless, a photographic record is necessary as a complement of documentation. I started with a 35mm camera and several lenses, using natural light or flash, depending upon the effects I wanted. Better equipment is planned when possible.

Infrared photographs. This technique should give good results on faded murals exposed to sunlight and should show details where the pigments are invisible to the naked eye, though they are still present in the plaster support. This technique is a particularly difficult to use, due to the remoteness of sites from an airport: the film, once exposed, needs quick development, especially considering tropical heat and humidity.

Studio Work In this stage, I prepare the copies for publication.

Reduction. I reduce the tracings of overly large dimensions to 1/3 size, by photographic means, or with the help of a 'clear room'. I photograph (on inversion film) the tracings and reproduce them by orthogonal projection on paper.

Color restitution in black-and-white. I codify the colors by means of the various shades of grey of Letraset adhesive films, which have constant tones and permit scientific reproductions. I choose appropriate graphic conventions to show erased, missing or obliterated parts, and doubtful tracings.

Color rendering. I reproduce the genuine tones according to the best preserved surfaces. Some missing parts are completed, based on documentation and with very diluted colors, clearly different from the original but without breaking up the whole chromatic harmony.

Physico-Chemical Analysis I also try to obtain laboratory analyses necessary to identify the pigments' composition, the adhesive substances used, mortar composition, etc. Until now, only a few blue samples are submitted for analysis.

Detailed Iconographic Analysis I am attempting this, based on the theory that the paintings represent a kind of language, a symbolic writing whose every detail is important (such as the proportions, colors, positions, orientations, perspectives, etc.).

Publication as a Catalog This is a major goal. The Mexican national Institute for Anthropology and History has asked to publish it.

The project began in 1975, and will last at least another five years. I have 23 registered sites yet to explore, and five private collections to visit in the U.S.A., Mexico and Germany. Additionally, some informants have told me about the possible existence of three unknown sites with murals in Central Yucatan. Much work remains to be done!

East to West Across the Sahara by Camel

Carlo Bergmann

American University in Cairo, Office of Student Affairs, 113 Kasr el Aini,
P.O. Box 2511, Cairo, Bab el Louk, Egypt

*West German, born May 4, 1948. Student of Colloquial Arabic, American
University in Cairo. Educated in West Germany and U.S.A.; Ph.D.
(Marketing) from University of Cologne (West Germany) in 1979.*

Since the days of my geography and anthropology studies at the University of
Cologne, I have been planning an extensive excursion on camelback through
the Sahara, my favorite field of academic interest during 1967-1969.

After a few years of work and after laying aside as much money as possible, I
now find myself in the position to put my project into practice.

In September, 1982, I left Germany with a heavy load of equipment and
settled down in Cairo, Egypt, for the following 9 months in order to study
Colloquial Arabic, an important prerequisite for a successful Sahara expedition. (Although the Colloquial Arabic is not spoken throughout the Arab
world, it is more familiar to North Africans than, for instance, the dialect of
Amman, Jordan, because of popular films, songs and other broadcasts
produced in Egypt and spread all over Arab countries.)

While studying the language, I have found enough time for testing essential
parts of my equipment. For that reason, I have performed some solo hikes in
Egypt's Western Desert, such as:

125 km from Lake Karoun and through Wadi Rayyan (3 days).

185 km from Farafra oasis via Ain Dalla to Bir Abu Minquar (4 days).

195 km from Dakhla oasis to Farafra oasis (4 days), following the footsteps of
the G. Rohlfs expedition of 1873/74.

Carrying equipment and supplies on my back during those hikes, I was also
able to test my ability to walk in the desert without assistance — a good
experience in case of emergency, e.g., loss of camels. Further equipment
testing will be performed throughout December, 1982, when I will be joining a
caravan travelling from Kharga oasis in southern Egypt to Wadi Atrun, Sudan.
After completion of the Arabic language course in May, 1983, I intend to travel
to the southeastern corner of Egypt and further on into Sudan in order to
investigate camel markets lying on the planned route of the expedition.

In addition to these preparations, and to obtain first hand information about
the water and vegetation situation, I am planning to pay short visits to Western
Chad, Bilma and the mountains of Air. Then, in August, 1983, I will return to
Germany for two weeks, for my last chance to stock up on supplies and to

purchase additional equipment before the start of the long hike in October, 1983. Up-to-date visas for the expedition will be granted by Sudan, Chad and Niger. There is no indication that Algeria, Mali and Mauritania will reject similar requests.

All available maps of the Sahara have been purchased and studied. After evaluation of additional information, I have chosen Berenice, Egypt as the starting point of the expedition. From there, the caravan will head south for Sudan and will cross the Nubian Desert to reach the Nile at Abu Hamed. From Abu Hamed, I will proceed to Berdoba, Chad, via Dongola (a stretch of desert known to me from previous excursions), Wadi Atrun and Wadi Howar. The itinerary through Chad will lead from Berdoba to Fada and further on to Largeau and Zouar, where the border to Niger will be crossed. In Niger, the caravan will pass the following places: Segedine, Bilma, Mount Fosse, Agadez, Iferoàne and In Assoua.

Continuing to Algeria, the objective is to reach Tamanrasset via In Ebeggi by the end of April, 1984. After a long summer rest, the caravan will resume travelling around October, 1984, following (in reverse direction) the footsteps of Geoffrey Moorhouse, through Aguelhok, Tombouctou, Néma, Tidjika and Nouakchott, intending to reach the latter in May, 1985. Short term adjustments and alterations to this project itinerary may be necessitated by dry wells and lack of camel fodder as well as by administrative restrictions.

The caravan itself will be of flexible size, with a maximim number of 12 camels and 4 men. The number of animals and men will depend mainly on the availability of camel fodder and water. Hence, a lack of fodder will require that a greater amount of supplies be carried along. Consequently, more camels and men will then be needed to handle the additional work. To keep the camel/men ratio as high as possible, I have purchased labour-saving appliances that will ease the procedures of camel-loading and unloading considerably. This equipment will be tested on the December, 1982 march from Kharga oasis in Egypt to Wadi Atrun, in Sudan.

All along the route from the Red Sea to the Atlantic Ocean, including the summer rest in the Hoggar Mountains, I plan to stay with my camels.

Due to the size of the caravan, the long travel time and the costly preparations and investigations, the expenses for the planned expedition are considerable. Before even covering the first mile, the costs will exceed U.S. $12,000. Each delay caused by illness, bureaucratic procedures, etc., and every customs payment will put stress on my limited funds. Therefore, winning one of The Rolex Awards for Enterprise would be a most welcome help, as it would enable me to buy essential equipment, to replace exhausted camels and to meet unexpected expenditures after having spent my own money on the first half of the journey.

Documenting Tibetan Medicinal Plants in the Himalayas

Tsewang Jigme Tsarong

Tsarong House, Upper Cart Road, Kalimpong P.O., District Darjeeling,
West Bengal 734301, India

*Stateless (Tibetan refugee), born September 30, 1945. Former Executive
Director of Tibetan Medical Center, Dharamsala, India. Educated In India
and U.S.A.; B.A. in 1970 from Valparaiso University, U.S.A.*

Due to the high cost and adverse side effects of modern orthodox medicine, millions of people all over the world are desperately searching for alternative forms of health care. It is in the light of the present state of 'crisis' of modern cosmopolitan medicine that unorthodox forms of diagnosis and treatment are enjoying public confidence of a kind that two decades ago seemed hardly conceivable. Even the World Health Organization has finally recognized that there is much to be learned from 'tribal' ways. A joint 1974 UNICEF/WHO report recommended that the services of practitioners of traditional medicine should, where possible, be utilised in primary health care, a recommendation accepted by the executive of WHO in 1975. In 1976, a working group was set up in Geneva to assist in the development of traditional medicine, and to seek to combine its knowledge and skills with those of the West.

In spite of the enthusiasm described above, it is most regrettable to note that no significant effort has yet been made to either preserve or develop one major area in human medical achievement. Consequently, it is the overall aim of my project to create an awareness, and encourage the development and study, of the ancient Buddhist medical tradition.

This tradition, popularly known as the Tibetan system of medicine, but technically known as *gSo-wa Rig-pa* or 'the knowledge of healing', forms an integral part of the four 'great tradition' systems of medicine that developed within the four main streams of human civilization, i.e., the Mediterranean, South Asian, Chinese and Central Asian civilizations. Yet, much to the dismay of those familiar with the Tibetan tradition, it is the least known and most neglected of all mankind's medical systems.

On careful examination, it appears that there are three main factors responsible for this unfortunate situation. First, after it was virtually wiped out of India during the 7th century Gupta era, the Buddhist medical tradition found fertile ground and nurture in a remote country that was 'cut off' from the rest of the world by 'inaccesible' natural barriers. Secondly, the deliberate isolationist policy of the then rulers of Tibet made it impossible to have friendly cultural, social, political and economic exchanges with other nations. Finally,

Tibet is presently colonized by Communist China and, therefore, the medical tradition is practised only by a few Tibetan refugees who neither have the necessary resources nor required infrastructure to make much of a global impact.

Notwithstanding the above, one can objectively point to the importance of the Buddhist medicine tradition both from academic and practical points of view. The very inaccessibility of Tibet has enabled her medical tradition to escape destruction. Consequently, to this very day, the tradition has not only the largest number of extant medical manuscripts but, also, an uninterrupted medical practice of well over 2,500 years! The tradition's integrated and holistic approach to health care, its vast knowledge of natural drugs and formulations, its emphasis on the psyche and the gentle method of therapeutics, etc., are all worthy of support and research so that mankind's suffering millions may be the ultimate benefactor.

I propose to document the Tibetan medicinal plants growing in the Himalayan ranges. Specifically, with the aid of a local physician, I first intend to concentrate on the plants grown in the Zanskar region of Ladakh, covering a route of some 550 kilometers over a period of about two months. The second exploration will be in Sikkim, where local help has already been offered, and where I have assistance from a Lachung monk-physician who has given me a list of 70 medicinal plants growing around his village. The third exploration will be in Bhutan, where established ties with the Royal Government will be of assistance. The fourth exploration will be done in Nepal, where I am in touch with numerous local physicians familiar with the area's herbal plants. Lastly, if permitted, I would like to go to Tibet, and cover all the areas where the former physicians used to collect their herbs.

As the medicinal plants flower only during the Monsoons seasons, I will be able to explore only one area per year, and that only for the two months. The duration of the project, therefore, is anticipated to be about five years. It is to be carried out in conjunction with my current preparation of *The Healing Art of Ancient Tibet,* a book on this medical tradition.

"Perie Banou Hall's Head"; the remarkable boat that completed a double circumnavigation of the globe without touching land.

The First Solo, Always-at-Sea, Double Circumnavigation

Jonathan William Sanders

20 Jukes Way, Glendalough, Perth, Western Australia 6016, Australia

Australian, born August 12, 1939. Master Shearing Contractor and Yachtsman. Educated in Australia.

I have been an avid sailor since my teen-age years, and have been racing in large boats since the sixties. It was while at sea during the 1979 'Parmelia Race' (Plymouth, U.K. to Fremantle, West Australia) that I first considered the idea of making a solo, non-stop voyage eastwards around the world. Then the dream began of making a *double* circumnavigation, under rigid rules of non-stop, one-man sailing.

Though I have sailed numerous yachts, it was the "Perie Banou" (which I co-own with my brother) that I chose for the attempt. A 34' yacht, designed by Sparkman and Stephens of New York, and built by Swarbrick Bros. in Western Australia, it had been launched at the Royal Perth Yacht Club in 1973. By 1981, I had logged over 65,000 nautical miles in 'Perie Banou', 15,000 of them solo. I knew the boat well, and was confident it would serve for the trip. There were no modifications required, such as strengthening or altering, for the hull.

As to the rest of 'Perie Banou', the expected rigors and needs of the trip did lead to a few changes and additions.

A heavier mast, with all external halyards, and steps for ease and safety in case of an emergency climb, was specially designed and installed. All of the standing rigging was replaced, virtually doubled in strength, and careful attention given to sails and other running gear.

The all-important wind vane steering system was carefully overhauled and strengthened where necessary, and a complete new outfit purchased as a spare, a precaution that very definitedly proved its value early in the voyage.

Crystals of suitable frequencies were fitted in the main radio transceiver, to enable contact to be made through Radphone stations throughout the world.

A small committee of friends undertook to provide 'cover and contact' for the voyage, as well as assisting in obtaining helpful sponsorships for victuals, etc.

After many months of planning, re-rigging, equipping and and re-furbishing, I set sail on Sunday, September 6, 1981, in the renamed "Perie Banou Hall's Head", from Fremantle. In brilliant weather, a marvelous gathering of small craft, both sail and power, escorted me down the river and out to sea on the first single-handed attempt to twice circumnavigate the globe, non-stop, via the world's great capes.

Just three days later, south of Albany, I was in the cabin when a large sea laid the yacht down, swamped the cabin, and carried away the vane of the self-steerer. An omen of things to come! I was able to make repairs the next day, when the weather calmed down a bit, but this report, so early in the trip, must have been disconcerting to my friends back home.

"Perie Banou Hall's Head" and I then settled into our long, but by no means uneventful sail. Rounding the tip of New Zealand, we encountered another storm, which led to a 120° roll, with the mast pointing down into the trough of a very large wave, and everything in the cabin, including me, hitting the ceiling. Once again, a major clean-up job, but then all was well up into warmer waters, as we crossed the Pacific on an average course of 42 degrees South. We rounded Cape Horn on November 13th, as the first solo Western Australian yacht to do so. Immediately we moved northward to find warmer weather, and then headed for our first re-victualing in Storm Bay, south of Hobart. Six hundred miles south-west of Cape Leeuwin, on January 26th, we were once again capsized, this time with a 360° roll. Having learned a bit about the consequences of such a happening, I was better 'stowed', and the clean-up didn't take as long as on the earlier occasions.

I had advised my friends of an expected arrival time and point in the vicinity of Maatsuyker Island, just off the mouth of the Darwent River on February 10th, and would have made it almost on the dot, except for becoming becalmed for 12 hours in the area's flukey weather, and having to amend the time to about 19 hours later. It was a heartwarming rendezvous, with Commodore Des Cooper of the Royal Yacht Club of Tasmania arriving in his 58' flagship 'Lotus Eater', full of replacement supplies, accompanied by a small flotilla of other craft, including a police launch filled with press photographers and reporters.

"Perie Banou" and I were drifting about 400 yards offshore, in a calm sea that was perfect for the needed transfer of supplies. I was able to pass over a packet of letters, and receive many in return, along with a needed new radio set (to replace the transmitter that had gone out in one of the capsizes) and new solar panels. There was a chance to talk a bit with the press people while police frogmen inspected the hull underwater, taking time out to catch a couple of lobsters which were cooked on board 'Lotus Eater' and handed over as a most welcome change of diet. After about an hour, in which all of the rules had been stringently observe ('Perie Banou' moving at all times, at no time anyone boarding her, and at no time me leaving her), the wind freshened a bit, I tightened sail and was on my way again.

Crossing the Pacific once again, we made excellent time, averaging about 120 miles per day without incident until a mid-Pacific scare. The same cyclone that did much damage in the Tongan islands, with winds over 100 knots, shredded and completely ruined a mainsail, even though it was fully reefed. After a few calm days of cleaning up, with radio in excellent condition and communications working well, we once again approached Cape Horn.

This time, rounding on April 13th, we had to take some extraordinary measures, as 'Perie Banou' and I were running before wild and breaking seas in gale force winds of up to 80 knots, with very low temperatures producing snow or hail. At one point, it was necessary to trail a one hundred foot bight of line astern to prevent over-running. After two-and-a-half days of this, calmer weather was a welcome relief.

We passed some 80 miles south of the Falkland Islands, where, in spite of the conflict going on there, I saw no ships or planes. Having nothing to report relieved my friends, who had been concerned about the possible dangers arising from man-made hostilities as well as those of the sea. Our course took us northward, where we became virtually becalmed at the edge of the Sargasso Sea on June 1st. In nine days, 'Perie Banou' and I covered only three hundred miles, although it was a pleasant change to be able to move about the boat without having to hold on. Then, on north-west of the Azores, heading for Plymouth and another re-victualing under the same rules of no boarders, no getting off the boat for me, and continuous sailing by 'Perie Banou'. On June 28, 1982, we sailed into Plymouth Sound and were met in the lee of Drake's Island by the Harbour Master's launch, carrying Commander Lloyd Foster, Secretary of The Royal Western Yacht Club of England, ably assisted by Brian Meegan of Royal Perth. The exchange of letters and supplying went smoothly, along with brief talks with the press, and within an hour, 'Perie Banou' and I were on our way once again.

We headed down the coast of France, past Portugal and Spain and North Africa, making good time, and having the chance to enjoy reading the letters and papers put aboard at Plymouth. On July 22, some 600 miles north of the equator, the good conditions eased off, with glassy seas and little breeze signalling our approach to the "Doldrums". With a view to avoiding the delays, we shifted westward, coming down off the coast of Brazil, and then swinging eastward to pass north of the island of Tristan da Cunha. Once again rounding the Cape of Good Hope, I planned to hold a course of about 40 degrees south latitude to head straight into Fremantle, but gale winds forced us south. In all, we were to undergo four more 'knockdowns' of 100° to 120° in rough and uncomfortable seas. However, by September 20th, I felt able to advise my friends by radio that we expected to be in Fremantle, some 4,360 nautical miles away, on October 31st.

'Perie Banou' and I made excellent time, and I appreciated a few days of light wind that gave a chance to re-varnish the woodwork topsides and generally tidy things up. On Friday, October 29th, a spotter plane picked us up about one hundred nautical miles south-west of Rottnest Island, and we knew our friends were getting ready for our arrival. As the 'festivities' had been planned for Sunday, October 31st, at eleven o'clock, I eased sheets, and we just loafed along, by this time being met by the vanguard of what was to be hordes of welcoming boats.

Arriving at Fremantle, we tacked into port amidst a veritable armada of well-wishers. The finishing gun was fired at 11:02, on October 31, 1982, four hundred nineteen days, twenty-two hours and ten minutes from the beginning of our voyage of 48,510 nautical miles; twice around the world, non-stop and solo.

Balloons in the service of archaeology, over Luxor's Valley of the Kings.

Mapping Egypt's Valley of the Kings: High Tech Archaeology

Kent Reid Weeks

American Research Center in Egypt, 2 Midan Qasr el Doubbara, Garden City, Cairo, Egypt

American, born December 16, 1941. Associate Professor of Egyptian Archaeology, University of California, Berkeley, U.S.A.; Director of the Berkeley Theban Mapping Project. Educated in U.S.A.; Ph. D. (Egyptology) from Yale University in 1970.

In spite of its fame and great archaeological importance, there exists no complete and accurate map of the Theban Necropolis at Luxor, Egypt. The Valley of the Kings lies here, as do scores of temples, thousands of tombs, ancient villages and shrines, covering a ten-square kilometer area that has never been adequately studied. The lack of such maps or plans is a serious problem for scholars. Worse, without detailed architectural and topographic studies, it is impossible to protect and preserve these monuments. The increasing number of tourists in the area, increases in theft and vandalism, and recent environmental changes, are all threatening these ancient sites.

A detailed archaeological map is a necessary first step if plans are to be made to preserve and protect these monuments. It is the goal of this project, working in association with the Egyptian Antiquities Organization and other agencies, to make such a map available.

Our map is being made using the most modern surveying techniques available, by a small staff of trained engineers, architects and archaeologists. Many of the methods that we have developed will be of value to other archaeologists around the world, and our project has been called one of the most innovative and important ever to be conducted in field archaeology. It is one of the most complex and ambitious, as is shown in the five annual reports (The Berkeley Map of the Theban Necropolis, Years 1978-1982) that we have produced on the work done to date.

To make our map of the Necropolis as complete as possible, we are utilizing a number of new techniques to locate hitherto undiscovered tombs and other archaeological features, both on the ground and below it. For example, once we have planned and sectioned known subterranean passageways, we can electronically survey the bedrock between them, thus identifying chambers or shafts without ever using pick or shovel.

To locate features in the rugged cliffs that cut through the Necropolis, we are making use of hot-air balloons. Floating slowly a few meters away from a cliff face, we can identify tomb entrances, concealed caches, graffiti and the like.

Last year, two features, a tomb and a Christian hermitage, were found using the balloon technique.

Because of these new techniques, we will not only be able to provide maps and plans of all archaeological features to the Egyptian Government to assist in its plans for the Necropolis, but we also will be able to trace the history and development of the area; to show how its archaeological features articulate with the geography; and to explore the region further without creating the problems that more traditional archaeological work normally generates.

Because of its rich history, the Theban Necropolis has long attracted both visitors and scholars. It was until the beginning of the 19th century, however, that maps in modern form began to be made of the area. While some were sumptuously produced, all were at very small scales — from 1:10,000 to 1:250,000 and more, and they only schematically indicated the position of the more obvious monuments. From 1900 to 1960, much progress was made, in part due to the active archaeology undertaken at Thebes, and maps on a scale of 1:1,000 were begun on the entire Necropolis (though the project was not completed). With aerial photography, more information became available, but a complete map still did not exist.

In spite of nearly 170 years of Theban mapping, then, there did not yet exist an accurate and archaeologically complete map of the Theban Necropolis. The questions were: What should any new map of this important area include?, and, Why is the preparation of a new map of special importance now?

Any new map of the Necropolis would desirably have a scale large enough to allow all significant archaeological features to be plotted easily and accurately, so that the close relationship between those features and the terrain could be studied. Sufficient geographical detail would be needed to permit the easy and accurate location of both hypsographical and subterranean features. Such items as paths, cairns, caves and fissures, or such modern improvements as roads and power lines would be included, along with a grid system to permit precise location of all elements.

Subterranean elements, such as tomb shafts and chambers, or natural breaks in the bedrock should be noted in detail. The Theban Necropolis, honeycombed with such features, would need a three-dimensional cartographic description. Such a map would provide archaeologists with a master key to the monuments of Thebes.

The Berkeley Map project began with the season of 1978, in an effort to make the most complete map of the area ever undertaken.

In 1978, during a three-month field session, the basic grid network, covering the entire Luxor area, was set out and defined, as the foundation for all future work. Elevations were run for all points on the major and minor traverses, and for many intermediate points as well. As an example of what the expected end-result map sheets might look like, two grid squares at the entrance to the Valley of the Kings were topographically mapped. This mapping was done at a scale of 1:250 and published at 1:500, with 1-meter contour intervals. Also, maps, sections and dimension schedules were prepared for seven tombs in the Valley of the Kings. Tomb plans were drawn at a scale of 1:100 for publication at 1:250.

During 1979, the project completed the targeting for aerial photography, and obtained complete aerial photographic coverage of the entire Necropolis in two runs; one at 3,000 feet, and the other at 5,000 feet. The grid network was extended north and south to the limits of the Necropolis, survey monuments were established in archaeological areas, and ten major tombs were planned and sectioned. Additionally, a comparison was begun with all former plans, including the ancient plans for the tomb of Rameses IV, with the project's own measurements.

In the 1980 season, we obtained a full series of oblique aerial photographs of the Necropolis, mapped and planned all accessible tombs in the East and West Valleys of the Kings, completed the Necropolis-wide grid network, and plotted surface archaeological features for the Valley of the Kings volume of the Theban Atlas. Also, toponymic studies of the Necropolis were undertaken, and we began experiments with computer graphics for the preparation of axonometric tomb drawings, and experiments with topographic sheet design and layout.

1981 saw the mapping of all accessible tombs in the Valley of the Queens and adjacent wadis, the mapping of all surface features of archaeological interest in the 3-square-kilometer area between QV and KV on our maps. Computer programming was developed for the preparation of tomb plans, elevations and axonometric drawings; the Necropolis-wide traverse was extended to the southern and northern limits of the archaeologically-relevant West Bank, and work was continued on the toponymy of the West Bank.

In the project's fifth season, 1982, all topographic and architectural work in the Royal Necropolis was completed. All tombs in the several wadis lying at the southern end of the Necropolis were mapped and re-numbered. Use of ropes and rock-climbing equipment enabled us to inspect and plan several hitherto inaccessible cliff tombs. Hot-air balloons enabled us to take a series of oblique aerial photographs of various parts of the Necropolis and to explore cliff faces for features of archaeological interest.

In 1983, we are planning to devote our resources to the publication of the Valley of the Kings "Atlas" volume, after which we intend to return to the field in 1984. Our work with the rented hot-air balloons (we had two on a short term lease) was so successful that the project has now purchased our own balloon. That, coupled with new electronic sensing equipment, will allow us to move into what we confidently expect to be an even more intensive search for the hidden secrets of one of the world's most magnificent archaeological sites.

Volcanoglaciospeleogy — Exploring the world of "fire and ice" in extraordinary underground caves.

Exploring Volcanoes Under Ice

Gérald Favre
1261 La Rippe, Switzerland

Swiss, born March 8, 1952. Geologist and film-producer. Educated in Switzerland; Hydrogeologist Diploma (1976) from University of Neuchâtel.

This present project includes the exploration and the study of a very special environment, utilizing the sciences of volcanology, glaciology and speleology. The best conditions for this kind of work are found in Iceland, under the Vatnajokull glacier, and I wish to organize an important expedition to this area in the summer of 1984. The expedition would extend further our research into one of the most unknown and fascinating subjects in the world; volcanoes under glaciers, to be explored with speleological and diving techniques.

After having spent 17 years exploring caves in limestone rocks, I decided to enlarge my field of investigations to the relationship between volcanic rocks and ice. I wish to push my studies of this subject further, and to bring the remarkable world of volcanoes under ice to a greater public through filming it.

When I first began to research the world of caves in ice, I had no idea of the implications that would emerge. With my group, over a period of two years, I investigated the most important alpine glaciers in France and Switzerland, between Chamonix and Zermatt (mer de Glace, Argentière, Trient, Les Bossons, Aletschgletscher, Bisgletscher, Gornergletscher, etc.). During these observations, we were able to discern the very important role played by thermal energy and the ice tectonic, causing the formation of an under-ice waterflow and caves in the ice. We then searched for the more powerful energy relationship between geothermic energy and glaciers. The best place to find such relationships in the European area is in Iceland, with its under-ice volcanoes.

A reconnaissance trip in 1980 gave us the certitude that the chosen area and subject were quite unusual, so we organized a speleological and cinematographic expedition to Iceland in 1982. During this trip, we had excellent results and we now wish to return to push the exploration and, if possible, to solve the question of the origin of the geothermic energy under the Vatnajokull glacier. In a party of scientists, cineastes and speleologists, we hope to focus on the two areas of Kverfjoll and Grimsvotn, where huge under-ice gallery systems have already been found.

Iceland is situated just on the Mid-Atlantic Rift, and is a country with permanent volcanic activity. Large parts of the country are still covered by glaciers; in some areas, volcanic activity coexists with an ice cover, and nobody knows exactly what happens in these special places. From 'outside', only some of the effects can be observed during the 'Jokullhaup' (the big water flow that sometimes matches the size of South America's Amazon, at 100,000 cubic meters per second) at some of the 'solfatares' (hot springs) situated near the edge of the glacier, as in Kverkfjoll. Under the Vatnajokull are several areas known as potentially active; Kverkfjoll, Grimsvotn, Bardarbunga and Oraefajokull. To the South-West, the Myrdals glacier has an active volcano and an under-ice geothermic area, the Katla.

The Kverkfjoll caldeira, or collapsed volcano, our main area of investigations, is situated on the northern flank of the Vatnajokull, and can be reached between June and October by a desert track with 4-wheel drive from the North.

Grimsvotn is situated very close to the center of the Vatnajokull, and can be reached only with a snowmobile for a heavy expedition, or by skis for a light party. This under-ice volcano is directly in relationship with the southern coast during the Jokullhaup. Its under-ice lake, formed during several years of buildup, empties in a few days, leaving open galleries under the glacier for a distance of 40 kilometers that drop from an altitude of 1,500 meters. The exploration and study of these galleries could be one of the most fantastic challenges of our time, and is possible only with very well trained people and sophisticated equipment.

During our two previous trips, a number of new techniques were developed for use in such a special environment, and we had to adapt our traditional caving techniques to the glacier's exploration and to the volcanic activity. Progression in and under the ice is based on a synthesis between mountaineering and caving equipment, with which we had experimented previously in the Alps prior to going to Iceland. In 1982, we were able to solve, in a limited way, the problem of lack of oxygen and excess carbon-dioxide. This enabled us to push the exploration of the 'hot river' in Kverkfjoll to a distance of 400 meters, with a closed oxygen system and bottles of 'chaux sodee'. To go further, we must now develop or find a more sophisticated breathing system that will allow up to 10 hours capacity.

During our two expeditions, we were able to make a 16mm documentary film while exploring several areas:

a)   The "Hot River". This runs out from the front of the glacier near Kverkfjoll, with a large, 50-meters deep pothole and has been explored to a distance of 400 meters; further exploration will require special equipment. In the surrounding soil, very hot water (35°C) springs exist.

b)   The "Upper River". This is the most significant discovery, with a length of 2 km under the glacier, and a difference in altitude of 400 meters, wich gives it the deepest actually known under-ice system in the world. No connection has been found to date with the "Hot River", but the exploration is not finished. All along the under-ice river, several areas of solfatares warm the water of the river and melt the ice over the 2 km distance now known. Several pitches and numerous lateral chambers, decorated with ice-concretions and crystallizations, need more study and investigation.

c)   Several ice-caves. Not far from the surface, and noted in the upper part of the glacier, these have already been studied by Icelandic geologists and glaciologists, though not in the dark, and underwater as was done by us.

We have prepared all the advance items for our planned 1984 expedition to the two above mentioned areas (Kverkfjoll and Grimsvotn). About ten people are organized for an expedition that will explore, carry out scientific studies and make a film. The feasibility of the project is certain, but it depends largely on obtaining financial support. The results of the expedition, to be distributed via films, a book for the general public, and publication in scientific reviews for specialized subjects, will bring new information and knowledge of volcanology and glaciology to light, and focus attention upon the relationship between geothermic energy and its potential utilization by people.

Beneath the waters of Truk Lagoon, a sunken ship's gun turret; mute testimony to lives lost in war.

Search For Lost Ships in Truk Lagoon

Klaus Peter Lindemann

Leuschnerstrasse 34, 67 Ludwigshafen, West Germany

*West German, born November 13, 1939. Business executive with
BASF AF, West Germany. Educated in Business and Economics.*

On February 17/18, 1944, U.S. Task Force 58 attacked Truk Lagoon on a
major scale with carrier-borne aircraft. As a consequence, the defending
Japanese forces lost almost all their airplanes and ships, on this "Gibraltar of the
East". The military value of the base was reduced dramatically, and the
garrison of 20,000 soldiers was isolated. The American advance continued
with the capture of Eniwetok and the Marianas without interference from what
had been a major aircraft and naval threat from Truk.

The operation, code-named "Hailstorm", had never been written up in any
detail. Even authoritative books in the U.S. (S.E. Morison: "History of the US
Naval Operations in World War II") and Japan ("Boueicho Bouei Kenshujo
Senshi Shitsu") make only short references to it. This project's objective is to
increase knowledge of this aspect of modern history by searching for wrecks,
recording eye-witness accounts, obtaining information and processing this
information.

Background

Truk Lagoon is located in the Pacific, within the Eastern Caroline islands, 565
miles (910 km) southwest of Guam, at 151° East and 7° North. It is the remnant
of an ancient shield volcano, which has eroded and is now partially submerged.
The outer coral rim, showing the original size of the volcano, is roughly
pear-shaped, 30 miles long and 40 miles wide (50 by 63 km). The water depth
outside the rim is abyssal, while inside the lagoon the bottom undulates, varying
between 10 feet and 250 feet (3m to 75m).

Truk's modern history began with Spain annexing large areas in the Pacific,
bringing Christian faith to the islands and pacifying the natives. When Spain
withdrew from the Pacific, these possessions were sold to Imperial Germany in
1899, who administered Truk until World War I. Japanese forces seized the
undefended positions quickly. The League of Nations later mandated the
islands to Japan.

In the years between the wars, Japan considered the Mandated Islands as
very important forward positions and closed them to all outside traffic and
view. Their development was shrouded in total secrecy. However, in Truk,
military installations were not constructed until shortly before World War II,

ODYSSETRON:
The First Unmanned Marine Circumnavigation of the Earth

Bryan Leigh Rogers

4391 24th Street, San Francisco, California 94114, U.S.A.

American, born January 7, 1941. Interdisciplinary sculptor/engineer/ educator; Associate Professor of Art, San Francisco State University; Chair, Editorial Board of Leonardo, Journal of the International Society for the Arts, Sciences and Technology. *Educated in U.S.A.; Ph.D. (Chemical Engineering) in 1971 from University of California, Berkeley.*

ODYSSETRON is an in-progress environmental enterprise designed to accomplish the first unmanned, marine circumnavigation of the earth. It evolves over four phases:

1) Launching 100 free-floating, *information carrying* modules in the San Francisco Bay waterways (1980). These geometrically rationalized, anthropo-morphic modules carry graphic information which, in part, pertains to the time and location of their deployment.

2) Launching 64 free-floating, solar-powered *transmitting* modules in the Pacific Ocean (1983). Each module, carrying both solar-powered radio and audio transmitters, will periodically transmit information relating to its origin and to the overall project. Encoded in electronic music, this information will be audible and radio-receivable by all life forms in its range of transmission.

3) Launching 32 free-floating, solar-powered *communicating* modules along the Phase 4 route (1986). Each of these modules will be capable of transmitting as well as receiving information. These capabilities will be linked, making each module a communicator. By being deployed along the cir-cumnavigation route planned for Phase 4, these modules will test the appro-priateness of the route and herald the forthcoming robot.

4) Launching 1 solar-powered module: a self-propelled robotic marine navigating system with intelligence on board for *circumnavigation* of the earth on a prescribed course (1990). This robotic craft's activities cannot be control-led remotely, but will be monitored.

Each phase becomes more complex and influences characteristics of subse-quent phases. The first phase has been launched; the second phase module is in the prototype stage. Beyond their conceptual roles, the first three phases serve two practical roles: 1) developing a sympathetic social climate on the planet to maximize the chances of success of Phase 4, and 2) establishing a step-by-step build-up of the technical, social and financial resources required to carry out Phase 4.

ODYSSETRON is a comprehensive, long-term project inspired by Homer's Odyssey. In the fourth and final phase of this project, Odysseus reappears after an absence of three millennia as an electromechanical device, whose task is a marine circumnavigation of the earth. ODYSSETRON operates with a considerable degree of technical sophistication and is infused with numerous social, artistic and scientific concepts. Therefore, it exists in a cultural emulsion.

ODYSSETRON is my response to a paradox. I believe that humanity is ceaselessly searching for planetary unity, yet it appears that the social fabric of the entire planetary culture is increasingly fragmented. To approach a sense of unity, the planetary culture must transcend the local perspective. With this quest in mind, I have designed ODYSSETRON, a meta-level work which indirectly addresses several contemporary issues:

The Environment/Planetary Cooperation

ODYSSETRON is conceived as a global symbol to generate planetary cooperation and create environmental awareness. It bears no arms and carries no flags. If the quest of the small craft of Phase 4 is to be realized, many nations must cooperate with and assist it. The effect of such cooperation in a nonmilitary venture will be a step toward management of life-supporting global resources and co-evolution of life and the planet. ODYSSETRON is becoming a global event, an international endeavor whose spirit can capture imaginations across ideological lines.

The environment is linked to international cooperation. Like the earth and its inhabitants, ODYSSETRON is extremely vulnerable. It is conceived as a poetic meta-model for planetary cooperation and environmental awareness — a requiem for national isolation and planetary resource exploitation.

Artificial Intelligence/Robotics

The philogeny of ODYSSETRON's intelligence echoes that of the organism, progressing from simple to complex in organizational structure. ODYSSETRON evolves to Phase 4 to become an electromechanical intelligence, the first robot to re-enact, in essence, the myth of the classical Odyssey. In the act of circumnavigating the earth, it questions humanity's identity and opens territory for new identity quests.

Art/Science/Technology Interface

ODYSSETRON is a synergistic melding of art, science and technology, which have become increasingly separated from each other since the integrated Renaissance culture. ODYSSETRON addresses audiences without regard to traditional cultural niches.

In our search for unity, we probe for understandings, values and connections in ever smaller containers, not only in art and science, but in all aspects of a fragmented world civilization. ODYSSETRON reverses the microscope. It is a synthetic gesture, a poetic collage of philosophical concepts, political issues, and art/science events, potentially having an enormous impact on the planetary culture. It is a line of thought which I am casting through the evolving present into a quasi-predictable future. I intend to use the line to wind in that future, to which we have a direct — if uncertain — connection.

If a robot succeeds in performing the terrestrial circumnavigation, the assorted obsolescence that follows invites new modes for human odysseys. In response, the Grand Design Equation can chart unbounded inward/outward trajectories for the spirit of intelligence.

The majestic Green Sea Turtle (Chelonia Mydas) needs help from man in protecting its nesting grounds in a hostile world.

Locating and Conserving Sea Turtle Nesting Grounds in the Andamans

♛ Satish Bhaskar

Honourable Mention — The Rolex Awards for Enterprise — 1984
c/o Maj. P. Bhaskaran, AC (Retd.), "Jaitavana", Kannamkulangara, Trichur 680 007, India

Indian, born September 11, 1946. Principal investigator in World Wildlife Fund-India project on sea turtle conservation. Educated in India in Electrical Engineering at Indian Institute of Technology in Madras.

In order to locate and conserve hitherto unrecorded nesting populations of endangered and threatened sea turtles, I propose to spend eight months alone on the uninhabited island of South Sentinel (approximately Latitude 11°N, Longitude 92° 16′ 36″ East) in the Andaman Islands.

Probably on account of non-availability of fresh ground water, and its remoteness, South Sentinel apparently has been visited only rarely by humans. The island is part of the Indian Union Territory of the Andaman and Nicobar Islands, which is administered from Port Blair by a Lieutenant Governor appointed by the Central Government at New Delhi.

South Sentinel is a flat, roughly circular, coral island, about 6 kilometers in circumference. Lagoons make up about half of its shoreline, with the other half being comprised of rocky or sandy beaches. One of India's loveliest sand beaches occupies about 2 kilometers of the northwest coast, and it is here that sea turtles, particularly the Green turtle, come for nesting and sunning.

Based on surveys carried out by me on nearby islands, it appears likely that one or more other sea turtle species (Hawksbill, Olive Ridley and Leatherback) additionally nest at South Sentinel. Earlier reports by scientists visiting the island were related to stays during February 1973 and March 1974, months which are believed to be lean seasons for nesting Green turtles in the Andamans, based on my studies. I am therefore proposing to spend the period of 1 June 1984 through 31 January 1985 on South Sentinel, to cover the peak nesting season for Green turtles, and to incorporate the nesting seasons for any other sea turtle species visiting the island.

All necessary equipment, provisions and medical supplies will be carried to the island at the start by hired motor launch from Port Blair, to make the expedition self-supporting. Some freshwater will also be carried, but most will be collected from the rains, as both monsoons — the southwest and the northeast — will occur during the study period. Two tents are being taken to provide accommodation and shelter for equipment and records. No contact with humans, including members of the friendly Onge tribe who inhabit Little

Andaman Island about 25 kilometers distant from South Sentinel, is anticipated over the course of the 8 month expedition. Official permission to stay on the island will be obtained in advance from the Central Government.

The Green turtle *(Chelonia mydas)* is considered by many as the most valuable of all reptiles, and is known all over the world as the 'soup turtle'. Its shell is a dirty green and its fat a light green, accounting for its popular name.

Six of the world's seven surviving sea turtle species, including all five known from Indian waters, are listed in the International Union for the Conservation of Nature and Natural Resources' Red Data Book for threatened and endangered species. The study at South Sentinel will be aimed primarily at the conservation of sea turtles on this island, and will provide further information on the basis of which the island could be set aside as a sanctuary for these reptiles.

Since September 1977, sea turtles have been placed in Schedule I of the Indian Wildlife (Protection) Act, 1972. This stipulates that all sea turtles are totally protected, and that anyone found killing a sea turtle or collecting its eggs or dealing in sea turtle products is liable to imprisonment up to 6 years and a fine of a minimum of Rupees 500 (about U.S. $60).

Notwithstanding widespread knowledge of the fact that sea turtles are protected, the eggs of all species are actively sought for almost everywhere in the Andaman and Nicobar Islands. Apart from the colonization of nesting beaches by man and the hunting of adult turtles for meat, this constitutes the greatest threat to the future existence of the sea turtles.

Natural predators include Monitor lizards which can and do excavate sea turtle eggs, often fighting among themselves for the right to the spoils. Wild pigs may be another predatory enemy. Hatchlings are preyed upon by ghost crabs, rats and, in all likelihood, hermit crabs.

In order to provide information to support the recommendation to make South Sentinel a preserve, the data to be collected will include:

1)   The species and numbers of sea turtles nesting on the island each night.

2)   Those species predatory on turtle eggs and young, and the degrees of predation.

3)   Observations on nesting, re-nesting intervals, clutch sizes, hatching times and temperatures, courting and mating habits, weights and sizes of nesting females, hatchlings and eggs, and nesting habits.

The survey method will be similar to that used by me in earlier studies of sea turtles in the Andaman and Nicobar Islands. The sandy beach coastline is first surveyed on foot for evidence of turtle tracks, nests, excavations, shells, carcasses, and skeletal remains or egg shells. Sightings of turtles at sea, on land while nesting, or when emerging from nests as hatchlings are recorded. Clutches of turtle eggs are located by digging, and data on these are noted under the following categories: number of eggs in a clutch, track width, depth of topmost and deepest eggs, sand temperature at nest level, size of eggs (using vernier calipers), evidence of predation on eggs, and the nesting habitat (eggs are carefully replaced in a manually excavated 'nest' after data are collected). Additionally, turtle carapaces or carcasses found on shore are measured for carapace curved length and curved width, and skulls for head width.

As many nesting turtles as possible will be tagged in order to obtain data on migration routes. If nesting numbers warrant, as is considered to be likely, follow-up studies of shorter duration will be undertaken in later years, in the hope of revealing re-migration intervals and growth and recruitment rates.

Other data of natural history interest will be collected. Special attention will be focused on the Coconut crab, which has been the subject of the only in-depth study at South Sentinel so far. The occurrence of the Estuarine crocodile *(Crocodylus porosus)*, to date unreported from South Sentinel, will be looked into also.

After the study, a detailed report and list of recommendations pertaining to the conservation of the sea turtles at South Sentinel will be prepared and sent out to the IUCN, the Indian Government and to other organizations concerned with conservation.

Combined Production of Paper and Starch from Bamboo Culm

Anisio Azzini
Rua Sampaio Vidal 466, 13100 Campinas, SP, Brazil

Brazilian, born December 15, 1942. Scientific Researcher at Instituto Agronomico de Campinas. Educated in Brazil; M.A. in 1976 from University of Sao Paulo.

The economical importance of bamboo has been increasing in tropical and semi-tropical lands as a raw material to be used in the production of pulp and paper, mainly because bamboo is a perennial, fast-growing species that supplies a long fiber.

Although bamboo has long been known to have certain qualities that are important in the paper making process, a key factor inhibiting the greater use of this apparently obvious raw material has been the high starch content found in the bamboo culm. This project is concerned with the further development of a promising method of production which can combine the production of pulp for paper, and starch for other uses, in one process.

Although the high potential of the various bamboo species for the production of paper with attractive strength qualities is known, conversion of the bamboos to pulp by conventional cooking methods encounters problems, the most important of which is the starch present in the parenchyma cells of the bamboo culm. The presence of this starch during the cooking process reduces the pulp yield, increases the consumption of necessary chemicals, and contributes to problems in the liquor recovery system.

Because of this, we have developed a system involving the shredding of chips in water to reduce the starch content prior to bamboo cooking.

Our preliminary studies with *Bambusa vulgaris* have shown that our soda process showed an increase on pulp yield from 42.4% for non-shredded chips to 59.1% for shredded chips. The starch content was about 4%. With the species known as *Guadua flabellata* we had yields of 8.53% starch, 22% of parenchymatous fraction and 62% of fibrous fraction after the shredding of the chips. Given these initial results, it is our intent to study a variety of bamboo species, in order to determine which might most efficiently be used for paper making.

We will be studying ten species taken from the bamboo collection of the Instituto Agronomico de Campinas: *Bambusa vulgaris, B. vulgaris var. vittata, B. tuldoides, B. beecheyana, B. stenostachya, Dendrocalamus giganteus, D. latiflorus, Guadua augustifolia, G.amplexifolia,* and *Phyllostachys edulis.*

The bamboo culms will be converted into chips by a semi-industrial bamboo-chipper. The chips will then be shredded in water, resulting in fibrous, parenchymatous and water soluble fractions. The starch will be removed by decantation from the water soluble fraction.

The basic density and fiber dimensions of the bamboo culm will be determined for both shredded and non-shredded chip samples, according to well-known scientific methods.

The pulp will then be prepared from shredded and non-shredded chips, using a laboratory digester with 20 liters of capacity. The soda process will be used according to the following cooking conditions:

- Soda (As Na20) — 14% (over dry material).
- Ratio of dry chips to solutions — 1/4.
- Maximum temperature — 160°C.
- Time to reach maximum temperature — 100 minutes.
- Time at maximum temperature — 45 minutes.

Total yield, screening yield, rejects, kappa number, whiteness and strength properties will be determined according to ATBCP (Associacao Tecnica Brasileira de Celulose e Papel) methods.

Expected Results

The results anticipated from this project represent a new technology that could be important in bamboo utilization. The combined production of starch and paper is an advance on present bamboo utilization because, in addition to the production of an important food product (the starch), it is now possible to obtain pulp and paper from bamboo with much better results.

The bamboo starch produced by the process has a very small grain size (5 microns), and can be easily used as a food or converted into ethanol by fermentation. Based on our work to date, the bamboo starch produced by this process shows the following chemical characteristics: Moisture — 12%, Protein — 0.67%, Amylose — 23.98%, Fat — 0.22%, Fiber — 0.28%, Ashes — 0.57% and Glucose — 98.41%.

Steps in the process commence with the bamboo culm, which is processed by chipping into chips that are then shredded and screened. Fibrous fraction then leads to pulp and paper. The starch and fines (residues) in suspension are twice decanted, allowing the removal of the residues, and leaving a starch suspension. This is decanted a third time, resulting in starch that is dried and stored.

The successful development of this process to an industrial level could be of considerable help in helping to save our world's rapidly diminishing wood forest resources.

The Mt. Elgon Caves: Mined by Wild Elephants

Ian Michael Redmond

38 Lime Road, Ashton Gate, Bristol BS3 1LT, U.K.

*British, born April 11, 1954. Wildlife Biologist and Conservationist.
Educated in U.K.; B.Sc. (Hons) in Biology from University of Keele, Staffs.*

Since their discovery in 1883 by Joseph Thomas, the caves of Mt. Elgon in Kenya have intrigued visitors to the area. Anthropologists wrote about the tribesmen inhabiting some of them with their cattle, geologists mused about their peculiar structure, and hunters observed that elephants and other animals used them as subterranean salt-licks. Surprisingly though, no one made any detailed studies of the caves or the animals that use them until I began this project in 1980. It is my intention to thoroughly investigate the phenomenon of elephant geophagy (eating earth and rock), particularly in relation to the formation of caves.

In 1980, Col. John Blashford-Snell and I monitored Kitum Cave for five nights and built a path for tourists to visit it without using the elephant path. I was fortunate enough to witness elephant visits on three nights — an experience that fired my curiosity but which proved frustrating. No one could give me any hard facts beyond the observation that the rock appears rich in sodium sulphate, and that this was presumed to attract the animals. Later that year, I returned to Elgon independently and mapped Kitum Cave with a friend, and then stayed in the cave alone for another week to make further observations. I succeeded in photographing a herd of seven elephants in the entrance chamber — the first ever pictures of elephants actually underground. On subsequent trips to the area, in 1981 and 1982, more photographs were taken and more research done, not only in Kitum Cave but in 12 other caves in the area. It was during a trip in 1982, assisting a film crew to make a documentary for English television, that I was able to observe the actual technique of mining used by elephants underground. I also recovered fragments of rock from faeces, and discovered that 45% of dung piles contained such fragments.

A major roof fall in the inner third of Kitum Cave occurred in July 1982, and the elephants can no longer penetrate to the maximum depth. Where their old route is blocked, they are now beginning to mine the newly fallen rock, thus demonstrating how a high roof can be created. The film crew used infra red light and an image intensifier to film in the cave in total darkness. If available for research, this technique would enable observations to made in the cave with little or no disturbance to the animals being observed.

The caves so far visited are all variations on the same theme — a cul de sac extending horizontally into the hillside from a cliff or valley side, with varying amounts of fallen roof obstructing access. The development of Elgon's caves fits neatly into a five stage process, from a simple chamber through high roof with many rock falls, and finally to complete collapse. Examples of all stages were seen, but the time span for a single cave to pass through all five stages can only be known when we have ascertained the rate of excavation.

Observations in the field and laboratory analysis of rock, food-plants and water samples have yielded certain facts, which can be summarized as follows:

In areas of high rainfall, where soluble salts are leached from the topsoil, plants contain low levels of ions such as sodium and potassium. Herbivores in these area must visit salt-licks (exposures of mineral rich strata) to redress the balance by eating this earth or rock — a behaviour known as geophagy. On Mt. Elgon, a dormant, late-Miocene volcano, caves seem to be the best source of minerals available to herbivores since exposures of mineral rich strata above ground appear too hard to be exploited. Several species of herbivore penetrate into the dark zone of the caves in search of mineral rich rock, including not only the elephants but also buffalo, water-buck, bush-buck, duiker, black-and-white colobus monkeys and baboons.

Elephants cannot lick the rock, and so must loosen earth or prise off bits of rock with their tusks and then convey the particles to their mouths with their trunks (this is what is referred to as 'mining'). Rock is briefly ground between the molars and swallowed, thereby being removed from the cave.

Elephants prefer certain caves to others; different social groups visit the same cave on different nights, usually arriving at dusk. An 'elephant visit' might last up to five or six hours, but not all this time is spent mining rocks. They indulge in other activities, such as play, rest, bathing, rubbing on rocks, etc., in the dark and relative warmth of the cave (night temperatures on the mountain usually fall below the constant temperature in the depths of the caves).

Conventional theories of cave formation (speleogenesis) fail to account for the features common to most of Mt. Elgon's caves: the suggestion that elephant geophagy could be responsible for such huge caves is a radical departure from accepted geological thinking. Our early work has provided the qualitative data and suggested an interesting theory. More importantly, it has served to crystallize the questions that need to be asked next, and shown the direction for further fruitful research.

Our research aim includes studies relating the caves of Mt. Elgon to their past and present usage by elephants, a study of the local elephant population, and assessment of the extent of its dependence upon caves for mineral salts, as well as other aspects of the role the caves play in the environment of Mt. Elgon.

The exotic world of the tropical rain forest canopy, with its millions of little known species, is now being reached by research biologists with a unique, new "web" that gives vastly expanded access to "upper space".

Exploring a Virtually Unknown World: The Tropical Rain Forest Canopy

♛ Donald Ray Perry

Rolex Laureate — The Rolex Awards for Enterprise — 1984
247 "C" Bicknell, Santa Monica, California 90405, U.S.A.

American, born July 10, 1947. Tropical rain forest/canopy biologist at University of California at Los Angeles; freelance photojournalist and film maker. Educated in U.S.A.; Ph. D. in Tropical biology (1983) from University of California at Los Angeles.

The tree tops of the world's tropical rain forests possess up to 40% of all species of life on the earth. This is estimated to be several million species, and over 75% of these are unknown to science. Extrapolating from the known uses of canopy products in medicine, agriculture and industry, we can expect the canopy to be a vast, unexploited storehouse of biological material useful to mankind. Yet effective methods for exploring and tapping these reserves have not been developed. I propose to design and construct an advanced canopy research facility that would make it possible to thoroughly explore the complex aerial communities of tropical rain forest tree tops.

The canopies of the earth's tropical rain forests contain the most complex communities on earth. The problem with trying to explore this region is that the trees reach heights of up to 200 feet, comparable to that of a 20-story builiding. This, combined with man's fear of heights, has kept the canopy a hidden community. Over the years, a few biologists have tried to conquer these problems, but with little success. A quote from *The Forest and the Sea*, by the tropical biologist Marston Bates, illustrated the difficulty of canopy exploration: "In my (South American) forest, I was always confined to the trunks of the trees. I had no way of getting out into interarboreal space (the zone of the tree limbs) — I could be a poor sort of monkey, but I had no way of being a bird."

When I began studying tropical biology in 1974, there was no simple method for gaining access to the tops of tall trees. To do canopy research, I had to devise a method for climbing trees. Michael Emsley, the editor of *Biotropica*, a scientific journal on tropical biology, says that these methods have brought us to "the threshhold of a period of enlightenment."

Special tree-climbing techniques opened the door to a wealth of information, while withholding the most exciting regions. Many tropical rain forest trees have limbs that are too weak to climb. As many as 75% of the trees fall into this category. Even the peripheral regions of the trees which can be climbed are generally inaccessible.

In 1979, I took another step in making the canopy more accessible for exploration. I developed a method that was unparalleled in the access it provided to large forest volumes. The method is a web of 1,200 feet of ropes that are suspended from the tops of emergent trees, the tallest trees in the forest. The web provides three-dimensional mobility within the canopy from ground level to above the tree tops. A researcher no longer needed to be tempted to take that dangerous step out onto a weak limb.

Still, danger is the primary reason that canopy research is not moving at a reasonable rate. Says Michael Robinson of the Smithsonian Tropical Research Institute in a recent *Newsweek* article about my canopy research, "People don't do their best work when they are scared to death." In the same article, Terry Erwin of the Smithsonian Institution emphasizes the need for exploring the tree tops. He says of the canopy, "It's the last biotic frontier."

It is my objective to take canopy access past this final hurdle. The canopy web has become the prototype for a facility that most tropical biologists will find easy to use. I am now designing a web made of stainless steel cable, a rigid chair, and wireless controls. This will automatically lift an investigator from ground level to above the tree tops at the rate of a foot per second. Removing physical work, while strengthening mechanical elements, should do much to take the fear out of canopy research.

Tropical rain forests are disappearing at an astonishing rate. About 3,000 acres are being cut every hour. By the year 2000, it is estimated that unique forests in about 20 countries will have vanished. Increased research means increased awareness and increased protection. To increase research and exploration of tropical rain forest, the automated web and systems like it are urgently needed.

It is widely acknowledged that we are rapidly approaching the greatest wave of extinctions since the end of the Cretaceous period, when a whole community of organisms, such as dinosaurs and gymnosperm jungles, disappeared forever from the earth. I wonder, can we afford to let the most complex environment in the history of life on earth disappear without trying to explore it and save it from extinction?

Central American forests are one of the world's most threatened ecosystems. By 1990, population growth, slash and burn agriculture, and cattle ranching will have reduced these forests to relic patches that ultimately may be pushed to extinction. While the web's primary function is for exploring the canopy, it also presents an opportunity to aid in preservation. A site has been selected that is in the foothills of Costa Rica's Caribbean slope. It is within a tentative park extension that would link mountainous Braulio Carrillo National Park with a lowland rain forest scientific research station, Finca La Selva.

If the extension can be secured, Braulio Carrillo would become a unique forest preserve having several elevational feet of forest. In the tropics, where rare organisms can be restricted to extremely small areas, this means the preservation of a phenomenally diverse cross section of rain forest life. The complementary polarity between Finca La Selva and a sophisticated canopy research center in the adjacent foothills would tend to throw a scientific and protective web over the park extension. A bit more forest may then survive into the next century, allowing mankind to unveil its multitudinous secrets.

Arrangements for establishing the canopy facility would be made through the University of Costa Rica, and the Costan Rican National Park Service, where I have established contacts.

The importance of the web for an adequate exploration of tropical rain forests cannot be overestimated. It has already been mentioned that a large percentage of the earth's life resides out-of-reach and unstudied in forest tree tops. An intriguing line of research concerns pharmaceuticals derived from plants. New findings in anthropology and biology place the precursors to *Homo sapiens* in jungle trees as recently as four million years ago. Prior to that, our ancestors and those of the other primates, were associated with forest canopies for a period in excess of fifty million years. During this period, canopy plants and animals would have undergone intense coevolution. There can be little doubt that rainforest tree tops hold a key to solving and understanding disease.

Similarly, the canopy is a storehouse for a variety of foods which could prove useful to mankind. In certain forests where I have conducted research, edible canopy nuts abound. They are identical to the large cashew nuts of commerce. These nuts are inaccessible and are usually eaten by forest animals long before they fall to the ground. The result is that these forests are considered non-productive in comparison to corn and bean fields; thus, local politics favor their removal in order to capture some of the land's productive potential, despite the fact that the soil is usually poor. If the web design proved it could be useful for harvesting canopy crops, forests that are now threatened would gain a reprieve. This would not be a general panacea for worldwide deforestation, but some very specific areas might be saved without governmental or private subsidy.

Scientific investigation of the canopy would begin once the automated web is completed. Gary Stiles of the University of Costa Rica, a noted ornithologist, has requested that he be given research time on the web so that he can study the foraging behavior of canopy birds, something that is impossible to do without a web system. Birds are responsible for dispersing the seeds of a large number of trees and epiphytes, and are thus of vital importance to tropical rain forest dynamics. K.S. Bawa, a tropical botanist at the University of Massachusetts, has expressed interest in using the new web for studying the nature of canopy plant breeding systems. The list of researchers who could put the web to good use is very long. Their disciplines range from epidemiology and physiology to behavior and horticulture. Many scientists are waiting for an effortless means to study jungle treetops.

Rediscovering the Prehistorical Engravings of the Saharian Atlas

Micheline Marcelle Colette Debecker

5b Chemin Golette, 1217 Meyrin-Geneva, Switzerland

Belgian, born September 17, 1934. Designer, explorer and speleologist. Educated in Belgium.

My project seeks to 'rediscover' one of the oldest and largest art collections in the world, the prehistoric rock drawings of the Saharian Atlas. Objectives of the project include:

Description: it has become an urgent necessity to inform the public of the masterpieces in this 150,000 square kilometer 'museum'.

Explanation: it is possible to retrace the mentality of prehistoric civilizations by studying the traditions of today's population.

Location: relocating the drawings is equivalent to identifying the underground water reserves, long since forgotten, which may exist in the desert today.

Saving the future starts with the preservation of the past!

The Saharian Atlas, stretching over more than 2,000 kilometers from west of Morocco to the west of Tunisia, forms the northern barrier to the inexorable encroachment of the biggest desert in the world, the Sahara. Only a few thousand years ago, today's parched soil and picture of desolation was a green and fertile paradise where prehistoric hunters/herdsmen lived.

The pastures and abundance of wildlife made the Saharian Atlas the center of highland grazing-ground for numerous civilizations. And for 8,000 years, men engraved the symbols of their mysteries and concerns on the stones of this area.

In this vast area, dispersed over Algeria and Morocco, scientific researchers have located and reported significant finds for over 60 years: Les Pierres Ecrites (Flamand, 1921), Les Pierres Ecrites de la Berberie (Solignac, 1928), L'Art Rupestre Nord Africain (Vaufrey, 1939) and Les Gravures du Sud-Oranais (Lhote, 1971) are a few examples. The majority of the identified and reported sites are located close to the few tracks and dust roads crossing these deserts. Hundreds of others, less easily accessible, remain to be discovered.

In four previous sojourns in the Atlas, we have been able to familiarize ourselves with the kind of organization and experience necessary to find and explore hard-to-reach prehistorical engraving sites. As a result, even though it sounds impossible to explore every square meter of these areas, our practical knowledge of these countries quickly permits locating rocks which might bear engravings. Our most valuable help still comes from the Nomads who drive their herds through these regions when the heat becomes unbearable in the South.

We must not allow these several-thousand-year-old testimonies to disappear forever. Everyday we witness their destruction, through human vandalism (collecting them, sometimes shooting at them) and natural erosion (the often soft rock sections off through the stress of differences between day and night temperatures, and is susceptible to the ravages of water, frost, etc.).

Modern techniques will provide support for these relics of the past. An incredible and unforgivable hypothesis would be that nothing but 'phantoms' will be left behind for our descendants. It is essential to rediscover these engravings immediately, to describe them, and bring them to public knowledge.

Each site that has been discovered and studied will be documented as follows:

1) Sketching landmarks of the sites and marking the dust roads.
2) Entering information on the map of "I.G.N. Institut Géographique National".
3) Sketches representing the site.
4) Sketches of engravings, including descriptions and dimensions.
5) When possible, making rubbings of each engraving in natural size.
6) Color photographs and indication of color shades will be made, using the "Munsell Soil Color Charts".

A further objective of our research will be to provide an aid to the location of long forgotten water sources and wells.

General indications, such as the discovery of prehistoric troughs in the immediate surroundings of carving sites, enable us to confirm that engraving took place in the vicinity of water, locations which were at the time obligatory meeting places for prehistoric hunters and herdsmen.

The sandstone walls, on which most engravings are found, consist of oblique, almost vertical, compressed stratifications. The fissures filled with marl and gypsum alluvium, forming impermeable layers, in ancient times. During each rainfall, water filtered down, swelling the already existing subterranean water reservoirs, which were of impressive size. The reservoirs remained intact. Up to the 16th century, a river flowed in Reggane. Here also was the Lake of Timimoun; today it is dried out, but only two centuries ago, it was the size of Lake Geneva.

Deep under the desert there still exists a huge and nearly inexhaustible water deposit. This treasure could one day revive the area and its inhabitants, and form a natural obstacle to the pitiless progression of the desert.

The spotting and topographical recording of the prehistoric engravings pinpoints the presence of ancient water runs and at the same time designates the most promising locations for research into potential water deposits. These places where prehistoric men rested, and made their engravings, are milestones. They are privileged areas which offer modern techniques the best possibilities for the discovery of water.

While our repertory map will not constitute an exhaustive document, it will be a tool fully appreciated by geologists and hydrographic researchers, a further aid in helping to preserve these valuable keys to our past.

Solo by Boat Across New Guinea's Unknown Irian Barat

Miroslav Davidovic

48 Rue des Trois-Frères, 75018 Paris, France

French, born August 31, 1948. Artist, sculptor, photographer. Educated in Italy and France.

My project consists of an expedition, in an exploratory, ethnological and photographic capacity, to the last uncharted areas of the Indonesian side of New Guinea, an area commonly known as Irian Barat. These 'blank spots' (shown on maps, for example, as "Relief Data Incomplete; elevations rising in a northerly direction not believed to exceed 7,000-13,000 feet") are found at 136°30′ East and parallel 2°30″ and 3°56′ beneath the Equator.

To this day, this particular region has presented unprecedented difficulty to adventurers and explorers. A 1959-60 expedition, made by Pierre D. Gaisseau and Gérard Delloye, was the first to travel from South to North in the western part of New Guinea, and resulted in the Oscar-winning film "Sky Above and Mud Below". Returning in 1969-70, Gaisseau and a team of others were parachuted into the area of the 'Stone Age' people in the central part of the high mountains and continued to the Idenburg (Taritatu) River, where they were forced to leave all film equipment in making an escape. Since that date, no other explorer has dared venture into this 'cursed zone'.

I began my travels in this part of the world in 1976.

These led, in 1978-79, to my staying for four and a half months in a small village of cannibals and head-hunters precisely souht-west of Irian Barat as an 'adopted foreigner'. This gave me the chance to explore extensively, to make the acquaintance of many tribes, and to learn much about survival in the country. I brought back 2,800 slides and six hours of Super 8 film from this time.

In 1979-80, I explored the Australian side of New Guinea for the second time, in order to make a comparison of a two-year lapse in time on the cultural life in the region, taking 2,000 photos and four hours of Super 8 film. In early 1980, heading toward the Nomad river, I was evacuated to Wewak village in Northern New Guinea for treatment of a venomous snake bit, thus ending my attempt to make firsthand contact with the tribes of the Nomad river.

Recovered and once again ready, I am now planning an exploration of the Rouffaer River, which flows through Tariku and Memberamo, the rivers estuary on the Northern side of the island. On the banks of this river live unknown stone age tribes. These tribes consist of cannibals, hunters and head-hunters who fashion remarkable sculptures called 'Korwar' that represent men holding their shields before them. The particularity of 'Korwar' is that they serve as shrines for the skulls of the owner's ancestors.

If we look at a map, we see that numerous villages on the banks of the Tariku have not yet been surveyed and studied. This would be part of the purpose of my expedition, which will take place aboard a pneumatic boat, or simply on a raft constructed on the spot. The Rouffaer River descends from an altitude of approximately 4,000 meters to about 1,200 meters.

I will depart from the village Timika, following the Mimika River, and leaving the Kaokanao Mission to the North, in order to visit villages bordering this river that have not yet been studied. Then on to Enarotali, a little known region I visited in 1978, with its Kapauku and Ekagi pygmies, and then northward in the direction of Jimbe village, past existing villages that have not yet been studied. I will then continue my voyage alone, with the intention of filming and photographing flora, fauna, geological formations, the indigenous inhabitants, their cultural life, their art, dance, taboos, etc. If possible, I hope to film cannibalistic rites, funeral ceremonies, marriages and initiations.

After Jimbe village (Stone Age people), and from this point, I will continue to realize a dream: the exploration of Rouffaer River, then the river Tariku and at last Memberamo and the unexplored zone. When possible, I will then rejoin the Northern Coast at the Memberamo estuary and radio in my position.

This voyage should permit me to finish a manuscript devoted to my adventures in New Guinea. This socio-, anthropo-, paleonto-, ethno- and geological study, illustrated by photographs, drawings and engravings, will be my "Encyclopedia of the Island of New Guinea".

The purpose of this work will be to communicate the knowledge acquired on my expeditions, to try to share the marvelous moments spent on the Island, and to let the prehistoric world of New Guinea be known to the outside world.

I hope, on my return from this latest venture, to organize further expeditions in order to share my voyage with others.

The approximate site of the lost and mysterious "Chinguetti Meteor", reputedly the largest ever discovered.

Upper left: Piece of the Chinguetti meteorite recovered by a French officer, Captain Ripert, in the desert of Adrar, about 45 km. southwest of Chinguetti, in the Western Sahara, Mauritania. This specimen, weighing about 4,5 kg. is now in the Musée d'Histoire naturelle, Laboratoire de Minéralogie, in Paris.

Search for the Chinguetti (Mauritania) Meteorite

Christian Köberl

Institute for Analytical Chemistry, University of Vienna, P.O. Box 73, A-1094 Vienna, Austria

Austrian, born February 18, 1959. Cosmochemical/astronomical projects collaborator at University of Vienna's Institutes for Astronomy and Analytical Chemistry. Ph.D. (Astronomy/Chemistry) in 1983 from University of Vienna.

This project deals with the attempt to rediscover the so-called "Chinguetti Meteorite". The first note on this meteoritical mass was published by A. Lacroix in the Comptes Rendus of the Académie des Sciences at the Séance du Lundi, for 4 and 11 Aug. 1924 (pp. 309-13, and 357-60). It is there stated that Captain Ripert, a French officer, had made the important discovery of a large meteoritical mass in the desert of Adrar, in the Western Sahara (now part of Mauritania). Captain Ripert returned with a piece of this meteorite, which is now in the Musée d'histoire naturelle, Laboratoire de Minéralogie, in Paris. As the description of the discovery is highly graphic, I here provide a translated quote from the French of the above Compte Rendus:

"The specimen in question was recovered about 45 km southwest of Chinguetti, and to the West of Aouinet N'Cher. It lay isolated upon an enormous metallic mass measuring about one hundred meters on a side, and about 40 meters in height. This mass jutted up in the midst of sand dunes that were covered by a desert plant, the sba, and it had the form of a compact, unfissured parallelopiped. The visible portion of the surface was vertical, dominating, in the manner of a cliff, the wind-blown sand that was scooped away from the base, so that the summit overhung, and that portion exposed to wind erosion was polished like a mirror. The sand had accumulated against the opposite face and hid it completely, which rendered it impossible to evaluate its third dimension. The summit of the mass bristled with tiny needles, which the Mauritanians had tried to break off, but the malleability of the metal had prevented them from accomplishing more than bending. A few metallic blocks of lesser magnitude were scattered in the neighborhood."

This description sounds very interesting, and would also be a little bit improbable, if it were not for the piece recovered. This piece weighs about 4.5 kg, and is now in the Musée d'histoire naturelle in Paris. The piece does not fit the common class of meteorites, and has been classified to be either a meso-siderite (Lacroix, 1924) or a nickel-poor ataxite with silicate inclusions (Hey, 1966). A known fact is, this one meteorite has no relative among the known meteorites, as its chemical characteristics are very strange. The literature does not, in fact, list many analyses of this meteorite (Lacroix, 1924; Wasson et al., 1974), but the ones available clearly support the singularity of this object.

The scientific community quickly regarded this the discovery of the century, since this would be the largest meteorite ever discovered on earth—the largest now known are the Hoba mass (Grootfontein, South West Africa, about 60 tons) and the Cape York masses (58 tons, largest piece 30.9 tons). If the description filed by Captain Ripert was accurate, the Chinquetti mass would be larger than all other known meteorites by magnitudes. So, in October, 1924, a search was conducted by Lt. Bonnin of the area in question, but achieved no success. This search, and a few others, mainly by French military personnel, continued to yield no results.

In letters at the beginning of the 30's, M. Ripert, now an old man, continued to insist that the mass existed as described, and also said that the natives in the region called the meteorite the "stone that fell from heaven", and that Arab blacksmiths regularly collected iron from it for their craft working.

The scientific community was alarmed, and in respect of the unsuccessful attempts to rediscover the Chinquetti meteorite, the International Astronomical Union (I.A.U.) adopted a resolution urging the French government to make every effort to rediscover the mass. This resolution, which passed at the Fourth General Assembly of the International Astronomical Union, held at Cambridge in 1932, read (in extract form) as follows:

"...in view of the fact... that the great mass in the Adrar, North Africa, has not been rediscovered; that the study of (this) locality will have immense value for meteoric astronomy... it is the fervent hope of the Commission... that the French Government make every effort to rediscover and study completely the great meteoric mass in the Adrar, North Africa, as long delay may render this almost impossible due to shifting sands."

The subject was treated again at the Sixth General Assembly in Stockholm (1938), in the following statement:

"...Neither has anything further been heard about the recovery of the Chinguetti (Adrar) meteorite, the existence of which has been recently put in doubt. According, however, to a letter addressed to Prof. Bosler by Mr. G. Ripert, administrateur en chef des Colonies en retraite, who discovered the mass in 1916, and brought back the fragment analysed by Prof. A. Lacroix, 'the metallic nature of the rock cannot be doubted'. The possibility therefore still exists that it has been covered up with sand from the desert, and a magnetic exploration of the site is more desirable than ever."

A search conducted by T. Monond, a French desert scientist, led to the discovery of the Aouelloul-crater, a meteorite crater in the neighborhood of the area in question. There, a black impact glass was detected. Some trials have been made to evaluate the prime body responsible for that crater by extremely sensitive subtrace element (platinum group — siderophile elements) analysis (Morgan et al., 1975; Köberl, 1982) and have shown that the traces of siderophile elements detected in these impact glasses bear no clear similarity to the Chinguetti meteorite sample in Paris.

Further studies would be of great value. The first thing to do would be to rediscover the Chinguetti meteorite mass. Of course, that's easier said than done. Nevertheless, all previous searches were conducted by untrained personnel, and/or with bad (or no) instrumentation. It is proposed to search for the meteoritic mass now with the help of ground-based search (by Land Rover) assisted by natives, and with search from the air. This latter would involve planes equipped with appropriate instrumentation, such as magnetometers and infrared-sensitive cameras to record the thermal behaviour of the area, where a large metallic mass must have another thermal behaviour in the infrared and might be detected from false-color pictures. These can be processed by computer assistance, as there is a good two dimensional picture processing system installed at the Vienna Institute for Astronomy's computing center. With the proper support equipment, the search should not take more than one or two months to yield success, or to prove that a more expensive undertaking is necessary.

To date, no real modern attempt has been made to rediscover what would be the largest meteorite in the world. The successful carrying out of this search would lead to the greatest "re-discovery" of the century, and would provide a most valuable scientific contribution.

Surveying Underwater Hydrothermal Vents off California

Michael Gene Kleinschmidt
Post Office Box 2334, Van Nuys, California 91401, U.S.A.

American, born December 7, 1948. Explorer and diver. Educated in U.S.A.; in art, sculpture and oceanography.

The Southern California Continental Borderlands Hydrothermal Vent Survey is a geological mapping effort. This application describes my continued effort to locate and chart geothermally heated submarine sulfur springs that vent in the hazardous, high energy coastal waters of the East Pacific Ocean.

Poets sometimes romanticize the ocean as a benevolent mother. Any diver who's been dragged by her boiling surf across a rocky reef and hurled headlong into an unyielding sea cliff might refer to the ocean as a cruel, humorless bitch. I was, and sometimes still am, confronted by the serious task of surviving a physical punishment served up by this often vitriolic environment. Moderate amounts of bodily injury are a matter of course. The fact that I've logged over a thousand hours of bottom time diving in hazardous coastal waters without serious injury is a testament to my watermanship skill.

Physically difficult scientific investigations conducted in the unfriendly, rocky, subtidal heavy surf zones located along the Southern California coast have ignited my imagination. I've become fascinated by geothermally heated submarine sulfur springs that vent from the ocean floor at a number of rough water locations throughout the Southern California continental borderlands. It is impossible to repeatedly swim into a vent site without forming some questions.

Where does the water come from?

What is the white stuff that grows around the mouth of the vents?

What chemical compounds are exhausting from the vents?

Why is the vent water warm?

Why do the flow characteristics change just prior to local earthquake events?

The quest to find the answers to these questions has become a grand adventure. I would never have guessed that simple questions would turn out to be so difficult to answer. The scientists whom I queried in my search for the answers were not aware that the vent sites existed.

So we began.

The microbiologists, geochemists and geologists who've come to share my curiosity are completely landlocked. Most do not swim well and are not eager to dive around rocks, reefs and rip currents. As a result, a unique union has been forged — my access to the vent sites and their scientific expertise. A handful of researchers have donated some of their precious time to patiently school me in their field work techniques. This enables me to make meaningful observations, accurate measurements and effective collections underwater. After an intensive course in historical geology and structural mapping skills, my new vent site charts now exhibit a location's basic geologic structure and are forming the basis for more sophisticated investigations. I have made difficult vent water collections and soon I will collect and handle microorganisms.

My prototype vent study site is named White Point, because its shallow shoals are continuously swept by foaming white surf that makes it a distinctive navigational landmark. It is located on the Palos Verdes peninsula, near San Pedro, California. Enthusiastic accounts of my hot springs mapping efforts have helped White Point become designated as a research shore station and an ocean interpretative center by the Los Angeles County Department of Parks and Recreation Underwater Unit, and by the Greater Los Angeles Council of Divers.

As a result of the initial investigations, I can report that the water exhausting from the vent openings is heavily laden with hydrogen sulfide and carbon dioxide. The white matter that grows around the exhaust ports includes at least one strain of chemosynthetic sulfur oxidizing bacteria that is apparently new to science. The flow characteristics somehow affect, or are affected by the occurrence of seismic events in the Los Angeles area.

The procedure for determining where the vent water comes from, and why it is warm, reads like a twisted inside-out 'Catch 22'. To further understand the White Point vent system's hydraulic mechanism, an accurate structural model of the Palos Verdes geological history is needed. This doesn't exist. The development of this model requires an understanding of the continental borderlands tectonic history. The theories regarding the tectonic evolution of all Western North American are in a current state of heated revolution. However, a detailed geological map, similar to the work I pioneered at White Point, and a geochemical analysis of the hydrothermal vent systems located throughout the Southern California continental borderlands will be a significant step toward understanding the borderland's tectonic history. This will form the basis for developing an accurate structural model of the Palos Verdes geologic history, which in turn will tell us how the White Point vent hydraulic system works. Then we will know where the vent water comes from, and why it's so warm.

For about a year and a half, I have dedicated all of my energies and resources to the full time exploration of Southern California continental borderland hydrothermal vent sites. With help from others, I have been able to locate about 20 vent sites. I am currently compiling my journals and field notes into formal maps and reports for publication. My situation is unique. I am not a scientist. Rather I am an explorer whose only duty is to press on, whose only responsibility is to record my observations, and whose only limitation is courage.

Saving the Last Vestiges of the Languages of Tierra del Fuego

Piotr Klafkowski
Vardoasveien 59, L.412, N-50, 1385 Solberg, Norway

Polish, born June 20, 1951. Research Fellow at University of Bonn. Educated in Poland, India and West Germany; Ph.D. (Linguistics) from Adam Mickiewicz University, Poznan, in 1974.

This project is based on the unknown, exceptionally important manuscripts discovered by me in the manuscript remains of Martin Gusinde SVD (1886-1924) kept at the Anthropos Institut, St. Augustin, Bonn, and never before described. These documents contain, inter alia, several vocabularies of the three native languages of Tierra del Fuego, running to several thousand entries, collected at the time these languages were still in everyday use. Since Gusinde's expeditions (1920-1924), most of the Indians have died, and the languages are almost extinct, with the little of them that remains being corrupted and full of Spanish words. Out of the three languages, one is unique and unrelated to anything else spoken on Earth, while the relation of the other two to the languages of the South American mainland has never been explained. At present, there are very few Indians left in Tierra del Fuego who still speak their languages, and comparing Gusinde's records with the surviving living speech is the last chance of making such a documentation and saving these languages for the future. In no more than 5-10 years, it will be too late.

The three languages of Tierra del Fuego — Yahgan, Ona, and Alakaluf (also known as Yamana, Selk'nam and Halaqwulup, respectively) — are the southern-most languages in the world. Being isolated from the continental mainland for a considerable period of time, they may contain vital data for the linguistic history of that part of the world. More than that, they together constitute a unique whole, reflecting the life and thinking of a unique civilization that has no parallel in the world's primitive societies. The chief language of Tierra del Fuego — Yahgan or Yamana — is not related to any other tongue spoken on earth, and its disappearance without a comprehensive scientific study would be an irreplaceable loss.

When I discovered Martin Gusinde's papers, I immediately realized that here is possibly the last chance of making a linguistic survey of Tierra del Fuego that we may ever have. The languages have never been described extensively in print, and the only English dictionary of any of them has been unavailable for decades, being printed privately in 1933 in an edition of 300 copies. Recent travellers report that there are still some people, mostly old and very old, who speak their native languages in different parts of the island, but their speech is corrupted with plentiful Spanish borrowings. Except for three New Testament translations into Yahgan, and a few prayers printed in the other two languages — all these items being extremely scarce — no comprehensive lists of words or sentences have been compiled by trained linguists. The manuscripts in Martin Gusinde's archives give us the unique chance of going to the island with *some* tangible records to verify, and not only in the hope that the old Indians shall remember something. Confronting the last native speakers with the actual language records may release a lot of their memories, besides being an important psychological argument showing them that someone cares for their heritage. I must stress that it is *now* that the unique opportunity exists, while in five or ten years it will be lost forever.

As the spoken wisdom of the Fuegian Indians has always been regarded as nil, a number of sources — including works by Anglican missionaries who lived among them for decades — present them as barbarian savages of no culture in any meaning of the term. However, when a *sympathetic* researcher, Martin Gusinde, went to Tierra del Fuego and convinced the Indians he could and should be trusted, most extraordinary things emerged, including the revealing of a complex set of religious mythologies which the Indians had never before revealed. It is highly possible that, after the contemporary elders are convinced about the sincerity of the researcher, a lot of unique materials, ranging from folk medicine to traditional songs and poems to religious mythology may come to light. The countless photographs extant in Gusinde's collection are of extreme importance, as they offer the researcher a chance to look for the descendants of Gusinde's own informers and confront them with their parents' pictures, thus breaking the barrier of mistrust that is a great obstacle in any such work.

The project divides into two sections. The first is the duplication of the entire linguistic part of Gusinde's archives, transcribing his field notebooks from shorthand originals into normal writing, arranging the photographs and printing the negatives (of which there are hundreds) to get a complete set of pictorial documentation, and present this in a book. This portion also includes re-playing the phonographic wax cylinders Gusinde recorded of Fuegian materials, which are now kept in Berlin. The second part of the project will be to conduct a research trip to Tierra del Fuego, to locate the Indians who still speak the languages and to tape record every possible word, phrase and sentence in order to ensure that the languages will not be lost.

Circumnavigating the earth in a balloon requires high-tech accommodations for two people; this capsule was a forerunner to the currently planned attempt.

First Manned, Non-Stop, Round the World Balloon Flight

♛ Julian Richard Prothero Nott

Honourable Mention — The Rolex Awards for Enterprise — 1984
Endeavour 84, 49 Mill Lane, London NW6 INB, U.K.

*British, born June 22, 1944. Private consultant engineer. Educated in U.K.;
Masters Degree in Physical Chemistry from St. John's College, Oxford.*

My project is to make man's first ever non-stop flight round the world using a
balloon.

The flight would take about 2-3 weeks, flying most of the time at about
35,000 feet in the jet stream winds, taking off from Australia, circling the world
and returning to Australia.

This would be the first time a balloon has flown round the world carrying a
crew, the first time *any* aircraft has made the circumnavigation without landing
or in-flight refueling, the longest ever unaided flight (excluding orbiting space
craft), the first time Australia has been crossed by balloon and would set very
challenging new distance and endurance records for ballooning.

The project is a dramatic demonstration of new technologies, and indeed of
a new approach to technology in general; one likely to be more and more
important over the next decade in a world of diminishing resources.

Is It Possible?
The fact that the project can be insured against failure through Lloyds
aviation brokers may be the most useful way to judge its feasibility. Not only is it
possible to fly round the world by balloon, it has been done tens of thousands of
times by thousands of unmanned meteorological balloons of all sizes. The
record for these is 2 years, making 33 non-stop circuits of the earth. Scientists
would like balloons to be spread all over the globe, but they nevertheless
congregate in the jet stream south of the Equator; bad for the scientists, but
ideal for this project.

The Balloon
To get round the world, one must fly at about 35,000ft/11,000m, in the jet
stream, above the weather. In principle, all that is needed is a man-carrying
cabin attached to a balloon launched on this well-established route. One
advantage I hold in making this attempt is that I have already built two
pressurized cabins for balloons, using the most modern technology. (One of
these is permanently in the Science Museum in London, the other was used to
set the 1980 world altitude record of 55,000ft/17,000m.) I have flown in the

jet-stream several times, once in India at over 130mph/210kmph, a far more violent world than the 65mph/100kmph wind found in the Southern Hemisphere.

For this trip, the second crew member will be a cameraman, jointly selected by me and Central Television of the U.K. I anticipate no difficulty in making both a balloon and a cabin capable of handling this flight. The basic pressure structure of the 1980 cabin, big enough to hold two fully equipped men, weighed only 30 lbs/14 kgs. The 'round the world' cabin will be an enlarged version weighing only 100 lbs/45 kgs. Similar construction methods throughout will give a very low all-up weight, sharply reducing cost and greatly increasing the likelihood of success because the small balloon will be easy to launch and fly. The problems of endurance in small cabins have been solved, as demonstrated by submarines, spacecraft and the like.

The balloon itself will be basically white, to minimise solar heating, and nearly spherical, unlike the pear-shape of hot air balloons. It will be 60ft/18m in diameter, with the cabin suspended 30ft/9m below. The lifting gas will be helium which is totally non-inflammable.

Route and Timing

The flight must be done in the Southern Hemisphere. The weather is much simpler as there is so little land adding heat to the atmosphere (which can be seen by looking at the globe with the South Pole facing you). Balloons have never flown round the Northern Hemisphere in the same way, and clearance would be impossible, even with a Russian, a Chinese, a Vietnamese and an Iranian on board!

The take-off should be about 35°South. It is proposed to fly from Perth, crossing Australia and circling the world before crossing Australia again to land near Sydney.

The best time of year is between May and October, but balloons fly round the world consistently in every season. Four to six months is needed to prepare for the trip. Length of the actual flight should be about 15-20 days. The balloon will fly at high altitude to use the jet stream, but will descend every other day to about 10,000ft/3,000m, a point where it would easily be visible from the ground, filmed, etc. When over land, it will descend on several occasions to less than a thousand feet/three hundred meters from the ground.

Relevance to the 1980's

While the project has a great element of adventure, it is by no means mere adventure, but a dramatic demonstration of specific technologies and, even more significantly, of an approach to technology which will be far more important over the next decade.

Post-war technology, up to landing on the Moon, was typified by projects using vast resources, vast computers, ever bigger machines and ever bigger budgets. Such projects are less and less possible, and less and less acceptable, in an age when every resource is under increasing pressure.

This project is quite the reverse, exemplifying the way technology is being compelled to go. As Lindberg was at the forefront of the technology of his time, so this project is at the forefront of 1980's technology.

Crucial to success are small computers and composite materials, like Kevlar or carbon fibre. People are aware of, and nervous about the 'electronic revolution'. Small computers are invaluable to preparing this project, but far from being off-putting, the project shows how personal computers set the creativity of individuals free. By contrast, the great potential of composites, and how they can be used without major plant, is not yet widely appreciated. As well as saving a great deal of weight, giving better performance and saving fuel in aircraft, boats and vehicles, composite construction has low raw materials costs, produces valuable products, and can use automated processes, thus representing the sort of industry that the developed nations must pursue in the future.

The project is just as relevant to the developing nations, showing what can be done by the wise use of limited resources.

The balloon could conduct a unique air sampling operation. There is increasing concern about pollution in the Southern Hemisphere; acid rain has even reached Antarctica.

The project has no negative aspects; it is cheap, uses no fuel, does not pollute and derives power solely from the wind.

Safety

Modern technology means the flight can be very safe. Comprehensive survival equipment will be carried and advice has been taken from the world's leading authorities.

The balloon will operated with an air traffic clearance like any other aircraft, navigating and giving position reports to the same standard as any intercontinental jet. In addition to the conventional navigation done by the crew (Omega/VLF plus astro-navigation), the balloon will be tracked quite independently by the ARGOS satellite system, so that its precise position will always be known to the outside world. Reports will be issued every three hours for safety and tracking purposes. ARGOS beacons will also be carried in the survival packs.

A commercial flight watch company at Gatwick Airport in the U.K. will track the balloon throughout, exactly as they do for company aircraft operating around the world, and will also act as a general communications centre.

Searching for an Unknown Horseshoe Crab in West Africa

Tom Mikkelsen

Tryggevaeldevej 134, 2700 Copenhagen, Denmark

Danish, born June 17, 1947. Research associate in phylogenetic and comparative immunology, University Clinic for Infectious Diseases, Rigshospitalet, Copenhagen. Educated in Denmark; completing D.Sc. thesis.

Three preliminary underwater expeditions by the applicant, over a period of 10 weeks in 1982-1983, have produced indirect evidence that a hitherto unknown population of horseshoe crabs exists along the west coast of Africa. The aim of this project is to localize and identify this population and have it protected in cooperation with local authorities. The identity of the species will be determined by means of comparative examinations with the three Southeast Asian species, and the American species, Limulus polyphemus. All three Asian species will be collected off the northern coast of Borneo, where they are believed to coexist. Techniques for the comparative studies of blood, cells and tissue for all species include crossed immunoelectrophoresis, electron microscopy, clot formation of amoeboycyte lysates induced by Gram-negative bacterial lipopolysaccharide, DNA fingerprinting and standard morphological examinations concerning the external morphology of the different species.

All stages of the expedition will be filmed and photographed, including underwater filming of the horseshoe crabs in their natural surroundings, as I have done previously with respect to Limulus polyphemus in the Mexican Gulf.

In January, 1981, I was given a dried horseshoe crab specimen in the Canary Islands by a local tourist boat operator. It was the last specimen out of a total of approximately 400 he had purchased in 1976-77 from fishermen based in Las Palmas who had apparently caught them (frequently in lobster traps) along the West African coast, from Senegal to Sierra Leone. On three subsequent, preliminary expeditions to West Africa (January-February 1981, June-July 1982 and February-March 1983), I have amassed abundant, though indirect, evidence that a hitherto unknown population of horseshoe crabs does indeed exist in this area. According to Senegalese fishermen, horseshoe crabs are caught, occasionally, as far north as Dakar, but they become increasing frequent further south from the Gambia River estuary to Cameroon where they reportedly (in March 1983) are brought to the fish market every day.

A Las Palmas based skipper, who was catching lobsters in Gambian waters for the market in Banjul, caught several live horseshoe crabs in his lobster traps in December 1982.

At least two dozen registered fishermen in Banjul and the villages of Tanji, Brufut, Kanifing and Gunjur knew and recognized the crabs - generally considered an annoying by-catch with no commercial value.

Thus, although believed to have disappeared from the West African coast some 65 million years ago, indirect evidence is abundant that a population of horseshoe crabs exists in coastal waters from The Gambia to Cameroon. Whether this niche has persisted since the separation of the African continent from the North American continent, or the population has been established more recently will be determined by comparative scientific investigations of blood cells and tissue performed in The Gambia and at the University Hospital in Copenhagen.

Due to very fluctuating visibility, it has not been possible to dive into the Atlantic very often, so I have decided to catch the crabs by means of a dozen specially designed traps, which are being manufactured in Denmark. By the end of 1983, the traps will be shipped to Mr. Mario Anguis in Banjul. Mr. Anguis is the director of an EEC-financed fishing project in the village of Tanji, and the traps will be brought into operation immediately by his staff. Mr. Anguis and I have exchanged telex numbers in order to have a direct line of communications. A house for my laboratory equipment has been provided by Mr. Joof, Director of Fisheries, who will also assist me in recruiting local staff. Mr. E. Brewer, director of the Wildlife Conservation Department, has granted me permission to export a number of live crabs, and has asked me to assist in designing adequate protective measures to prevent the deplorable slaughter of horseshoe crabs known in other parts of the world, due to the formidable success of the Limulus test.

During the past 8 years, we have produced detailed information regarding blood, cells and tissue from Limulus polyphemus, including specific antibodies to the major fractions of the amoebocyte lysate. Due to lack of systematic research, little is known about the geographical distribution, general ecology and blood chemistry of the three South-East Asian species of horseshoe crabs. It is also my intention, therefore, to confirm earlier reports that the three species coexist off the northern coast of Borneo. I plan to have blood substances and a limited number of live animals flown to the University Hospital in order to perform comparative investigations similar to those designed for the West African crabs.

We know of very distinct cellular and subcellular differences, but little is known about the immunological and evolutionary significance of these differences. A thorough understanding is a prerequisite of producing the (semi-) synthetic amoebocyte lysate, which will restore peace for a fascinating 'living fossil', and retain an extremely valuable diagnostic aid in clinical medicine and scientific research.

A Private Search for Extraterrestrial Intelligence

Robert Hansen Gray

3071 W. Palmer Square, Chicago, Illinois 60647, U.S.A.

American, born March 7, 1948. Private consultant and university lecturer. Educated in U.S.A.; Master of Urban Planning and Policy Analysis in 1980 from University of Illinois at Chicago Circle.

The Small Search for Extraterrestrial Intelligence (SETI) Observatory is a private radio astronomical project that has the goal of detecting evidence — in the form of radio transmissions — of extraterrestrial life. A state-of-the-art radio telescope has been built and is currently being tested; programs to largely automate the observatory operations are now under development. This system will be used in a full-time search until detection occurs, or until other surveys have systematically covered the entire sky to an equal sensitivity, which will be several years at a minimum. The operation phase will require installation at a radio-quiet site and financial support, neither yet funded.

It is appropriate to undertake this exploration now because extensive theoretical work both strongly supports the likelihood of life elsewhere and specifies the equipment capable of detecting any interstellar radio broadcasts that may exist. Furthermore, very promising strategies for frequency selection (Dixon, 1973), search direction (Gray, 1977) and signal detection (Lord, 1981) have been developed. At least one signal strongly suggesting an extraterrestrial transmission has been observed, but not confirmed (Kraus, 1979; Gray et al manuscript). The Small SETI Observatory project is intended to implement these strategies and findings on a much faster schedule than larger institutional projects, such as NASA's program.

This paper describes the radio telescope and some of the methods that make a small scale SETI effort feasible. Detailed information on equipment, observing plan and budget have been prepared.

Major components of the radio telescope include a steerable dish antenna twelve feet in diameter, a 65 degree Kelvin low noise amplifier, a programmable receiver and a 256 channel HP 3582 FFT spectrum analyzer. A high performance microcomputer controls the instruments, processes acquired data, and records the observations on magnetic disks for later analysis.

The system is capable of scanning 1 MHz above and below the 1420.405 MHz Hydrogen emission frequency suggested by many writers as the most likely frequency for interstellar broadcasts. The multi-channel analyzer allows a very high sensitivity to be obtained with the small antenna: the system can detect signals of about 10^{-22} Watts per square meter, using a 1 Hz bandwidth and 16 seconds of integration time. This is within one order of magnitude of the best sensitivity reported to date (Tarter, 1981), and is adequate for the detection of a 1000 MW transmitter 50 light years away (assuming a broadcast antenna 50 meters in diameter).

Several strategies will be used to choose the most promising parts of the sky for earliest observation. A principal goal of the project is to extensively re-observe locales where interesting radio emissions have been briefly observed but not confirmed. These locales will be observed for hours or days, vastly increasing the probability of detecting signals that may not be broadcast continuously. The search program will also include locales that have been suggested as having high potential for detection. These include:

— The galactic center, due to its high concentration of stars.

— The galactic axis, du to possible astrophysical activities of advanced civilizations.

— Other galaxies, due to their small angular size.

The search plan includes some features not previously used, and many that have not previously been used together, in order to make detection and confirmation feasible in a reasonably short period. These features are:

1. Selection of celestial areas for exploration by strategy, rather than by proximity.

2. Use of a small, wide beamwidth antenna that can see many stars at once, rather than essentially one at a time.

3. Allocation of hours or days of observing time to individual target regions, in order to receive signals that may appear periodically. Most projects allocate only minutes per target.

4. Use of an extremely narrow bandwidth detector, wich will allow interstellar signals to be distinguished from local ones by measuring Doppler drift caused by relative motion.

5. Mobility of the radio telescope, which allows the observatory to be moved to a temporary site to confirm celestial origin, as opposed to a local origin, of any signals discovered.

The Small SETI Observatory has an excellent chance of detecting the type of interstellar broadcast considered most likely by many scientists. The essential equipment is in place and functioning due to the efforts and dedication of more than a dozen individuals. The intensity of operations, analysis, and systematic improvements will depend on additional funding. Unlike university and government funded projects, this one cannot be terminated by a policy decision or a cut in funding, and it can respond immediately to new concepts for the search for extraterrestrial life.

Painstaking attention to hand-made construction is preparing this ship to repeat the classic Magellan-Elcano voyage around the world.

Building a Classic Schooner to Duplicate the Magellan-El Cano Voyage

♛ Esteban Vicente

Honourable Mention — The Rolex Awards for Enterprise — 1984
Astilleros Isuntza, Lequeitio, Vizcaya, Spain

Spanish, born July 11, 1953. Mountaineer, canoeing champion, photographer and sportswriter. Educated in Spain; Physical Education studies in the Instituto Nacional de Educación Física.

I have designed, and am now leading and participating in the building of a classic wooden schooner, in which I and my team shall repeat the first circumnavigation of the earth, following the Magellan-El Cano route.

This ship is not a copy of the old original craft that made the voyage, but rather a schooner as such ships were built at the beginning of the 19th Century. The design is that known as a 'French Schooner' in many places.

The combination of such a ship and the original voyage is not such a contradiction as it might seem; after the voyage, we plan to dedicate the schooner to teaching how to sail large craft. With a ship built along the lines of those from the beginning of the 16th Century, such sailing instruction would be impossible.

We are building the schooner by ourselves, following the same methods of the old ships' carpenters that are nowadays almost totally lost. Commencing in 1979, I studied information about, and plans of, wooden ships in numcrous libraries. I learned technical drawing, made a model of the ship I wished to have, had it checked by a naval engineer, and assembled a group of young people with the dreams and determination to work on the project. In the beginning of the project, life was very austere, but now, little by little, the project has become known, and we are receiving publicity and some support.

Once the design and plans for the schooner were finished, and we knew the shape and measurements of all the pieces, we went to the forests to choose the most appropriate wood for each piece. We did this according to rigorous selection of natural curves of wood that matched the curves of the ship's pieces. We felled trees in winter in the waning or new moon, when the sap is in the roots, and carried them to the sawmill to be cut according to our plans.

Basically, the ship is a wooden schooner like its forebears, but a little wider for reasons of stability.

The total length is 24 meters, the beam is 7 meters, and depth is 2.95 meters. It has very seaworthy lines, and the scupper is long and slim.

The keel is formed by two pieces of irokowood, each 11.5 meters, joined with a union of 150 centimeters through which not even a razor blade will pass. The rest of the gallery of the ship is made with strong planks of irokowood, strengthened at the joints by strong pieces of oak, with the grain going in the same direction as the curves on the joints. All these pieces are further strengthened by 1.5 meter bolts.

The skeleton of the ship is formed by 45 ribs, all of them of oak and ash; woods that were used from the beginning of the history of naval construction. All of the ribs are doubles, each one being made of eight planks of 11-centimeter thickness. Each rib weighs between 400 and 500 kilograms. Making them has been our most difficult task.

In the beginning, we departed from our provisional plan by making a half model of the ship's hull. In this model, we corrected all the mistakes from the original plan (with much sandpapering) to obtain a correct and elegant shape. We then dismounted the model, and made a new sample plan drawn to actual scale in a loft we improvised on the roof of the hut where we lived.

Once we had drawings of each one of the ribs on an actual scale, we had to make a pattern in wood of each one. This meant looking through hundreds of planks around the shipyard to find the most suitable, so that the finished framework would have matching grains in the wood.

The deck foundation was made of 33 beams of irokowood and 12 of oak. Planks of irokowood 11 meters long were used to top the deck. The hull is covered with irokowood of 65mm thickness and Scots pine of 90mm thickness. All of this is connected to the ribs and beams by a total of 1,100 grams of square nails.

On the deck, there are three open hatchways. One at the bow admits light and provides the entrance to the forecabin, which holds 12 berths. A second, mid-ships, opens into a repair workshop. The third, in the stern, provides light to the quarter-deck, as well as access to the machine room, where a small motor of 140 horsepower will provide power for the ship when it is in harbours.

Two masts, each of 17.50 meters, descend through mast holes to base on the 'slops' that are just above the keel. Two top-masts, each of 10 meters, will be situated on the mainmastheads, bringing the height of the mainmasts to 24 meters above the deck.

Balustrades, turned out of engraved pieces worked over the last years, ornament the schooner's interior and exterior, and provide support and reinforcements of different parts.

We have now brought into being the fundamental structure of the hull and some of the interiors with our own work and the help of many open hearted people. We also have our rigging and tackle set. We hope to set out on our voyage around the world in early 1984 and return approximately three years later.

The Limnology of South America's Pantanal Region

José Galizia Tundisi

Universidade Federal de Sao Carlos, Via Washington Luis, Km 235, Caixa Postal 676, Sao Carlos, Sao Paulo State, Brazil

Brazilian, born May 2, 1938. Full Professor of Biological Sciences, Federal University of Sao Carlos. Educated in Brazil and U.K.; Ph.D. in Botany from University of Sao Paulo in 1969.

The 'Pantanal' region in Western South America (Latitude 18°S, Longitude 57°W) is an extensive ecosystem of wetlands, including a large variety of shallow lakes, ponds, saline lakes and rivers. This system has considerable value as a natural resource and it has a very high scientific and economical interest value. In order to increase the scientific knowledge of ecological processes in the scarcely studied tropical wetlands, it is proposed to conduct a two-year exploratory study of the Paraguai and Cuiaba rivers and their adjacent ecosystems. The basic idea is to understand scientifically the ecological structure of the rivers/lakes from chemical and biological points of view, and the interaction of the rivers/lakes in order to provide background data for conservation and protection measures to be applied to this vast unknown area.

The tropical wetlands have a great socio-economic significance. These ecosystems are exploited for water supply, fisheries, aquaculture and agriculture. On the other hand, tropical wetlands have been subjected to considerable human impact during the last ten years. Studies of regional limnology and ecology in different parts of the tropical belt are of fundamental importance, as they can provide basic information on key processes of functioning, on species diversity, and can provide an understanding of the biology of species that could be used for cultivation.

Of Brazil's three lake systems (the other two are the Amazon floodplain lakes and the Rio Doce Valley lakes), the Pantanal wetlands are the least known scientifically. The climate is warm and humid, with a long dry period. The region is a floodplain of some 156,298 square kilometers, and the main rivers responsible for the flooding are the Cuiaba and Paraguai rivers. Seasonal differences of 3-4 meters in water level are characteristic of this region. Since the flood plain has a low declivity, it is easily inundated during the period lasting from December to May. The low water period corresponds to a time of continuous drying lasting from May-June to October. The shallow lakes thus formed, and the permanent lakes, are thus connected to the rivers or to other lakes for a period of about six months.

After the flood period, the drying takes place, severing the connections between these bodies of water. The resulting shallow lakes, ponds, small rivers and saline lakes represent, therefore, an accumulation of biological material, including a complex food chain of phytoplankton-zooplankton-fishes, macrophytes, shrimps, birds and top carnivores. The understanding of the functioning of this community and its diversity, in the lakes and rivers, and its interactions with the basic ecological processes is the aim of this project. As it was demonstrated for the Amazon river, the connection of a lake to the river in a flood plain is of fundamental importance for enrichment processes in the lakes. The lakes therefore act as capacitors of biomass for the river, and the river provides nutrients to the lakes. This project is designed to pursue the quantification of this process.

The objectives of this exploratory study of the Pantanal region are, in order:

a)   To quantify the amount of living biomass (phytoplankton, macrophytes, zooplankton, fishes and birds) in a series of shallow lakes and ponds in the region.

b)   To quantify the process of interaction between the shallow lakes, the ponds and the rivers.

c)   To study the nutrient cycle in some shallow lakes and small rivers.

d)   To study the water chemistry and its relations with the geochemistry of the rivers Paraguai and Cuiaba.

e)   To determine the primary production of phytoplankton and macrophytes, and its relations with hydrological and climatological factors.

f)   With the information listed above, to try to understand the general mechanisms of ecological functioning of shallow lakes, ponds, saline lakes, rivers and the interactions of the community with primary ecological processes and forcing functions (such as water level, water temperature and solar radiation).

The area to be studied will cover the rivers Paraguai and Cuiaba, between the towns of Corumba and Cuiaba, from approximately 19°S to approximately 16°S. The region will be studied in four cruises from 1984 to 1986, during low water and high water periods. Continuous samples from a boat cruising the rivers will be performed, and selected lakes, ponds and small rivers will also be sampled. The field work is designed in such a way that all analysis will be made on board the boat.

This project is part of a larger project which was started by our group in Brazil in 1972. Not only do we wish to understand the mechanisms and functioning processes of natural lakes and reservoirs with the aim of increasing scientific knowledge of these ecosystems, but we also wish to provide reliable information that can be used for applied purposes. Among these latter are the conservation of freshwater aquatic systems, and their multiple rational uses, such as protein production, irrigation and the cultivation of aquatic organisms (algae, fish).

An essential part of these investigations is the use of the same methodology in the different freshwater ecosystems. The data thus obtained will provide possibilities to compare and to understand better the other systems under investigation, as well as making the results of this study more meaningful. The tropical ecosystems are highly complex and diverse; this complexity and the ecological processes can be better understood only if comparative research work is carried out in the various systems with the same methodology.

Our collections, methods of sampling and estimating biomass will follow the techniques and methodology described in I.B.P. manuals. A series of sampling stations will placed in the rivers and lakes. Measurements to be made include the following:

Climatology: Solar radiation (global), wind direction and speed, air temperature, relative humidity, data on precipitation from local meteorological stations.

Hydrography: Submarine solar radiation, water temperature, dissolved oxygen, pH and alkalinity, conductivity, nitrates/nitrites and ammonia, total N and P (dissolved and particulate) and sediment chemistry. A complete profile of each water mass will be obtained.

Samples and experiments for determination of biomass and processes will include: Determination of chlorophyll a, number of cells/litre of phytoplankton, number of indicated cubic meters of zooplankton, measurement of primary production, and determination of biomass of fishes, benthos, macrophytes and birds.

As a sideline, but also of importance, the work will be filmed and photographed, in order to produce materials for use in general education and teaching.

Reconstructing a Prehistoric Ship and its Mediterranean Trade Routes

Aleksandar Palavestra

Bulevar Revolucije 225, 11000 Belgrade, Yugoslavia

Yugoslav, born January 29, 1956. Teaching fellow in prehistoric archaeology at Institute for Balkan Studies of the Serbian Academy of Sciences and Arts. Educated in Yugoslavia; M.A. (Prehistoric Archaeology) in 1982 from Belgrade University.

The aim of the 'Liburna Project' is to reconstruct the prehistoric ship used by the tribes of the Eastern Adriatic for navigation in the Central Mediterranean during the first millenium B.C. The ship, as well as the trade routes linking Southern Italian and Central Mediterranean centres with the eastern coast of the Adriatic and Balkan hinterland, is to be reconstructed on the basis of existing archaeological data (primary and secondary).

Recent archaeological data indicate that Illyrian tribes from the Eastern Adriatic played an important role in the well-developed maritime traffic of the prehistoric Mediterranean. Their significance as 'cultural disseminators' has, until recently, been neglected. The Liburna project, involving an archaeological experiment in which the reconstructed Liburna ship would be put to sea, should shed light on yet another aspect of navigation and cultural dynamism in the prehistoric Mediterranean and Europe.

Even during the Bronze Age (2000-1100 B.C.), maritime trade was lively in the Central Mediterranean, with amber, precious stones and metals attesting to the contacts between the Aegeans, Mycenaeans, Italy and the Eastern Adriatic coast with its Balkan hinterlands.

Among those who took part in the Aegean migration (around 1100 B.C.), which completely changed the cultural and ethnical map of Europe and the Mediterranean, was a people who were later to become the Illyrian tribes of the Balkans. One of the tribes that emerged, well-acquainted with navigation, and as aggressive newcomers, was the Liburni. They soon became one of the most important maritime forces in the Central Mediterranean.

During the Iron Age, from the Aegean migration to the 4th century B.C., when the Greek presence grew in the Adriatic, the Liburni and their Lembus and Liburna ships ruled the Adriatic and parts of the Ionian sea. They became the undisputed masters of the Ionian Gulf over a period of several centuries, and many mentions of their sailing prowess occur in ancient texts.

In their cruisings, the Liburni performed an important function as the vehicles of cultural transmission. Studies, however, often neglect this point and merely credit these agile seafarers as successful pirates. Archaeologically proven cultural contacts between the peoples of the Central Mediterranean, the Balkans, the Appenines and the Alpine regions are attributed only to the 'more civilized peoples' of the Ancient world, i.e., the Greeks, Etruscans and Phoenicians. This is still an echo of the obsolete diffusionist theory of culture being carried from East to West, from 'civilized' to 'barbarian' peoples.

By the method of archaeological experiment, which increasingly is being used in modern archaeology as an equal research and analytical form of exploration, the Liburna Project will try to show that the Liburni were among the conveyors of cultural impulses and currents in the Mediterranean. This would alter the image of the 'primitive' peoples of the Eastern Adriatic and Balkan peninsula as being capable only of off-shore petty piracy, and show that the Liburni and other Balkan peoples were also ready and able, with their own cultural dynamism, take active part in the lively and frequent cultural interactions shaping the history of Europe.

Very little is known about the actual appearance of the ship used by the Liburni. Classical writers point up its exceptional navigational features, and rank it ahead of well-known Greek vessels. In Roman times, the Liburna ship became the main vessel of the naval forces, and Augustus won the battle of Actium with the support of the brave seafarers of the Eastern Adriatic.

In creating the Liburna Project, I have worked side by side with Branislav Misa Zivojnovic, a marine archaeologist and scuba diving instructor. The project provides for reconstructing the ship (10-12 meters long) on the basis of available archaeological and literary sources. Using the winds, the Liburna ship would sail from the Eastern Adriatic coast towards the islands and shores of the Adriatic and Mediterranean, which, according to archaeological and other sources, are known to have been within the sphere of interest of the Liburni. Our route links all the major islands of the Eastern Adriatic and goes as far south as Crete.

The Liburna Project should shed light on two important subjects:

First, the challenge of constructing a ship which differs considerably from all familiar types of vessels in this region. The technique of fitting the ship together is of particlular interest, as we have seen from hydroarchaeological findings in the Liburnian port of Nin.

The second challenge is to confirm the existence of the Liburnian naval and trade routes in the Adriatic and Mediterranean. Archaeological evidence exists. What is missing is theoretical and practical proof of the routes. The Liburni did not sail in desperate search of the new world. These were sober undertakings motivated by the desire to acquire and affirm their own power and knowledge.

It is this that the Liburna Project should prove.

Mount Everest altitude 8880 m. called the roof of the world is the highest peak in the world. Close to the towering peak of Mount Everest lies the body of an Englishman, Mallory or Irvine. Finding it, and the camera believed to be with it, might prove that the world's tallest mountain had been first climbed in 1924.

The Search for Mallory and Irvine: First to the Top of Mt. Everest?

♛ Thomas Martin Holzel

Honourable Mention — The Rolex Awards for Enterprise — 1984
52 Lang Street, Concord, Massachusetts 01742, U.S.A.

American, born October 26, 1940. President of Arcturus, Inc., an engineering company. Educated in U.S.A.; B.A. (Economics) from Dartmouth College (1963).

"British climbers Mallory and Irvine were last seen 250m below the summit of Mt. Everest in 1924, 'going stong for the top'. The body of one of the two lies on a snow terrace at 8,200m. His camera holds the answer to the greatest mystery of 20th Century exploration: Did the pair reach the summit 29 years before Hillary and Norgay?"

So I wrote in 1970. Ten years later, Japanese climbers wrote me that the body of an English climber — who could only be Mallory or Irvine — had been spotted on the 8,200m snow terrace.

My ambition is to recover this camera and so finally solve this great historical mystery. But there are two challenges: A special Search Expedition must be mounted, which I have started; and a special oxygen system must be built to allow time at extreme Everest altitudes to conduct the search. I have designed such a system, which must now be built, tested and then used by me in "The Search for Mallory and Irvine".

The Background

Wearing primitive, but effective, oxigen equipment, George Leigh Mallory, 37, and Andrew Comyn Irvine, 22, were last spotted at 12:50 pm on June 8, 1924 some 250m below the summit of Mt. Everest, "going strong for the top". It was a climb from which they never returned. Their disappearance left unanswered the burning question of whether or not they were the first to reach the summit of the world's highest mountain. Although Mallory and Irvine were revered by their contemporaries, as well as subsequent generations of climbers, expert opinion at the time held their success in reaching the summit unlikely. It is this negative assessment that remains in the history books.

In 1970, the authoritative climbing journal 'Mountain Magazine' published my study — the first comprehensive analysis of the Mallory and Irvine climb — which strongly suggested that one of the pair had indeed reached the top. Photographic proof would be found, I predicted, in the camera of young Irvine, whose body must still lie on the 8,200m snow terrace of Mt. Everest's north face. Publication of the theory touched off an explosive debate in the pages of the London Sunday Times.

It was not until 1980, after having read my study, that the Japanese Alpine Club passed on a report to me about a Chinese climber having discovered a body at 8,100m that could only be that of Mallory or Irvine. The body had been left in place, and no search was made for the camera or other clues. Since this announcement, four Everest Expeditions have evinced great interest in trying to help solve the Mallory and Irvine mystery; the Japanese in 1980, the French in 1981, the Dutch in 1982 and the Americans in 1984. Each would be stymied by two practical difficulties:

1. Everest Expeditions are either large and highly nationalistic, or, recently, small and underpowered. With large expeditions, the political pressures to reach the summit become overpowering, negating the best of intentions to conduct a subsidiary search effort. With small expeditions, the chances of reaching the top are so slight that the climbers never get to the point where they can spend the precious extra effort to look around for evidence.

2. Present mountaineering oxygen equipment is all of the open-circuit type which is barely adequate to let the climbers exist at the extreme Everest altitudes. Even with it, they must struggle up and down the mountain as quickly as possible to avoid life-threatening 'high altitude deterioration'. Week-long sojourns at 8,100m, necessary to conduct a search for the body and its camera, are not possible with present oxygen systems.

I have addressed these two problems in the following ways:

The Mallory and Irvine Search Expedition

The most practical way to recover the 8,100m high body and camera is to launch a north-side Mt. Everest Expedition exclusively for this purpose. This expedition would remove the enormous pressure to reach the top and let all efforts be concentrated on a search party.

The Chinese Mountaineering Associations (CMA) has invited me to apply for the next available Everest date — the spring of 1987 — which I have done. Permission is expected to be received in the near future. Commandante J. C. Marmier (of the French Everest attempt of 1981) has agreed to repeat his role as the Expedition's climbing leader. The American Alpine Club has agreed to endorse the Expedition. Because of the extraordinary public interest in the Mallory and Irvine mystery, financial support for this venture will be readily obtainable.

The Holzel-Chemox Rebreather (HCR) Oxygen System

To solve the second part of the equation, I have designed a unique closed-circuit chemical-oxygen system specifically for extended use at high (terrestrial) altitudes in sub-freezing temperatures. It promises to overcome all the extraordinarily troublesome difficulties that still beset the most modern mountaineering oxygen systems, and will offer several highly desirable additional features. (No existing closed-circuit system is available for mountaineering use.)

To appreciate the severity of the oxygen problem, it may be helpful to examine briefly the problems of conventional mountaineering oxygen systems.

Present-day mountaineering oxygen systems are all open-circuit systems. Oxygen, contained in vessels at pressures of 3500-4500 psi (238-306 Atm) is metered into a face mask and mixed with outside air to be breathed by the user. The quantity of oxygen his lungs can absorb declines with altitude. A clear example of the rapidly declining ability to work (climb) with increasing altitude can be seen in Climb Rate Charts.

Lower work ability translates to lower climbing speeds, less load-carrying ability, lesser ability to generate body heat, poor sleeping and recovery, reduced appetite, vastly increased water-loss (due to excessive panting), and the concomitant need to carry more fuel to melt water. All these result in the need for more intermediate camps, more carries and more time. In addition, the physiological problems resulting from the use of conventional systems is not their only weakness. Up to 50% of such systems have failed in use, or suffered reduced function or leaked, above 8000m. The biggest problem stems from the pressure reduction valve, which freezes up with frightening regularity. Once a mechanical failure occurs, it is not practical to field-strip the precision regulators in an attempt to fix them.

My HCR system is designed to avoid all the above problems. Here are its features:

1. The HCR is a closed-circuit oxygen system, meaning that the user receives at least as much alveolar oxygen tension at any altitude on the mountain as he does breathing air at sea level. This should permit climbing speeds nearly equal to sea-level rates (less weather difficulties).

2. The HCR is a pressureless chemical system, meaning there are no high-pressure regulators or tanks, and no metal valves of any kind. The entire system is made of silicon rubber that can be squeezed by hand to clear it of ice. The HCR can be field stripped in minutes without tools.

3. The chemical source of oxygen means that user exertion rate determines the quantity of oxygen flow. No use adjustment or calculation is required. The same chemical that generates the oxygen (potassium superoxide — KO_2) also absorbs the user's CO_2, thus eliminating the need to stock filters.

4. Since the user 'rebreathes' his own air (which is cleansed of CO_2), he no longers loses water vapor to the outside. In addition, the exothermic reaction contributes significant water vapor. This reduces to half the six liters/day/man of water required for hydration, as well as the fuel needed to melt it.

5. The high oxygen concentrations of the HCR system should keep the user in his normal physical and psycological state. He can enjoy mountain climbing once again, instead of enduring it.

Though the HCR system is heavier than conventional open-circuit systems in terms of weight vs. duration of oxygen production, the real measure of efficiency is weight per 100 meters of elevation gain. By this standard, the HCR offers a revolutionary advance. The HCR system will permit very fast ascents of Himalayan peaks — probably 2 to 3 times faster than the open-circuit speed. The system will also enable climbers to live at extreme altitudes for extended periods — certainly as long as is necessary to search the 5 acres of the 8200m snow terrace.

Finally, once standing on the wind-blasted slopes at 8200m on Mt. Everest, whether successful in our search or not, I will have done everything humanly possible to solve this great mystery. Regardless of its outcome, the effort will have been a score of years passionately well-spent.

First Woman-Man Team to a High Mountain: Cho Oyu, 8153m

Thomas Erhart Bubendorfer, Jr.
Pechstrasse 18, 5600 St. Johann/Salzburg, Austria

Austrian, born May 14, 1962. Alpinist, language student. Educated in Austria, U.K. and Italy; languages being studied at Universita per gli Stranieri in Florence, Italy.

It is Ruth Steinmann's and my plan to reach the summit of Cho Oyu (8,153 meters) in Nepal on our own, without the aid of sherpas or other alpinists. After establishing our base camp, we will proceed on our own, carrying all the necessary equipment to higher camps without help from porters and using no artificial oxygen.

In the Himalaya region, the highest mountain range on earth, and therefore the most interesting one for alpinists, there are fourteen peaks known as the 'Eight thousanders', all over 8,000 meters. Each of them has been climbed.

In the beginning of the exploration of the Himalayas, alpinists in huge groups 'attacked' the 'Eight thousanders', financially supported by nations and governments. Tons of equipment and artificial oxygen, and dozens of people, used to reach the summits by whatever means possible. One chose the easiest ways.

Gradually, men began to understand that, with the support of all this technical help, all problems on mountains could be solved. Expeditions got smaller and smaller, existing with less and less material, and trying new and more difficult ways of reaching the summits.

A further step (the last, it then seemed) was achieved when Mt. Everest was climbed in 1978 for the first time without artificial oxygen, and then again two years later, solo, by Reinhold Messner. Yet again, in December 1982, another 'first' was recorded by a Japanese alpinist who accomplished the first Everest ascent in winter; tragically, he was killed in a storm while descending from the summit.

With these last ascents (some people called them the 'final ones'), there seemed to be no new way of climbing left. New, very difficult ways, yes. But nothing, it seemed, remained to be discovered, nothing unknown that might challenge the spirit of alpinists.

Yet, there is one last possibility, another undoubted 'final one'; a woman-man expedition to an 'Eight thousander'.

There have been large women-expeditions to high peaks. Women, as part of large groups of male alpinists, have even reached — with artificial oxygen — Mount Everest. Yet only once before in the history of Himalaya climbing has a two-man team reached the top of an 'Eight thousander' (1975, the Austrians Peter Habeler and Reinhold Messner).

Never before has there been an attempt by a woman and a man alone on such a high mountain. One knows nothing about the reaction of two such team members in difficult and dangerous situations, the reaction of two people, male and female, who have lived together in one small tent for weeks and months, one hundred percent dependent upon each other; living together under continual stress, physically and psychologically.

For us it is an immense challenge, to go to Cho Oyu, not knowing how we shall get along together during all that long, and often hard, time on the mountain. We both know that the chance to reach the summit in bigger groups, with sherpa help, is a rather good one. Ruth Steinmann has already climbed to a height of 8,250 meters on Lhotse, one hundred meters higher than the top of Cho Oyu. I was on Pik Korshenewskaja, 7,105 meters, in the Russian Pamirs, at the age of eighteen. So, we believe, in a normal expedition group, we would both have a fair chance of reaching the top.

Now, for the first time, the problem of attacking one of the Himalaya giants will not be the technical difficulty, the height, the cold nor the wind. It will be the fact that we two, Ruth Steinmann, 47, and me, Thomas Bubendorfer, 20, will not be able to use the experience of any previous woman-man teams; we do not know whether we shall be able to tolerate each other's faults and mistakes, or if we can cooperate enough to get along in the face of all the difficulties of such a challenge — once we are alone on the mountain.

We have a big drive, and it is, of course, very important for us to reach the summit of Cho Oyu. Above all, however, we shall not be human climbing machines, simply chasing for success, but rather two people, with all of our feelings and fears and problems, going into one of the world's most severe regions, trying to fulfill — successfully or not, it does not matter too much — the fantasy of Cho Oyu by a woman-man team.

We shall leave Europe around the 20th of March 1984, and plan to return by the end of May 1984.

The Oral Art and Literature of the Sahara's Pastoral Ghrib

Gilbert Jules Marie Claus

126 Wevelgemstraat, B-8630 Gullegem, Belgium

Belgian, born June 15, 1944. Cultural anthropologist. Educated in Belgium; studies at State University of Ghent.

In the northwestern Tunisian Sahara, there live a nomadic people, the Ghrib, about whom our knowledge is very scarce. From my first visit to this area in 1973, and my initial acquaintance with these people, I have been fascinated by their culture and concerned about its potential loss. This project reflects my deep concern that their valuable history must be saved before it is too late to do so.

Even the origin of the Ghrib people is unsure. During my investigations, some Ghrib claimed that they originated from Saguiat l-Hamra, the utmost southern part of Morocco. Their migration should have taken place between the 14th and 15th centuries. At that time, they would have spoken a Berber dialect, though they subsequently became Arabic-speaking and converts to Islam.

They were then all stock-breeders, their life and their work based on goat-, sheep-, and dromedary-breeding, all connected with caravan trading and 'razzias', or raiding expeditions. They became experts in dromedary breeding, and renowned as real connoisseurs in this field. Their fame was widespread, far beyond the borders, and it was not surprising to hear that Libyan and Algerian tribes were purchasers of fine Ghrib dromedary offspring.

The Ghrib tribe is, in fact, a confederation of nine subgroups, or clans. Without anticipating the results of my research, still underway, I must say that I am under the impression that not all clans are of the same origin as is alleged by each of them. Analyzing the collected data, I see particularities in the fields of vocabulary, kinship terminology, manners and customs, dromedary property brands and lineage that, little by little, have convinced me that in former times there were fewer than nine clans, and that the origins of some will prove to be different from what they now pretend, perhaps going back to Saudi Arabia or Yemen.

The Ghrib society is changing very rapidly from nomadism to a sedentary life, a transition directly menacing the Ghrib culture and the collective memory of the Ghrib community. Indeed, sometimes I have the feeling of collecting the last breaths of the oral tradition, this amazing reservoir of centuries of memories quickly becoming empty. In the field of oral tradition, the first commandment and sacred duty of every anthropologist confronted with a rapidly shrinking body of oral tradition is to collect by any means what still can be saved, because later on his data will be the only source material available for posterity. Preserving the Ghrib oral tradition — music and spoken word — is my real aim. It is soul-destroying to think that the passing away of an old man or woman may be equal to the burning out of a library.

To pursue my goals, I have made many trips to the Tunisian Sahara. During all these periods, special attention was paid to the Ghrib oral literature and the Ghrib popular beliefs and usages relating to the subject of birth.

Until now, the result of my research may be called successful and fruitful. I have collected 52 songs (poems) — some lasting more than 20 minutes — attributed to 29 poets, of whom 26 are no longer alive. In other words, these are songs traditionally transmitted from father to son, or to another relative. Many times I have been the witness to the loss of a story or a song, told by an old man who died before passing his art on to younger people, who today tend to disregard this form of expression.

In addition to the songs, I recorded some fine ethnographic and linguistic documents about social events, e.g., a very complete and detailed story about Ghrib marriage ceremonies and practices (in the Ghrib tongue), supported by original recordings (12 hours) made during the marriage of my 'brother' dromedaryherd, which took place on July 4, 1980 and lasted until July 11. When I write of my 'brother dromedaryherd', it is because I had had a regular dromedaryherd, whom I engaged in Fall 1975 to look after my own herd (which now consists of 8 dromedaries and about 20 goats and sheep). The relationship between this informant, his family and myself became, over the years, so close that they consider me now as one of them, in the strictest sense of the term. In the summer of 1980, when my dromedaryherd's brother contracted a marriage, it was a unique opportunity for me to become intimately involved and investigate the matter to the very bottom.

Although I had attended many weddings before, I had never been able to observe and study such an event in such detail. As a result of my relationship with the family, it was possible for me to record and photograph many things never before recorded.

The final aim of my research is to compile and publish an anthology with recordings of the oral literature of the Ghrib tribe, and to publish a book: Ethnographic and Linguistic Documents of the Pastoral Ghrib. Afterwards, all the collected data and material could be used for a linguistic study on the lexicography of the Ghrib tongue, grammar, etc.

The investigations will include both the worlds of music (instruments, melody, voice, performances, songs of different natures) and the spoken word (proverbs, riddles, poetry and stories). There is much to preserve, and time is limited.

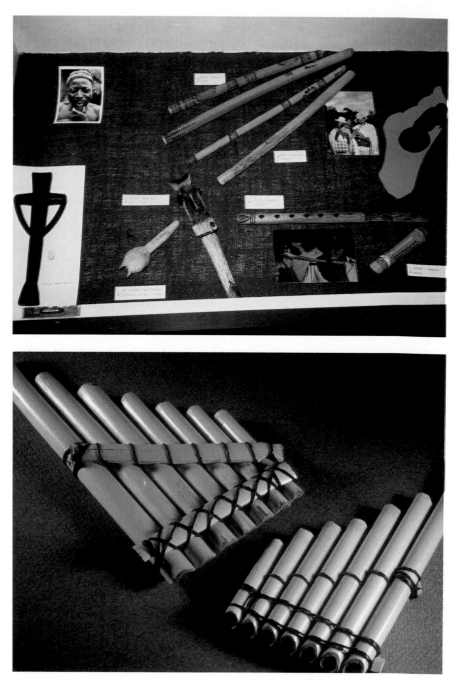

Displays of man's oldest musical instrument – the flute.

A Museum for
The Flutes of the World

♕ Charles Tripp

Honourable Mention — The Rolex Awards for Enterprise — 1984
3 Rue de Brasse, 90000 Belfort, France

French, born March 14, 1921. Researcher/animator of music; lecturer on music (organs, flutes). Educated in France; Certificates for Modern Literature (1950) and History of Music (1951) from Paris University.

The objective of this project is to collect all the types of flutes now existing in the world (about 1,000) and present them in a completely animated form (including a collection of slides, books and audio systems). The project is taking nine years (1980 to 1989), culminating at the Universal Exhibition of Paris. The end result will be the creation of the Museum of "The Flutes of the World", for the year 2000 A.D. My present collection now numbers some 400 flutes gathered over seven years.

As a musicologist and teacher, one of the passions in my life is to bring music to as wide a variety of audiences as possible, and in particular, to children. The musical heritages of the peoples of our world are richly filled with both sounds and instruments and stories; a collection of attributes I have been able to combine in presentations of music that seem to reach deeply into the minds and souls of those who see my animations.

Some years ago, recognizing that peoples around the world have all sought to make music with the wind instruments we know as flutes, I began to collect these universal symbols and tools of humanity's musical urge. I have learned that flutes are 'endangered' as a unique form of human artifact, due to the homogenizing influences of our increasingly global community.

My project, therefore, is concerned with gathering, before they disappear, all the flutes of the world; the flutes, whistles, and bird calls that belong to all peoples and ethnic groups. In number, there are about a thousand, representing an extraordinary mixture of craftsmanship and skills. Made of many different materials and equipped with a great variety of mouthpieces, they are played in many different ways. They use different scales, have different tones, possess unique forms of ornamentation and produce quite distinct kinds of music.

It is my objective to continue my research into traditional flutes, and display my collection with a view to creating a museum of world-wide significance, designed not only for exhibition, but also for studying; a museum of sound and a meeting-place as well.

What is New in My Project?

It appears that nowhere in the world is such a museum to be found, a museum where people can see, study and listen to all the flutes existing in the world.

The Originality of My Project

This lies in the fact that this museum will enable the public to discover our planet — the earth — in its variety of peoples and ethnic groups through an instrument we all share in a common heritage; the simplest and most widely used — the traditional flute.

The Museum as a Symbol of Creativity

It will show and discover the inventiveness of humanity, through the apparent simplicity of traditional flutes. It will explore the world through the amazing variety of traditional flutes from several points of view. One example of these is the materials used: bone, reed, bamboo, wood, metal, terracotta and horn. Another example lies in the type of mouthpiece and the way they are played: duct flute, vessel flute, notched flute, collar flute, water flute, Pipes of Pan, transverse flute, nose flute, oblique flute, double flute, ocarina, toy flute, flutes without a mouthpiece or holes, whistles, bird-calls, cuckoos.

Human Interest in the Museum

As the third millenium is drawing near, we need to bring traditional flutes out of oblivion and rescue them before they disappear, because they constitute part of mankind's cultural patrimony.

As the second millenium is drawing to a close, the discovery of man as craftsman — homo faber — from prehistory up until now through technological investigation into the most universal instrument, is a meaningful pursuit of the history of mankind.

The museum will enable the discovery of peoples through the shepherds' traditional flutes, which are a heartwarming symbol suggestive of the fight against wars and racism. Similarly, the cultural identity of various peoples and ethnic groups in the world will be seen in the panoramic presentation of all the flutes existing in the world; to each people corresponds a certain type of flute, to each flute corresponds a certain type of music. Further, the museum will aid the discovery of humanity's music by enabling the study of traditional flutes, which constitute a living and picturesque expression of our artistic sensitivity.

Meaning of this Project

In our modern, conditioned, standardized world, the nomads and shepherds are tending to die out. The discovery of men still faithful to popular art, thanks to a museum of "The Flutes of the World", will stand as an oasis of freshness and fantasy for the year 2000. This project pays tribute to:

— People's creativity, their sense of music characterized by their choice of particular scales of music, their sense of improvisation, their freedom in ornamentation and the distinctive tone of the various flutes.

— Craftsmen's esthetic senses, which are often revealed in the way flutes are decorated; they are painted or braided, engraved, inlaid, etc.

— Shepherds, musicians without status or title; never having studied, they draw their inspiration from the wind. As Claude Debussy says, "Nature is their conservatoire."

An Urgent and Necessary Enterprise

A slow disappearance of the traditional flute is to be feared. Its causes are numerous; wars, political and other disruptions, industrialization, and urban-

240

ization. As a result, there are changes in the traditional lives of the peoples of the world; a smaller number of shepherds and country craftsmen.

It is sad to note — even while admiring the progress of science and technology — the disappearance of a flute such as the whistle of the knife-grinders of Spain (Pan's Pipes, made of box-wood). Yet this instrument has been in existence for about 2,000 years, for it was already known by the Romans. It has scarcely been played in Spain or Portugal for the last 10 or 20 years. The 20th century will have seen it disappear. It still appears in a sculpture of a shepherd of the Cathedral of Chartres: the eternity of stone!

The plan for the Museum is based on assembling 45 show-cases, with the story of flutes being presented along these lines:

Introduction: In the beginning, the Wind, the Reed, the Breath, the Shepherd; The Wind organ; Whistles for the hunt, as signals, for entertainment, for work; The magic of the flute.

In Asia: Japan (Shakuhachi, and the mendicant monk "Komuso"); Korea (Taegum, and the calm of the mountains); China (The membrane flute); India (The flute of Krishna); plus Central Asia, South-East Asia and Indonesia. *Curiosities:* Nose flutes, flutes for ceremonies of initiation, whistles for pigeon's flight and horn-flutes.

The World of Islam: The ney; Middle East, North Africa and other countries. *Black Africa*

In America: North America; South America (Kena and Sikus); Flutes of baked clay (zoomorphic and anthropomorphic flutes).

Technical Aspects: Flute shapes, sizes, mouthpieces, the false flutes, woods from which flutes are made, and decoration (sculpture engraving, painting).

In Europe: Prehistoric bone whistles, Pipes of Pan, A story 2000 years old (the Pan Pipes of the Spanish knifegrinder), shepherd flutists, the word 'flute' in Europe, Carnival (the Basel fifer), organ with pipes, and regions.

Children's Corner: Hans-the Pied Piper of Hamlyn, whistles, bird-calls, cuckoos, nightingales, ocarinas, spring flutes, slide flutes, toy flutes, bird-flutists and children's flutes.

To complete my collection (which now has flutes from 55 countries), I must find 600 flutes for the creation of the Museum. This requires the creation of a vast network of contacts and informants throughout the world, journeys of discovery to distant countries, and the purchase of authentic specimens from craftsmen who are getting scarce, and of course, lasting patience!

Tracing the Greenlandic Sagas in the New World

Frederick N. Brown

P.O. Box 11692, Phoenix, Arizona 85061, U.S.A.

American, born April 15, 1929. Manufacturer/overhauler of machines and devices for manufacturing products. Educated in U.S.A., emphasis in design.

After six years of extensive study and research, I believe that I have achieved a sufficiently high correlation of my suppositions to indicate that one or more of the specified locations described in the Greenlandic and Icelandic Sagas describing Lief Erickson's "Vinland" may have been discovered.

I am aware of scientific excavations of the site at L'Anse Aux Meadows in Newfoundland, Canada, which have been accepted as Norse in origin. The site has been offered as that of Vinland, but some scholars cannot accept this as, the Sagas being the instrument of discovery — or indication — of a Vinland, it appears there is little correlation between that site and the descriptions given in the Sagas.

The basis for the belief of the existence of a Vinland lies in a number of references in several Sagas — or Epics — still extant in translations in both Iceland and Norway. Some of these are in the form of family chronicles, and the two main ones that concern us are "Flatayjarbok" (from Greenland) and "Hauksbok" (from Norway). Though they are folkloric, indistinct and difficult to translate, it is thought that parts are generally valid and factual within certain limits. Some of them tantalizingly describe voyages to a land which seems to have been North America, though no scientific proof has ever been found which correlates with the vivid descriptions given in these Sagas.

The subject has been studied and theorized by numerous persons before me, and my work has been to study these various theses and translations. There are many specifics that are agreed on by most as having high probability. Among these is the belief that there were at least five expeditions between 986-1035 A.D., of which the first was not planned, which saw or settled on land thought to be North America. Three sites are described which were settled or lived in for various times over the span of 20 years or more. The three sites are identified as *Straumney*, an island with strong currents; *Vinland*, in two forms, being specifically the camp or home of Lief Erickson, the Greenlander, or generally, the land in its entirety; and another place named *Hop*, which was one locality of the third and major colonizing attempt led by an Icelander named Thorfinn Karlseffni. These three sites, together with the two other places on the route to them, named Helluland and Markland, are described at greater or lesser length, and it has been thought by many that there is enough information in these descriptions to locate the precise places.

I undertook to do the same, but with the emphasis on *Hop* as, the descriptions being more complete and with the further help of there being two separate descriptions of a single battle, there were possibilities of success in locating this particular settlement. The Sagas are vivid and detail much; there seemed to be enough to justify trying to find *Hop*, if it were impossible to find Vinland.

A sizable portion of previous studies have theorized that the possible location of Vinland might be on the Southern New England Coast. If it can be accepted that Lief Erickson managed a voyage to Cape Cod and around it, then the Norse seamen of the Saga traveled an East/West coast and found an island which is north of a land (mainland) and which has a "sound" to its own north. And north of this sound is a fiord (sic); in or adjacent to this fiord is a river which has peculiar characteristics, according to the epics. These include: a sand bar across the mouth allowing entry only at high tide; extensive white sand beaches; forest growing right to the beaches; a long tidal 'fetch'; many other sand bars; navigability for a distance upstream where it enters a 'lake'; and it is on, or near, this lake where Lief founded his camp. The river is described as "coming down the land into a lake, and then into the sea". I believe this description is as cryptic as it is clear, and identical in most translations of the Sagas.

For several reasons, one accidental, in comparing a number of translations, I found I was confusing the description of Vinland with the description of Hop, and, on investigation, I found they were vaguely similar. I speculated on the possibility that Vinland and Hop might be the same place and I discovered a number of points in the Sagas indicating that this might be so. I resolved to investigate the possibility that Lief Erickson's camp (or Lief's Vinland) and the area later described as Hop might, in fact, have been one and the same place.

Reviewing some 33 specific elements of reference from the Sagas, I proceeded to search the coast line of Southern New England, using topographic and nautical maps for a site that might match all or some of these factors.

I did find such a place.

I was able to go there on a number of occasions and to explore and photograph. I found that all 33 items I searched for fell into place; every single one. Additionally, I found and photographed a number of old sites evidencing ancient man-made activity.

To date, I have spent some six years on the project, reading countless translations, and developing my theory in a progressive way. I wish to pursue the project to the point where the thesis may be presented to the Nation of Iceland. The tale and the Sagas are a Scandinavian treasure, and Icelandic scholars and peoples should be informed and able to participate in the study. I should like the site to remain undisturbed until it can be scientifically explored.

Way of the Vikings: Sailing and Navigating the High Seas

Olaf Tormodson Engvig

Bygdöynesveien 33, Oslo 2, Norway

Norwegian, born June 2, 1937. Writer, photographer, explorer and expedition leader. Educated in Norway.

Four years ago, my daughter (13) and I sailed the 115-year-old Viking-type square-sailer "Hitra" from Hitra along the Norwegian coast to Oslo and further to York in England by the southern route of the North Sea. The voyage was made in support of the Viking excavations in York, but also in order to do research on Viking migration routes and the sailing and maneuverability of the Viking ship.

Our voyage, with no map or modern navigation equipment, confirmed Viking merchant Ottar's description of the route in terms of places, distances and time of travelling. Navigation was simple; land was on the port side most of the time. We landed and asked for information on the place and what we needed to know to continue the voyage in the best way. The enterprise was successful, and we reached York after 38 days of sailing and rowing from Oslo.

To continue the enterprise, I would like to take one or two intermediate sized ships of the Viking type on the northern route across the high sea. The plan is to try to find the islands out in the ocean without maps and compass, or any other modern items not available to the Vikings.

The experiment will hopefully give useful information on many aspects concerning ocean crossing in the old times in our part of the world. To the best of my knowledge, such an experiment has not been tried or recorded in modern times. Until now, quoting the limited information on practical navigation and sailing given in the Sagas has been the rule when it comes to discussion on how the Vikings managed to find small islands out in the ocean.

The attempt intends to gain practical experience on how men in the old days could keep a fairly straight course and be able to reach certain places of destination well below the horizon, several days off the coast. There are many beliefs, and several theories on the subject, all full of pre-suppositions. Practical testing and experimenting with the subject have never been carried out in our part of the world, where fog or bad weather is quite common, even in the summer, and where sailing at high latitudes in the season gets no help from the Guiding Star due to daylight.

I believe the time has come to try to gain basic information on the subject of ancient navigation just by collecting know-how from the material and experience of a voyage of this type. The expedition should do away with much of the speculation and guesses on the subject.

The original Viking ships still keep their secrets of rigging and sail. No one today really knows how the rigging was made, its details arranged and how it was worked. The same goes for the sail, the handling of the sail, the rig and the boat itself. All these details of highly skilled craftmanship, still unknown to us, are very important for the proper handling of the boat. Most of the experiments with desk-construction of the rig and sail on Viking-ship replicas have resulted in very poor performance of the ships.

To eliminate as many as possible of these questions, I have decided to use two traditional boats of the Atlantic type from the coastal area. Boats of this kind have been used in Norway from the days of the Vikings in an unbroken line up to the turn of the last century, when engines and decked boats were introduced. Details in the methods of their construction insured that each new generation had to do exactly as the older ones before them. An example of the faithfulness of copying is seen in one of the boats I intend to use. The boat is original and was constructed almost 20 years before the very first Viking ship was found and excavated, the intermediate sized boats from the Gogstad find in 1880. It is interesting to note that some archaeologists were disappointed in this first excavation, because the ship looked so much like a traditional open boat.

"Hitra" does better than 10 knots under sail under very good conditions. In case of no wind, and propelled by all 10 oars, it does 2-4 knots. It doesn't have to sit for days waiting for wind, or getting set off course by currents. The Vikings did row when the wind failed.

We plan to start from the west coast of Norway in the summer. The course will be fixed and tried to be kept to on the basis of bearings from the sun, which should give us an idea of direction. As the sun travels fast, adjustments will have to be made often. Due to the summer's long light, we will unfortunately have little or no aid from the stars on the expedition.

The first goal is the Shetland Islands. The crossing and experience gained will probably give so much knowledge that we can then decide if it will be natural to continue south to the Orkneys and Scotland, or further on to the Faeroes and Iceland. Both of these routes were Viking paths.

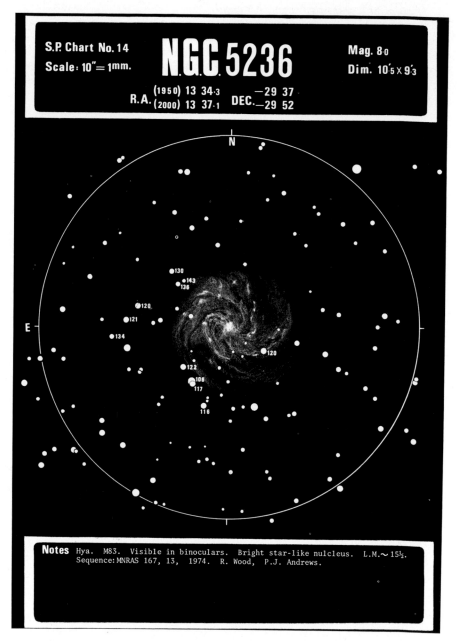

S.P. Chart No. 14
Scale: 10" = 1mm.

NGC 5236

Mag. 8·0
Dim. 10´·5 × 9´·3

R.A. (1950) 13 34·3   DEC. —29 37
(2000) 13 37·1   —29 52

N

E

130
143
136
120
121
134
120
122
106
117
116

Notes Hya. M83. Visible in binoculars. Bright star-like nulcleus. L.M. ~ 15½.
Sequence: MNRAS 167, 13, 1974. R. Wood, P.J. Andrews.

There is still an important place for the amateur astronomer in unlocking the secrets of space. Star maps such as this will aid in the rapid location of supernovas.

The Visual SuperNova Search Charts and Surveillance Program

Gregg Donald Thompson
7 Strover Court, Springwood, Brisbane 4127, Australia

*Australian, born September 28, 1949. Working Director of an Office
Stationery Company.*

The Visual SuperNova Search Charts comprise 225 charts of over 300 of the
brightest galaxies, meticulously prepared to illustrate the visual appearance of
the field stars surrounding these galaxies so as to enable quick visual discoveries
of bright supernovae (SNe).

Today, astrophysicists believe that if we could clearly understand the causes
and processes involved in a supernova explosion/implosion of certain stars, it
would greatly enhance our knowledge of stellar evolution and its further
implications. To achieve this requires detailed observations of many bright
SNe over their observable lifespans. Of the 500 discovered since Zwicky found
the first in 1937, relatively few have been bright enough to enable detailed
observations. Many of those that were bright were discovered photographically after they had faded into obscurity, and only recently have a mere few been
briefly recorded during the short time in which they rise to maximum
brilliancy.

Investigators cite a continuing need for the early discovery of bright SNe to
promote research. Professional astronomical searches have been the exclusive
tool for this purpose; yet, as early as 1939, some astronomers envisioned a place
for competent laymen to directly aid their efforts. The expectation throughout
the last forty-four years has been that persons with small telescopes could
efficiently close gaps in the coverage of formal programs. That this long-
standing invitation to amateur astronomers remains open today is slightly
anachronistic, but nonetheless practical. Opportunities for amateur contributions to astronomy have been steadily closing as the science rapidly evolves to
meet increasingly complex problems. Such contributions must necessarily be
carefully prepared and executed.

My friends and I believe that, after eight years of preparation, we have
arrived at a well-developed and practical response to the need for a nonprofessional search for SNe in nearby, bright galaxies.

The Visual SuperNova Search Charts Project and Visual Surveillance
Program were conceived to facilitate the discovery of bright SNe. It was
considered that serious amateur astronomers could discover SNe *if* a suitable
set of reference charts could be created. These charts would need to depict
those stars visible in our galaxy through which we look to observe other

external galaxies. In this way, a SN occurring in another galaxy could then be detected, appearing like a 'new star' in the field.

Why not use existing photographs as reference material? Stars that are superimposed on bright regions of a galaxy are often hidden against the background of the galaxy. HII regions and other distant galaxies can often appear starlike, creating confusion. The outer regions of galaxies where SNe often occur are usually cropped. Scale and orientation are seldom given, and if so, vary greatly in most publications. Most pictures of galaxies are long exposures which record far more stars than can be seen visually, thereby confusing observers. Photographic magnitudes of stars can diverge considerably from their visual appearances.

Why not search for SNe photographically, as has been done by several professional searches conducted over the last 50 years? These searches, for the most part, concentrated on clusters of galaxies so as to enhance the probability of discoveries. These clusters, however, are mostly distant, causing their SNe to appear faint, in which case only the portion of the light curve near maxima is recorded and spectra are seldom obtained. Bright SNe do not promote such limitations. We also consider the following factors to be of significant advantage with our visual approach:

1.   The eye can more readily detect a SN occuring over a bright section of a galaxy than can photographic emulsion.

2.   A visual observer can detect a SN brighter than his limiting magnitude almost immediately, whereas a film may not be developed and inspected for many days after.

3.   Far more bright galaxies can be searched visually than is possible to photograph and check in the same period of time.

4.   Vastly greater numbers of observers can participate visually as compared to those possessing the necessary equipment and expertise for photography.

5.   A visual searching program incurs negligible costs compared with photography.

Our reference charts, now nearing completion, include all galaxies in both hemispheres listed in the Revised New General Catalogue to magnitude 11.5 and brighter, except for a few dwarf galaxies. Some 12th magnitude galaxies involved in clusters have also been included. This encompasses 255 charts comprising 345 galaxies. Deliberately, no discrimination of galaxy type or orientation was applied as has been the case with most previous searches. It is hoped an unbiased selection of nearby galaxies may present useful statistics in future years.

Most SNe discovered with the use of these charts will be brighter than magnitude 15 at maxima; a range that is more accessible for detailed study by professional astronomers, not only in the visible region of the spectrum but also in the now important UV, IR, X-ray and radio regions. Being fairly bright, SNe will allow the monitoring of the light curves for longer periods.

We considered it to be most important that the charts should feature a common scale if the measurements of the position of a SN were to be quick and accurate. To achieve a common scale, our galaxies were copied from the 48″ Schmidt plates with a macro lens. 20x25 cm. prints were then made to a scale of 10 arc seconds per millimeter for all galaxies except M31 and M33, which are copied to a larger scale. This scale gives a convenient chart diameter of half a degree, being comparable to the field of view in most telescopes up to 40 cm.

aperture at low magnification.

The prints act as a base on which the arduous task of plotting all the stars visible to a particular limiting magnitude are traced. The limiting magnitude is typically between 14.5 and 16. Telescopes from 20-30 cm. aperture are used to make the tracings. Short exposure photographs taken by J. Child in the U.S.A. with a 46 cm. telescope also assist in the accurate plotting of stars on overexposed portions of our enlargements of the Sky Survey Plates. It is planned to publish some of his excellent pictures with the charts.

The photographic magnitudes of the stars are adjusted to visual at this stage. I then transform the tracings into finished artwork, which in turn is taken back to the telescope for final checking. This time-consuming process can need to be repeated several times in some cases. This alone consumes 10 hour per chart on average. The work is exceedingly difficult, being both physically and mentally taxing, requiring much experience in observational astronomy, so it is understandable that it has not been attempted previously on this scale.

Over the past years, we have continually updated the charts as further experience and information was gained.

Upon completion, the 225 maps will be printed in the style of the photographic masters, and published in a single volume. The semi-translucent waterproof paper on which they will be printed allows a remarkably close approximation of the galaxy and its field stars when placed on a simple dimmer box. The charts will be supported by an illustrated handbook of approximately 50 pages, which will give a detailed explanation of the charts, tables of data on each galaxy and companions, and suggestions for the development of serious observing programs.

Because of the delicate nature of the printing, it will be necessary for me to supervise the making of the reversed, combination line and halftone negatives. Commercial bids for this approach $10,000. The probability of sharing the printing costs is also likely. Presently, we lack any subsidy to pay such costs. Our team is in desperate need of a 40 cm. telescope to enable more efficient searches to be made, in particular for the confirmation of faint suspect SNe. This, too, is currently beyond our financial resources.

Two American publishers have already expressed interest and view the charts as a unique contribution to astronomical literature. To our delight, many amateur and professional astronomers have already requested that they may purchase the charts as soon as they are published. We expect to be able to offer them, with the 50-page handbook, for sale at about $35.00.

If funds become available, we will be able to realise the goals of our long endeavour; to publish the charts in order that they may be the basis for a program of regular surveillance of all our bright galaxies, with the view of enhancing our knowledge of the processes that create nature's most singularly energetic phenomena — a supernova outburst.

A New Treatment for Migraine Headaches

Sylvester Theodore Gunaratna

Physiotherapy Institute, 365/2 R. A. De Mel Mawatha, Colombo 3, Sri Lanka

Sri Lankan, born September 10, 1908. Physiotherapist in private practice. Educated in physiotherapy in England.

I have developed what I believe is a guaranteed treatment for migraine problems, based on professional experiences I have had with patients suffering badly from this illness.

I had for some time treated patients with various physiotherapy complaints before 1975, when I went to America to attend upon an American diplomat whom I had met in my clinic. He had previously consulted with specialists in several countries regarding a neuritis problem he had had from birth, and told me he believed nothing could be done for him. I assured him I believed I could cure the problem, although I didn't know how long it might take. After one week of using western massage techniques, plus an eastern medicated oil, he felt much improved. On his return to America, he sent me plane tickets to come to attend upon him there.

When I was attending upon this diplomat, I had an American lady come to me, wishing to know if I might be able to help her; she had suffered from migraine headaches for five years. I had never treated a case of migraine before, but I wanted to help her. Guided by intuition, I massaged the back of her neck (over the nerves leading to the brain) by hand, using the medicated oil, under the head of an infra red lamp. Then more massage using a powerful electrical vibrator. After twenty minutes of this, her headache had disappeared. I repeated the treatment for four days, and she appeared to be cured. A fortnight later, she visited to inform me she was free of the migraine, and to invite me to lunch. When I returned to Sri Lanka after the three-month treatment of the diplomat, I was very keen to test the treatment for migraine.

I found a local lady who had had this complaint for over twenty years, and she came round completely after four visits. After further tests on others, I realized I had perfected a cure for migraine. In Madras, I met a doctor who had a nursing home there, and spoke to him about my treatment. He said he was also a sufferer, and if I could cure him, he would be able to bring many patients to me. I did cure him, and thereafter I worked daily in his nursing home for a couple of hours each day, handling and curing patients with migraine problems. More recently, on a holiday in Sweden, I attended upon many friends of my host, all of whom were so pleased with my treatments that they showered me with lavish gifts. As a result, all I can say is that I have a cure for migraine and am prepared to prove it. Here is what is involved.

The medicated oil consists of mustard oil as the base. First, heat the oil on a low fire until it is fairly hot. Then, take a handful of garlic, nicely cleaned and slice it into the hot oil; this releases its properties into the oil. When the garlic turns brownish colour, put a handful of cloves in the hot oil. This immediately doubles in size. Then you allow the oil to cool and filter it to remove the pieces of garlic and cloves. The oil is now ready for use. This recipe was given to me by a learned monk; while in London, I used it in the curing of bad cases of arthritis.

The treatment for migraine involves the application of the oil, with the fingers, on the back of the neck, in the area with the nerves leading to the brain, while an assistant holds an infra-red lamp to heat the area where the oil is being applied. Keep massaging the space between the head and the shoulders for a few minutes, and then apply an electric vibrator until the oil dries up on the neck. Repeat this three times, over a period of about twenty minutes, on four successive days.

I would love to pass on this treatment to the world, as it would benefit people everywhere. It is difficult, however, as the West does not believe in medicated oils, which is the saddest part of it. Even when I went to America, it was with the use of this oil and massage that I cured people who came to me. The Western technique of massage is undoubtedly the best, but coupled with the Eastern medicated oil, wonderful results can be achieved.

I have been using my treatment since July 1975, and I am so confident of it that I would like it to tested so that all migraine patients, everywhere, could be cured. Any physiotherapist can do this job.

A Search for
Natural Satellites of Asteroids

Paul Maley
15807 Brookvilla, Houston, Texas 77059, U.S.A.

American, born May 21, 1947. Space Shuttle Flight Control Engineer, at the NASA Johnson Space Center, Houston, Texas. Educated in U.S.A.; graduate studies in Computer Science, University of Houston at Clear Lake City, Texas.

The objective of this project is to attempt to resolve the five-year-old controversy regarding the possible existence of natural satellites orbiting some of the larger asteroids. Under the project's brief, two-man expeditions would be undertaken to locations where rare eclipses (occultations) of stars by certain asteroids would occur. If more than one occultation was simultaneously observed for a given case by the two independent observers, this would provide primal evidence that the asteroid under study had a companion. Such a result could:

— influence the theory of the origin of the asteroid belt.

— be beneficial to planners of missions of spacecraft to rendezvous with an asteroid.

— verify the potential nature of a hitherto undiscovered sub-population within our solar system.

The author has been conducting an on-going project for the last two years, with limited resources, to observe occultations of certain stars by asteroids. To do this, I have used portable optical, timing and recording equipment. Only 34 such events have ever been observed, according to available data. Astronomers in 14 countries have reported limited observations. Of these, some 50% have also had a secondary occultation not associated with that of the asteroid itself. This circumstantial data may or may not be due to satellite material in a natural orbit about the asteroid.

Conventional techniques have failed to resolve whether such satellites exist, due primarily to limitations imposed by the earth's atmosphere. The occultation technique, which led to the discovery of the ring system of Uranus, has been used by me to successfully observe 4 of the 34 events mentioned above. Attempts on the others have been unsuccessful due to poor prediction quality on the part of the few astronomical observatories that provide this data worldwide.

I have developed a mobile approach to this system of research, which allows me to carry my equipment almost anywhere in the world, a degree of flexibility other astronomers do not have.

The project calls for the selection of particular occultation events involving asteroids large enough to support a satellite on a prolonged basis. This will allow me and another competent observer to position ourselves as close to the center of the path of visibility as possible. We would be spaced several kilometers crosstrack and downtrack from each other, so that both of us could obtain time-correlated timings of any occultation(s) seen. This method will allow verification of authenticity of any secondary events and would also permit relative velocity between the asteroid and any possible satellite to be calculated.

Equipment used will include 13 to 20 cm telescopes, micro-cassette tape recorders, WWV time signal receivers and digital quartz watches. Locations chosen for observation would be those where geodetic coordinates could be obtained for sites selected. I would use Landsat and other earth resources satellite imagery that I have access to, in addition to local geodetic survey maps.

By observing a statistical number of such occultations, it should be possible to determine whether satellites of asteroids do exist. One good set of coincident time-consistent data should provide enough proof of this. By having a two-man observing team with enough experience, dedicated to continuing observation of asteroid occultations, and with current proven equipment already in hand, such data could be gathered within a two-year time frame, if enough good quality prediction data are available.

Photographic confirmation, using a star trial technique perfected by me in a South American occultation in 1979, would be used for occultations of stars brighter than 8th magnitude. The most recent direct observation attempt to resolve a satellite reported in connection with the asteroid (9) Metis has been unsuccessful. This further points out the need to observe asteroid occultations to build up evidence that has been accumulating very slowly, based on reports of random events observed to occur in various parts of the globe by both professional and amateur astronomers.

My qualifications for carrying out this project include: 22 years of observational astronomy experience; 9 years optical satellite tracking experience with the Smithsonian Astrophysical Observatory; 5 years optical tracking experience for North American Rockwell; 12 years lunar occultation experience; coordinating solar eclipse expeditions to Africa, South American, Indonesia, Mexico and Canada; having papers published on astronomical topics in 7 countries, and being the first person to obtain a photographic record of an asteroid occultation (Guyana, 1979).

The Thimpu Market in Bhutan. One of a series of paintings of the people and places of the 'Altiplano', the nearly inaccessible high plateau cultures of the world.

Peoples of the High Plateaus; The Surviving Mediaeval Cultures

Carl Robert Berman

672 Kearsarge Circle, Charlottesville, Virginia 22901, U.S.A.

American, born July 5, 1915. Retired industrial designer. Educated in U.S.A.; Philadelphia Textile Inst., New York University.

This project brings into focus 'people' who inhabit the high plateaus throughout the world. It is compiled by me, as an artist, who journeys to these areas, lives with existing mediaeval families, and paints their portraits and their environment. I am involved in their daily lives and they relate to me as an interested participant.

Because most of these areas are distant from the industrialized parts of the world, the cultures, customs, costumes, and philosophies are much as they have been for centuries. Much of this is good: many qualities that they have, we have lost. I want to bring together this visual knowledge with appropriate commentary in book form. Those who have seen the paintings have been fascinated with the people and their surroundings: I hope to bring the same information to a larger audience.

It was in 1935 that the people of the high plateaus first drew me into their cultures with such intensity that, to this day, I am still fascinated by them and concerned with their life styles.

At that time, I was in Lima, Peru, examining pre-Colombian textiles at the National Anthropological Museum. My field of industrial engineering encompasses cotton and other textile fibers, and I am always eager to study the earlier examples of superb workmanship.

From Lima I travelled to Cuzco and from there began a series of treks to outlying areas in which dwell the Quechua, and to the South, the Aymara. Returning to base from each trip, I spent weeks and often months, painting the people I had visited and sketched, and their environment, silent and almost hostile in its austerity. I believed then, and still do, that photographs do not tell the story properly, that the artist catches and records subtle nuances that somehow elude the lens and the rigid field of the camera.

As the years passed, I explored the highlands of Guatemala, the State of Chiapas in Mexico, and Ecuador, Colombia and Venezuela. Their peoples, their family groups, the steady response of body and mind to the challenges of the altitude, the cut and texture of their clothing (often times unchanged in style for many hundreds of years), their attitudes reflected in their faces and stance, are the subjects of my paintings.

It is significant that my background in agriculture affords me a clearer and fuller insight into the different ways of subsisting that the 'people' have. So, underneath and intermingled with the oils I apply to the canvas is a working knowledge of the trials, tribulations and triumphs that these cultures experience in their daily lives.

In my continuing search, I have travelled to Asia, again to the high plateau, to visit with the Tibetan groups with whom I have long felt a kinship, so close are they in many ways to the tribes of the Altiplano of South America. North of India, I visited the Ladaki, Kashmiri, Sikimese, Nepalese and Bhutanese — all members of the 'people'. I lived with them and continued to record them on canvas. I also travelled among the tribal groups of the Central Indian Highlands, who fill an important place in our history. These 'people' were indigenous to the area before the arrival of Hindu, Bhuddist or Aryan civilizations. To live with, and communicate with, these mediaeval and ancient 'people' is to return to the original family of us all. The fruit of such an encounter is tranquillity.

After many years of recording the 'people' in paintings, after endless travels to their secluded areas, after years of sitting in huts, hunkering around a fire, gossiping about varieties of vegetables, fruits, animals and varied philosophies, the startling realization came to me that they are not inhabitants of some vague dream world, but that they are existing, vibrant members of our family of man. Many of the foods we eat, the spices to season that food, words in our modern vocabularies, fibers for our manufacture, and philosophies that underlie our civilizations, come from them over the millenia. So few of us know this branch of our family, know that they smile and joke, sorrow and despair, and are as involved in everyday tasks as we are.

I want to bring the 'people' to a wider audience by way of my paintings and commentaries in book form. The vignettes of their lives are possibly the last glimpses we can ever have of the way we were. I need help and funding to accomplish this.

The record is not complete, however, and I am anxious to fill in some of the remaining pieces in this far-reaching panorama. In the area of India north of Delhi, and stretching to the Nepalese border, there lies a corridor of high plateau peopled by many groups, each with its own tribal characteristics.

Using Chandigarh as a base, I plan to paint the 'people' of the area from there to the Kulu-Manali Valley, including Simla and the Dalhousie area. Completed, this would encompass a very broad spectrum stretching from the Nilgiri Hills in Mysore, northward along the spine of India, Ladakh, Kashmir, Nepal, Darjeeling, Sikim and into Bhutan.

Underwater Excavation of World's Oldest Sea-going Ship

Michael Mensun Bound

19 Williamson Way, Littlemore, Oxford OX4 4TT, England.

British, born February 4, 1953. Post Graduate Archaeology student, Lincoln College, Oxford University. Educated in Falkland Islands, Uruguay, U.S.A.; M.A. (Classical Archaeology) from Rutgers in 1979.

About 18 months ago, I brought together a team of talented archaeologists, technicians and divers to start a programme of maritime research and excavation. Too many vital maritime sites and important new findings were being lost to divers. A site and its information once lost is gone forever. The work we have started so far has been exceptionally successful.

The most important project is our discovery of, and work on, a 7th century B.C. ship of possible Etruscan origin at Giglio Island, North Italy. This is the oldest ship ever to have been explored by archaeologists. The site contains vital new information on ancient trade, ship building, mining and metallurgy. The cargo consists of beautifully hand-painted fine ware pottery worth a fortune. The wreck had also been carrying copper, lead and possibly iron ingots. In the bay area around the wreck, we discovered a previously unnoted Etruscan industrial zone, consisting of mines, smelting works, habitation areas and burial grounds. Already our work of last year has opened up an entire new chapter on the Etruscans.

The project began as the result of a visit I paid upon Alexander McKee (a professional historian, now famous for having discovered the Mary Rose) regarding another maritime project. Learning that I was a pottery specialist also involved with underwater archaeology, McKee told me of a site he had seen being looted twenty years ago, on the northern Italian island of Giglio. He believed the site was of more than ordinary importance and insisted I meet Reg Vallintine, who was in possession of McKee's notes and photos on the site and who had long attempted to bring it to the attention of archaeologists. When I met Vallintine, I was immediately struck by the importance of this wreck. If it had survived unlooted, it would be, without doubt, the most important wreck to come down to us from all antiquity.

Work began at Oxford, trying to trace people who might have been on the island twenty years ago and who might have more information and photographs. Through our research, we learned that the ship must have been carrying an assorted cargo of fine wares, the complexity and importance of which had never been seen before in archaeology. By dint of much detective work with helpful owners of pieces that had been recovered many years earlier, we were sure the ship came from the 7th century B.C. At this early stage, we had no idea how to obtain a permit from the Italian government to do the work, where we would find the money and diving equipment needed, or, indeed, how we would set about obtaining the University's approval so that the effort could become an official 'Oxford University Project'. By summer 1982, however, all these problems had been overcome, a team had been assembled, and we were ready to go to work in Giglio.

The Site

This we found after a few days searching, using divers' log books giving key underwater landmarks we were able to follow. Once we knew we were in the area, our electronic detection equipment led us to the precise spot. Almost as soon as we began removing the sand, wood and painted pottery began to appear. It was more than we had ever hoped for. Obviously, once the earlier looters had taken what was visible, they had thought there was nothing left, and the extreme depth and hazardous nature of the site protected it thereafter. The site itself is an offshore reef, some 150 meters from shore. The reef rises to within a few feet of the surface, and the wreck lies at the very base of the reef, at about 50 meters depth.

The Pottery

Samples removed from the site this season not only confirmed our findings at Oxford, but expanded them considerably. We now know that not only was the ship carrying Etruscan, Corinthian and Phoenician pottery, but also Laconian (i.e., from the city of Sparta) and Ionian (modern Turkey) pottery, as well as amphorae from the island of Samos. Some of this fine ware pottery was decorated with birds, lions and other animals in coloured paint. Not only will art historians and students of pottery be excited by this, but also Etruscologists and scholars who study the mechanics of ancient trade. The pottery is of outstanding archaeological value, not just because of the way it will help us date some of the less well known types, but also because it proves for the first time the sophistication of trade during the Archaic period. With pottery from 6 very distinct localities about the Mediterranean, the ship would have had to have been working through a system of brokerages. She could not possibly have picked up this cargo by herself.

The Metal Ingots

Copper, lead and perhaps iron, these add a new and very important dimension to the ship. Around the end of the 8th century B.C., the wealth of the Etruscans and their standards of living rose dramatically. This was reflected in their tombs, which became incredibly rich with exotic imported goods from all over the then civilized world. On the surface, this trade seems, however, to have been very much one directional: goods were pouring into Etruria, but only a few native vases were leaving. What, then, was it that the Etruscans used to pay for all these imports? What did they have that others needed so badly? They had timber, which we know they exported; oil also went abroad, and their wines were held in high esteem by other nations. These, however, were all commodities available elsewhere. What the Etruscans had that other nations did not have was an endless supply of iron, copper and lead. The ingots were not a mere appendage to the trade our vessel represents; they were its very essence.

The Hull
The wood we saw last summer was in excellent condition. If a significant portion of the ship survives, we will raise it for conservation and exhibition.

Security
The Italian police are watching the site 24 hours a day, but with such a fortune in pottery sitting there on the bottom, security remains a major worry. For this reason, the Italians have asked us to keep the site quiet until June 1983, when we are due to resume work.

The Etruscans and the Sea
The Etruscans left no histories of their own. All we know about them is what the very biased Greek and Latin authors had to say. Archaeology, up until now, has really only been able to concentrate on their tombs, and this has given a somewhat one-sided view of the Etruscans as a rich and sumptuous nation. But, when we look carefully at the ancient authors, we see from their asides that they were unanimous on one point; that the Etruscans were 'Lords of the Sea'. The much respected Greek historian Euphorus (quoted in a later work) states that before the time of their earliest colonies in Sicily (c. 735 B.C.), the Greeks would not venture on the western seas for fear of the Etruscans. For about 300 years between, let's say, Homer and Pericles, the Etruscans ruled the seas. Greek mythology tells of Etruscan pirates stealing the baby Dionysus, while later historic accounts tell how they stole the statue of Hera on the island of Samos, raped the women of Brauron near Athens, and even quarrelled with the Phoenicians over islands in the Atlantic. If the Etruscans could speak for themselves, they would probably say that their rule of the sea was the most important achievement of their civilization, and certainly that which had most influence on world events.

Now, for the first time, there is an opportunity for archaeologists to explore not just one of their ships, but one of their ships full of the most important product of the day. We could not wish for a better document on the Etruscans. Its position, in the heart of Etruscan waters, gives it added importance. And its date is also significant; the Etruscan star was soon to go into decline, and with it their fleet. Our wreck, in other words, is not just the oldest sea-going ship in the world, but the one Etruscan vessel ever found, and which dates to the absolute peak of Etruscan power and influence.

We expect to work three more years on this site, with a field season of two months on Giglio each season.

Quest for a
Surviving Dinosaur

Sydney Wignall
Taklakot, Craig Road, Old Colwyn, Clwyd. LL29 9HN, U.K.

*British, born October 16, 1922. Historian/Author/Explorer. Educated in U.K.;
History read at Liverpool University.*

I have for a number of years studied the subject of 'Loch Monsters', or more
correctly, 'large unidentified fresh water dwelling carnivores', with the intention of positively identifying the species. In two already completed phases of
this project, I have made four sightings and photographed one swimming 7
meter length creature, wich confirms for me the validity of my personal
hypothesis as to 'when and why' these creatures appear on or close to the
surface of Loch Ness and other fresh water bodies in various parts of the world.

Large carnivorous aquatic creatures have been seen in Loch Ness since the
middle ages. They have also been reported in at least nine other lochs in
Scotland and in fresh water lakes in Ireland, Iceland (the fabled Skrims),
Japan, Canada, Sweden, and Soviet Russia. All these habitats have several
factors in common. These lakes must either be, or once have been, connected to
the sea, they must possess an all-year-round temperature for the main body of
water of no more than 5°C, and must have an adequate population of fish such
as salmon, eels and trout.

The Loch Ness Monster (so named by the British press) came into international prominence in 1933. A road building program along one side of the loch
involved cutting down trees that had previously denied a view of large expanses
of the loch, and dynamiting sections of cliff faces into the loch. The latter
explosions apparently propagated acoustic effects underwater which resulted
in more than 200 sightings in that year; of these, reasonable scientists regarded
at least 80 as genuine.

Over the years, visual and photographic observation from land or boats has
given way to a more scientific approach via the use of sonar apparatus for
recording the passage of the creatures underwater, and photography using
high speed cameras coupled to high intensity strobe lights in turn coupled to
sonar equipment, which triggers off both strobe and camera whenever a
creature passes within range of the apparatus.

I am of the view that those who seek to identify the creatures (there have to be
colonies, not single examples) are now taking themselves into a cul-de-sac due
to their close involvement in using science and technology. Sonar graphs
depicting a large moving object underwater, swimming at a depth of, say, 200
meters, might impress the scientist, and they do impress me. They do not,

however, impress zoologists, who must verify evidence and declare that such a creature does or does not exist.

I believe I have correlated sufficient data to have come to a point of being able to predict the appearance of the creatures with a fair degree of accuracy. The factors involved include water temperature above and below the sub-surface thermocline, the movement of salmon into the lochs, meteorological conditions at sea near to the lochs and over the lochs and the effect of the latter on the thermocline movement. The creature's means of communication, which must be acoustic, also forms part of the theory.

In winter, the creatures exist in a state of semi-hibernation, living off eels that colonize the deeper water, and occasionally rising close to the surface for a change of diet, i.e., the growing young salmon and also trout. The main mass of the loch water does not exceed 5°C in winter; above the thermocline, the water approaches only 5.5°C up to the surface.

The thermocline is always in a state of slow rotation in winter.

The creatures, lying at great depth, have no means of ascertaining where the loch surface lies, due to the fact that acoustic underwater signals cannot pass through water layers, i.e., thermoclines. Thus, the creatures can only 'bounce' their sonar signals off the underside of the thermocline. Under Force 4 wind strengths, the thermocline begins to rotate at a faster speed, coinciding with ingress of salmon into the loch. The thermocline becomes wedge-shaped, and its underside begins to oscillate. I believe this is the 'trigger' which brings the creatures up to and through the thermocline, where they are able to feed. The salmon continue to move through the loch, and the large unidentified creatures may occasionally, accidentally or by design, break surface immediately following the cessation of high winds and disturbed thermocline.

Putting this theory into practice, I observed a *two-humped* animate creature which surfaced 100 meters from my boat for about 20 seconds, on the first flat calm day after a storm. Later observations from two small aircraft led to the sighting of a creature (estimated 7 meter length) swimming just below the surface, and to movie film of what appeared to be a large creature (about 6 meter length) lying on the loch bed in about 3 meters of water.

I am currently developing a new design for a two-seater, twin-float, ultra-light aircraft from which not only aerial observation can be made, but which can be landed on the loch surface by the pilot. This will allow me (I am an experienced diver/cameraman) to drop into the water for the first 'eyeball to eyeball' footage of what could be a species of dinosaur, perhaps the plesiosaur, that did *not* become extinct at the upper end of the Cretacious Period, 65,000,000 years ago.

One of the last available color photos of the magnificent "Shwesandaw" Pagoda in Pagan, built by King Anawrahta in the 11th century, before it was badly damaged by earthquake.

Preserving the Ancient City of Pagan in Photographs

♛ Jimmie U Lwin Aye

Honourable Mention — The Rolex Awards for Enterprise — 1984
Paganian Private Enterprise, "Utopia", Yondan Yet, Pagan, Socialist Republic of Union of Burma

Burmese, born July 13, 1955. Tourist guide for government corporation. Educated in Burma; B.Sc. (Marine Biology) from Moulmein College, Rangoon Arts and Science University in 1977.

The ancient capital city of "Pagan" flourished in the central lowlands of Burma in the late mediaeval period, becoming a unique cultural and architectural center; during its "golden age", Pagan was widely renowned for its 5,000 masonry Buddhist temples, pagodas and monasteries.

In 1287, the army of Kublai Khan invaded the city, forcing the Burmese mission to flee to Khan Balik — the Beijing. Fortunately, however, the occupying and pursuing Tartars did not destroy all the great monuments in Pagan. With the abandonment of the capital by the Burmans, Pagan declined rapidly in importance and population. Seven centuries later, nevertheless, the extraordinary grandeur of the city, with its gold-leaf covered spires, intricate and gracious buildings, and still impressive wall murals, continued to delight and inspire visitors.

On July 8, 1975 a massive earthquake inflicted terrible damage on this ancient and lovely city. Most of the superstructures of the immense monuments were toppled to the ground, and there was an irreparable loss of many of the rich and illuminating wall paintings that provided much of the historical knowledge we have about this impressive culture.

My project is concerned with saving, photographically, that which can still be documented in Pagan, before the rapid weathering of these now-crippled buildings and monuments erases all traces of this ancient world's art. In this goal, I am fortunate to have a very large collection of photographs of Pagan taken before the terrible earthquake, showing many of the buildings as they were before the damage occurred. My father had begun to document the buildings with photographs in 1956. Many of these old photographs are now in poor condition, but they still show the great monuments of Pagan as they were before the earthquake.

While some of the major buildings of Pagan have been documented by archaeologists who have come to study this ancient site, there are over 2,000 monuments in Pagan for which only limited data are recorded. Within, and on, these buildings, there are thousands of artifacts; terra-cotta plaques and wall paintings that have yet to be recorded, located and preserved. If we do not make the effort to preserve these historical treasures, they will soon be gone, leaving no trace of our ancestry for our children, and nothing for the study of those scholars who wish to learn more of the history of this great period.

There are officially 2,217 temples and pagodas identified in today's Pagan. Dr. Lwin Aung, lecturer in Architecture, states that there are 1,280 monuments, survivors of the earthquake, left with considerable structure for study. A French architect, Pierre Richard, says more than two thousand buildings have no data recording. As an example of the magnitude of the project, there are 1,472 different terra-cotta plaques described in the temple Ananda alone. It is precisely these unrecorded ruins and decorative elements that I wish to compile in a photographic record. So, please imagine the numbers of pictures I wish to take.

It is my objective to photograph all of the wall-paintings and plaques, and to assemble this collection in a centrally available location for the convenient use of those people who need it, or wish access to it.

To date, the equipment I have had available to do this job has been limited, and I do not now have the capability of reproducing and restoring the already fading and aging older photographs.

For recording the interior wall-paintings, I need powerful electronic flash units, plus certain special lenses, for correction of perspectives and work that must be done sometimes at very close distances. While I should like to have better equipment, I have begun the work with that which I now possess, as will be seen in the hand-bound book submitted with this application. My concerns are two-fold; that I find a way to preserve the pre-earthquake color photographs, which are now beginning to fade, and that I complete the record of the wall-paintings before it is too late. With both black-and-white and color photographs, I hope I can catch the meaning of architectural function and expression in our ancient temples and pagodas.

My final ambition is to present these photographs in book form, where colors could be preserved in multiple copies, and these treasures could be seen in many different places. In doing this, I would be providing a convenience for scholars in their extensive research, by reducing the amount of field work that would otherwise be necessary. The matched pictures will also be of assistance in further preservation and restoration work at the site. Such books would also show our descendants the former glory of Pagan. By allowing Pagan to become better known, the books would help encourage tourism to our country as well.

Experiencing Nature for the Disabled: A Guide to Ways and Means

Sandra Peehl

228 O'Connor Street, Menlo Park, California 94025, U. S. A.

American, born June 20, 1957. Previously education assistant/coordinator of student programs; currently disabled by multiple sclerosis. Educated in U.S.A.; B.A. in International Affairs from University of Colorado.

I propose to formulate a guide to aid persons with physical disabilities who desire an outdoor experience, but are deterred by lack of available information by which to judge accessibility and feasibility of contemplated ventures. Disabled people often share the increasingly common desire to escape crowded living conditions and to explore the ever-diminishing natural world, yet are faced with numerous obstacles in their attempts to do so.

The first of the many bridges-to-cross is often in the mind — although disabled persons formerly enjoyed outdoor activities such as skiing, hiking, camping, birdwatching, etc., they may feel that they now must forego these hobbies. We hear of disabled folks who still climb mountains on their stumps, or raft the Colorado with oars clenched between their teeth. Certainly such people are admirable and respected, yet for everyone of these exceptional types, there are undoubtedly a thousand other disabled people watching these fantastic exploits on TV and wishing that they had enough courage just to attempt a simple picnic in the park.

The second problem is one of practicality. *Could* a person in a wheelchair go camping, or is that just a pipedream? And if he made the attempt, would he have to bring an army of friends and family along to help? Would they enjoy anything in return, or would the whole works be confined to one corner of a 20 acre RV parking lot between the rest rooms because that's the only place accessible to a wheel chair? Not much of a wilderness experience — and a sure way to lose friends!

By writing through the eyes of one who is disabled and has actually experienced outdoor situations, I hope to research and compose a guide that will help circumvent these obstacles. Throughout the progression of my disease of multiple sclerosis, I have been confined during some periods to a wheelchair, and currently walk slowly with the aid of forearm crutches. The 'personalized' viewpoint of my publication should be advantageous in several respects. First, I hope to prove inspirational to many of us "ordinary" types who wouldn't skydive even if we *had* all our physical faculties! By considering my impressions regarding the 'fear factor' and physical difficulties, the reader could gauge whether his desires concur with what could realistically be gained from a particular outdoor experience. Was it really not so scary after all, or was it in fact mostly unpleasant, because the wheelchair ramp built to the bald eagle's nest had a fifty-foot drop-off? Some want an opportunity to do most or all for self and face the challenge of independence, whereas others (or at other times) need a less stressful or taxing situation, more to provide relaxation than challenge.

My second goal is to provide more detailed information about handicapped nature trails, camping sites, sports activities, outings geared toward the disabled, etc., than is currently listed in public access guides. While these guides do supply the basic information that access is provided, they usually give no further details regarding utility or ambiance. As many a disabled person has discovered, even hotel rooms or restrooms specified for the handicapped are not always what they're cracked up to be, because of poor design. I will attempt to assess the situation in each case, and determine if it is truly accessible for the wheelchair-bound, or whether the ability to walk at least a short distance, or up a few stairs, would be immensely helpful; if a certain amount of help from another person should be available; whether special equipment would be necessary, etc.

Finally, I hope to considerably expand the listing of potential outings and adventures beyond those listed in current handicapped access guides (which are mainly, in fact, limited to camping sites as far as outdoor facilities are concerned). Even though a particular trail may not be designated 'handicapped', it is often accessible to a wheelchair, at least during certain seasons. During a fall dry spell, for instance, we found that a hard-packed dirt trail through the Sacramento Wildlife Refuge was perfectly fine for a wheelchair, although of course this would not be so if it were muddy.

This proposed undertaking would involve extensive travel through-out the state of California, which was selected as the focus of this project for several reasons beside the fact that it is my current residence. Its many parks, campsites, wildlife refuges, and nature trails in diverse habitats are attractive year-round to both residents and national and international visitors as well. It is likely that many disabled persons may undertake a trip to California alone, or in conjunction with friends or family. In addition, California seems to have more than its share of 'activist' disabled, who have already made great progress in educating the state with regard to its obligation to make many activities — educational, cultural or job-related — available to the handicapped. We have stumbled upon several community or local projects for the disabled that a visitor or even a permanent resident might easily overlook, as these projects are not often widely advertised.

A guide such as I suggest would be beneficial to the wider dissemination of this valuable information. The Palo Alto Wetlands near my home, for example, recently completed a wheelchair-accessible boardwalk to provide excellent bird-watching for those inclined, or just bay-watching for the rest.

Since California does seem to be in the forefront in providing for its disabled residents, a popular guide to this area could also serve as a prototype to stimulate people in other areas to find out what is available and, if it doesn't seem like much, to try to get more. Perhaps similar guides could eventually be compiled for other parts of the the country, or for other nations.

Because of the diversity available in California, I will be able to describe different types of activities that would appeal to a wide range of interests. My sister Donna, who will help me travel, and I have already discovered a wide range of accessible activities. We recently went on an expedition to watch the grey whale migration, a popular spectator sport in California. I first called several advertised expeditions to ask about accessibility and provisions for a wheelchair-bound person. Many did not seem at all comfortable with the prospect of dealing with a disabled person and if they knew of an organization that dealt with such 'problems', they tried to steer me to them by phone. Eventually, I found a trip sponsored by the Oceanic Society, an organization which sounded very confident and experienced with such dealings. The crew members helped every step of the way (and were especially concerned that *I* see the whale, even if everyone else missed it!) and gave me a worry-free, enjoyable trip. If a guide similar to the one I propose had been available, I would have suffered far less frustration trying to find the right people to help me. We plan to try future expeditions offered by the Oceanic Society, such as to the Farallon Islands off San Francisco.

By searching out, experiencing and evaluating many outdoor activities for the disabled, I know that I would contribute information appreciated and utilized by the handicapped as well as their friends and families. In recent years, particularly since the designated International Year of the Disabled (1981), great strides have been made to brighten, expand and enhance the lives of the handicapped. Most of these accomplishments have occurred as the result of the insistence of handicapped groups and individuals like myself, who have a very personal reason for such growth. Successfully completing and enjoying even one outdoor activity provides a great boost for one's morale and encourages further growth and enterprise in all individuals.

My sister Donna, with whom I reside, will help me throughout this project. Since I can no longer drive, she will be my "wheels"; her flexible schedule in her position as a lab director will enable her to travel with me. She has always been active in naturalist activities, and since the onset of my disease has encouraged and aided me to do as much as possible outdoors. Donna has traveled extensively in California, and is now re-examining her knowledge of the state's attractions for those that might prove accessible to the disabled. Because of the tremor in my hands, Donna will be responsible for note-taking and light typing; her experience in publishing original scientific articles will also prove useful in editing. I anticipate that the project will run from May 1984 to May 1985.

The Paleolithic Art of the Cave of Nerja (Málaga)

Lya Monique Dams

171 Avenue Latérale, 1180 Brussels, Belgium

Belgian, born 6 November, 1926. Free-lance interior architect and decorator. Educated in Romania, Belgium and France; Ph.D. (Human Sciences and Letters) from University of Toulouse in 1980.

My project's objective is to publish my report on the Paleolithic cave-art of the cave of Nerja, in Málaga, Spain. I wish to obtain help in undertaking a comparative study of the signs and symbols of this cave and those of other caves, in order to establish a pattern in what may well be a primitive system of notation, or a mnemonic device; a forerunner of an archaic system of writing.

Having always felt that for a better understanding of our future, we must know more about our past, I have worked as a part-time prehistorian for the last 25 years, using part of my earnings as a interior architect for prehistoric research. I have done this work on my own, without grants or subsidies, with my husband doing my photographs. In this way, I have been able to publish a number of papers, and complete my study of the Andalusian cave of La Pileta. For my Ph.D, I did a complete corpus of the Levante rock-art of Eastern Spain, taking a period of seventeen years of my spare time.

I have now completed the study of the cave of Nerja, in Málaga, as regards its Paleolithic art. This is a very unusual cave, as it contains a very high percentage of signs and symbols versus the usual animal figures. Another year of work will be necessary to develop all the implications of this accumulation of symbols, in which I am inclined to see the roots of a notation system akin to a primitive form of writing.

Professor Marshack of the Peabody Museum has already pointed out a system of notation based on notches occurring on bones and artifacts. I believe this to exist also in the cave of Nerja, where the repetitive symbols indicate a pattern.

I would like to be able to publish my findings from the Nerja cave as a first step. Secondly, I intend to undertake a study of the signs and symbols of the other ornamented caves, in order to find out in what way they may or may not be related to Nerja. I shall start this phase of my project in 1984; otherwise I shall be too old for the physical difficulties that are involved when working in some of the caves. Up until now, I have taken advantage of cheap gasoline in Spain, cheap charter fares for flying to Málaga, and cheap accommodation. Now, prices have gone up and it is very difficult to manage research out of my earnings alone.

In order to do the comparative work, I will need to spend time in Northern Spain, in the Pyrénées and in Central France. I should like also to be able to include the cave-art of Calabria and Sicily, which I have not yet been able to investigate, and which include very interesting symbols.

In my writings about Nerja to date, I have illustrated the symbols I believe to be of importance to this study. At Nerja, in the section of the cave known as the "Organ", there is a series of red paintings tucked away within the folds of the rock. Completely hidden from a normal view, their location suggests that they were made with no original objective of providing visual or esthetic appeal; this supports my view that they were supposed to be hidden, as a means of communication. For these symbols to have been made in the first place, in places difficult of access, the cave-artists must have been strongly motivated to express themselves in such a mysterious manner. Throughout the cave, there is an abundance of such symbols. Although I know about 100 caves, and have worked in several, I know of no other Paleolithic art-cave containing such an unusual accumulation of these symbols.

I believe this large quantity of abstract designs form an alternative iconographic system, where the symbol replaced the animal image with a majority of specialized ideograms. If we consider signs as a substitute for the representation of an animal or object, and symbols as the means of expressing an abstract fact which cannot be formulated more exactly in the absence of writing, Nerja may contain the earliest beginning of written records. It may complement the system of expression and communication necessary to the group life of nomadic or semi-nomadic hunters or fishermen, as the cave is located on the seashore.

I intend to arrive at a better understanding of this system by detailed comparisons with other caves, including the location and positioning of symbols, a subject to which little attention has been paid until now. To date, my collection includes hundreds of drawings of these symbols, made by me, of which many have been photographed as a prelude to the publication that would bring these symbols to the attention of scholars for further study.

Antarctica's looming and forbidding Brabant Island; never explored until now. Adventure and scientific research are being combined in an expedition that is registering numerous 'firsts'.

The Joint Services Expedition to Brabant Island

♛ Kenneth William Hankinson

Rolex Laureate — The Rolex Awards for Enterprise — 1984
Royal Air Force College, Cranwell, Sleaford, Lincolnshire NG34 8HB, U.K.

British, born April 19, 1952. Royal Air Force Flight Lieutenant, Instructor at Royal Air Force College. Educated in U.K.; B.A. (Hons) from Liverpool University in 1973.

The purpose of our project is to place 24 young men on Brabant Island, British Antarctic Territory, in 3 teams of 8, for a period of 16 months. The aims will be both adventurous and scientific. They include:

1.   Becoming the first party to over-winter in tents in Antartica.
2.   Making the world's most southerly canoe journey.
3.   Making a complete survey of the biology of the Island, with particular emphasis on the role of krill.
4.   Making a complete geological survey of the area.

The Joint Services Expedition (JSE) to Brabant Island is probably the most ambitious amateur expedition launched in Antarctica in recent years. It is intended to combine exploration and adventure with worthwhile scientific investigation; these activities should greatly increase our knowledge of what is the largest unexplored Island in Antarctica and probably the world.

Brabant Island lies to the West of the Antarctic Peninsula at 64°S; it was first visited by de Gerlache's expedition of 1898, which included Amundsen and Dr. Cooke, who stayed for 3 days. It would appear that only 3 visits have been made since; all brief, by helicopter-borne geologists. Although surrounded by scientific bases of several nations, the costs of visiting the Island have prevented scientific exploration despite its considerable interest. The Island, therefore, forms an unexplored link between the relatively well-studied areas of Signy and Elephant Islands to the North and Anvers Island to the South.

The size, military training and past experience of the Expedition mean that we will be able to explore the area and surrounding sea using the most adventurous methods available. Furthermore, the need to keep costs to a minimum means that traditional means of movement, with the exception of sledge dogs, will be used where possible and tents will be used for accommodation.

The adventurous aspects of the Expedition are as follows:
a.   To make a complete survey, on foot, of the Island
This will entail about 2,500 man-days of travel across the Island, of which

only 1% is snowfree, on skis. Much of the area is inaccessible unless snow-climbing techniques and equipment are used: the dangers from crevasses, avalanches and the weather will be considerable.

b.   To make first ascents on all significant peaks of the Island

The Island is particularly mountainous, and rises to over 8,000 feet (2,460 m); the mountains appear to present a wide range of technical difficulties and many of the ridges are surmounted by ice mushrooms. Several of the Expedition are members of mountain rescue teams and mountaineers of some ability.

c.   To make the world's most-southerly canoe journey

If finance permits, a canoe journey will be made around Brabant Island and through the Le Maire channel; otherwise-inaccessible bays and islands will be visited. As far as is known, canoes have never been used at this latitude before.

d.   To become the first party to overwinter in tents in the Antarctic

The only accommodation available, apart from tents, is an insulated hut 8ft x 8ft (2.5m x 2.5m), which will be used for the storage of scientific equipment. A range of tents will be taken and the expedition will become the first to deliberately spend the winter in tents on the Continent.

e.   To commemorate past explorers

Two of the leading polar explorers of the 20th century, Cooke and Amundsen, landed on Brabant Island in 1898. It is intended to place 2 plaques, one donated by the Norsk Polar Institute, the other by the Explorers Club, at the site of their landing.

The scientific aims of the Expedition are less closely defined, but any research in this remote area is most valuable and unlikely to be carried out by other bodies. The area therefore, though barren, could be particularly interesting. The main area-related projects are:

a.   To prepare a complete geological, geomorphological and glaciological survey of the Island

Brabant Island appears to be a "hinge-point" in the structure of the peninsula and may well hold the key to a better understanding of the formation of the Antarctic Peninsula and the Bransfield Trench. Although 2 graduate geologists have been recruited from within the Armed Forces, some of the work will be beyond their ability. A young research geologist (who has yet to be selected) will be recruited from Nottingham University, under the supervision of Dr. Peter Baker. Other work will be done on geomorphology and glaciology of the Island with particular reference to glacial recession.

b.   To study the feeding habits of Crabeater Seals

Although the Crabeater is the world's most common (17,000,000) seal, little is known of its feeding habits as it is difficult to catch in the summer, and little work is done in the winter. An understanding of these seals is particularly important, as they feed mainly on krill and may be affected by recent proposals for large scale harvesting of this resource. The Soviet Union has been investigating the harvesting of krill for the past 20 years, and recent British interest in the area is likely to have a similar direction. It is intended that the feeding habits of the seals will be checked at regular intervals throughout the Expedition and the stomach contents preserved. This study will also provide useful evidence of the population patterns of fish and krill in the area.

c.   To prepare a complete Environmental Survey

Brabant Island is untouched by man's influence; it does not, therefore, have the disadvantages of disturbances of study areas by the human waste, garbage dumps, vehicle tracks and noise that are associated with permanent Antarctic stations. The Expedition will study the complete ecosystem with particular reference to botany, terrestrial invertebrates, penguin colonies and visiting seabirds. These studies, whilst not remarkable in themselves, will join together to produce a comprehensive picture of the Island before disturbance, and will thus neatly complement the study of Anvers Island made in 1956-58.

The exposure of large numbers of men to adverse conditions for a long period provides particular opportunities for studies not related to a specific area. The most important of those planned so far is by Surgeon Lieutenant H. Oakley; this will be a study into the effects of prolonged exposure to cold. Surprisingly, little is known of the effects on men working in the cold for long periods as most of the institutional scientific work in Antarctica has been based in warm, hutted accommodation. Lt. Oakley will study the psychological and physiological stresses of such long exposure to cold and isolation on the party, with the main areas being; the relationship between long term exposure to cold and changes in the body's metabolism; effects on the individual of exposure to various regimes, particularly cold wet and cold dry; and the effect of isolation on small groups and individuals. Recent commercial interest in exploitation of polar areas suggest that such knowledge, and associated work on clothing efficiency, will be most valuable.

Throughout the 16 months, it is intended to maintain a balance between the demands of discovery and science; every man will both explore the Island and produce worthwile scientific results.

The dates of the JSE Expedition to Brabant Island are from December 1983 to April 1985. Team members will travel out and back as follows:

Britain — Ascension Island: Routine Royal Air Force flights
Ascension — Falkland Islands: Routine troopship passages
Port Stanley — Anvers Island: HMS Endurance
Palmer Base — The Hump: Skidoo/sledges over Marr Ice Piedmont
Anvers Island — Brabant Island: Inflatable boat over Dallman Bay.

On Brabant Island, each seasonal party will work in two mobile groups, in radio contact and meeting regularly to interchange team members. Weekly radio contact with Antarctic Bases using Plessy PRC 320 radios will be our only contact with the outside world.

An International Treasury of Bird Names

♛ Michel Desfayes

Honourable Mention — The Rolex Awards for Enterprise — 1984
Prévan, 1926 Fully, Switzerland

*Swiss, born December 14, 1927. Independent winegrower. Educated in
Switzerland; Diploma from Lycee-college, Sion in 1947. Self-taught
ornithologist and linguist.*

My project is a compilation of over 100,000 bird names in the dialects of
Indo-European and Semitic languages, as well as in the ancient extinct
languages. The purpose is to find out more about the semantics of bird names:
Why is a bird called "duck", "falcon", "eagle" or "heron"? Or why is a bird called
"bird", "ptits", "ave", "vogel" or "uccello"?

Research in the semantics of bird names requires a good knowledge of birds
in the field and a feeling for the spoken languages of the people who named
those birds. I have been privileged to spend my youth in the countryside, in
daily contact with nature, where I gained an intimate knowledge of birds and
their lives. I have spoken the dialects of the people around me. My interest in
linguistics has never abated: on the contrary, it has grown to be a devouring
passion which is now occupying my whole life. This combined interest in the
field of ornithological and linguistic sciences puts me in an ideal and rare
position to undertake this kind of research.

My project is the result of a lifetime interest in both ornithology and
linguistics. My first notes on bird names were taken in my college years. In
1966, I started to put together my notes methodically. Thanks to a stay of twelve
years at the Smithsonian Institution in Washington, D.C., where I worked as an
ornithologist, I have been able to search through a vast amount of literature in
the Library of Congress and in the Smithsonian Libraries.

My undertaking is a compilation of over 100,000 bird names and their
variants in over 30 languages (not including some 8,000 names in the Spanish,
Portuguese, Creole and Amer-Indian languages south of the Mexican-
American border). In addition to the European languages, I have also included
the Iranian, Caucasian and Hamito-Semitic languages. The reason for doing
so is that the area covered by those languages encompasses the Western
Palaearctic Region, a zoogeographical entity over which most European birds
are distributed.

Whenever possible, I have collected bird names personally. Outside the
Valais, I have thus collected names in the Basque, Catalan, Albanian, Farsi and
Baluchi languages.

The publication will be divided in two main parts. The first is a compilation by languages, starting with the westernmost Gaelic and ending with the easternmost speeches of Afghanistan. The second part is composed of appendices where the semantics of bird names will be found. Birds have been named after their characters;; their denominations may be acoustic, chromatic, or morphic, etc., in origin. Each of these categories is represented by several bases or roots under which are assembled the denominations of things derived from a particular base. The names in the first part are referred to these bases by an arrow.

This method has also permitted me to systematically assemble thousands of Indo-European color names under their various roots, an endeavor never attempted before, which will throw new light on the etymology of color names as well as on many things or animals named after their color (or their shape, the sound they produce or the movements they perform).

Another benefit provided by this project will be to make available to all linguists and ornithologists the denominations of birds, things, colors, etc., in those languages written in non-Roman letters, such as Arabic, Armenian, Georgian, Greek, Persian or Russian.

I expect the project to be ready for publication in 5 years.

I believe the scientific value and the importance of this undertaking justify its presentation to The Rolex Awards for Enterprise. Any award, if merited, would go toward the computerizing of the material and the publication costs.

The Snailshell Book
of Memory:
A Poetic Environment for Man

Saul Kaminer Tauber

40 Boulevard de la Paix, 92400 Courbevoie, France

*Mexican, born July 8, 1952, Painter/sculptor, with architectural
background. Educated in Mexico and France; Architecture degree from
Universidad Nacional Autónoma de México in 1976.*

This project could be described as the physical creation of a philosophical
environment where Man can mingle with the ancestral memories of the natural
world; it is a journey of initiation, passing from the instinct as a life force to
creativity, allowing each person to discover areas of poetry in himself, and
realizing his capacities for play and imagination. It will be a space where Man
can express himself as a sum of all ages and all memories.

Since this space will be a philosophical environment where the memory of
the species intermingle, this project must be given a magical perspective aiming
at a society where Man would be led by his imagination, where organic and
inorganic elements merge to demonstrate Man's relationship with the world of
minerals, animals and plants.

The snail is a creature that is both organic and inorganic; it has been a shape
pregnant with the meaning of the origins of life from the most pre-Colombian
and Celtic civilizations right up to the present day. Therefore, we take the
shape of its shell, and its use of space as our starting point for an open book, in
which each page shall be a structural feature with an ambiguity of shape that
will lead to various different interpretations and the generation of new
meanings. The structural features (pages) can become nerves, hieroglyphics, a
plant-like texture, a mesh of wires, a shady arbour.

— A snailshell is a shape with constant infinite evolution.
— It has several interconnecting levels giving the idea of a whole.
— The spiral shape can be construed as a shape that evolves in time,
each stage being necessary to the next.
— The snailshell is an analogy, an allusion to the creation of life and its
development, as if it had an animistic cosmic meaning.

The *Snailshell of Memory* will be built as a small mound. The shape of the
snailshell will be traced and built as a totemic path in ceramics, with relief work,
engraving, and the use of different colors, so that the level and the texture
change continually. The path will be partially covered with sand and terra cotta
figures, to give the impression of an archaeological dig. The shape descends
gradually into the mound, where the shell is suggested by the structural
elements previously noted, including the shapes of skeletons and the pages of a

book, so that the visitor has the impression of an unexpected encounter with the Universe. "This journey is a series of symbols, each symbol is a vaster space waiting to be given a name." (Kenneth White, La Figure de Dehors, 1982.) Here the unexpected should be taken to mean a dimension of poetry and beauty. Like an encounter with something unforeseeable which could be a potential form of reality, there is a constant flow of exchanges between the conscious and the unconscious. Here Man is considered as One: a being who is not split up into childhood, adult life and old age, but a person who has a continuity of memory and who lives each cycle or age of his life simultaneously.

This is why the snailshell will lead us to the central moment of creation in this space, where the arbour of pages, the structural elements, join together in a center piece that is the synthesis of natural elements and their mechanics as well as the ideas formerly mentioned.

In this project, natural elements are necessary:

— earth, as a primary element, source and nourisher of life, where the snail will be dug into shape.

— water, also a primary element, used for acoustic textures, trickling under the totemic path — its pressure will create movement.

— the wind, to create movement in the structure, and when harnessed, to energize other elements.

— the sun will be the central element, dictating the orientation of the snailshell, to form a sort of cosmic table following other prehistoric models.

— we shall also use elements of shadow, the heating capacity of the sun and other potentials.

We are attempting to put forward the idea of a space where the possibilities of inter-identification are maximal, where Man is in the center of the Universe and sees phenomena with the greatest possible depth and breadth, and where he can see an infinite network of inter-relations. Gaston Bachelard expressed a similar idea (The Moment of Intuition), "The practice of an essential simultaneity where a person, however lost or unfulfilled he may be, achieves his own unity."

Work on the project began about six months ago; we are in the research stage, and are making the designs for the structural elements that will be the pages of the Book in this spatial environment. We are looking into materials and construction techniques and experimenting with small models, trying to mobilize our center-piece: we aim at a plant-like swaying in this structure. The next step is construction of a proper scale model of the whole spatial environment, together with an audiovisual presentation of same in order to promote it. We would like to build the model in a life-size construction.

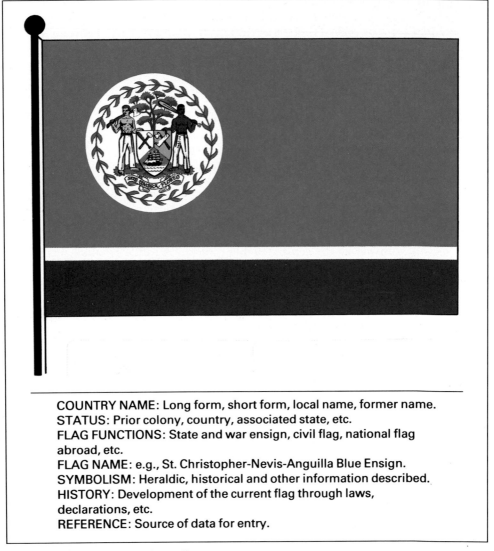

COUNTRY NAME: Long form, short form, local name, former name.
STATUS: Prior colony, country, associated state, etc.
FLAG FUNCTIONS: State and war ensign, civil flag, national flag abroad, etc.
FLAG NAME: e.g., St. Christopher-Nevis-Anguilla Blue Ensign.
SYMBOLISM: Heraldic, historical and other information described.
HISTORY: Development of the current flag through laws, declarations, etc.
REFERENCE: Source of data for entry.

Top: A proposed flag for Belize.
Bottom: A development prototype outline for the information on each flag to be included in 'CORPUS VEXILLORUM MUNDI'.

CORPUS VEXILLORUM MUNDI:
The Shorthand of
Human History

Whitney Smith

3 Edgehill Road, Winchester, Massachusetts 01890, U.S.A.

American, born February 26, 1940. Executive Director, Flag Research Center. Educated in U.S.A.; Ph.D. (Political Science) from Boston University in 1968.

CORPUS VEXILLORUM MUNDI is no less than a graphic statement in permanent form of the past millennium of human history — expressed through the premier symbol of the nation-state, the national flag. In brief, CORPUS VEXILLORUM MUNDI will present all the national flags (more than 3,000) of all nations, existing now or in the past, since these symbols of national self-identity first began to develop a thousand years ago. Presentation (both text and illustrations) will be optimal in accuracy, comprehensiveness, usability, and excellence of design and presentation. Distribution will be world-wide to appropriate archives and scholars.

Ultimately, every exploration is valuable for what it allows men and women to learn about themselves. Whether we turn to the stars, to the natural world around us, or to records of the past buried in the ground, the real adventure is always within our own minds. My project — the CORPUS VEXILLORUM MUNDI — is an intellectual enterprise, that will both inspire our imagination and expand our understanding of the world in which we live.

Remarkably, flags have been overlooked as a source of knowledge about humankind. For a quarter of a century, I have striven to overcome the traditional attitude that flags are no more than pretty bits of color, "the floral border in the garden of history", as one writer unhappily put it. In point of fact, national flags in origin and use provide a unique statement of self-image for all organized political groups. Flags — the shorthand of history — chronicle the rise of empires and great leaders, the assertions of ideology, and the philosophy of great religions. They incarnate history, territorial claims, political and economic systems, and the hopes of every society on the globe.

It has been my constant objective to bring proper recognition to the worth of vexillology, the study of flag history and symbolism. I left Boston University after six years of teaching, determined to use my experience and knowledge as a professor of political science toward the goal of collecting, preserving, organizing, and disseminating information on all aspects of flags. In the intervening years, I have established the world's largest library of flag documentation. I have published a journal (THE FLAG BULLETIN) which for 22 years has championed the cause of scholarship in the field of vexillology.

I have helped to organize professional associations in several parts of the world, most notably the North American Vexillological Association and the umbrella group to which it belongs, the International Federation of Vexillological Associations. I have helped to plan several of the International Congresses of Vexillology held every other year since 1965. I have written ten books and numerous pamphlets and articles, published in many languages, which are recognized as the leading sources in the field of flag study. I have also built up a network of correspondents with hundreds of vexillologists in all parts of the world for data exchange.

All of this has been designed from the very beginning as a prelude to the two prime requisites of vexillology, ones which I intend to realize in my lifetime. One is the transformation of the Flag Research Center collection into a permanent, non-profit, funded institution associated with a university. The other project is the realization of the CORPUS VEXILLORUM MUNDI. Although in the planning stages for many years, it has never been brought to fruition because, so far, the resources have been lacking to do it in the way in which I feel it must be done.

Just what would CORPUS VEXILLORUM MUNDI consist of? Stated in the simplest possible terms, it is the presentation in both graphic and verbal form, in as complete detail as possible, of *all* the national flags of *all* countries that have existed since national flags developed a millennium ago. For the first time in history, it will be possible to find in one place, in an organized and comprehensive form, the designs, colors, symbolism, dates of usage, official standing, and other information concerning any flag — presented in a "user-friendly" format where further references, cross-cultural comparisons, and statistical analyses are not only possible but easy.

This collection will protect the past achievements of vexillology by integrating the disparate work of hundreds of scholars. It will spread that knowledge as widely as possible to those institutions and individuals around the world most likely to need it and make use of it.

Perhaps most importantly, CORPUS VEXILLORUM MUNDI will make vexillology truly an auxiliary historical science by creating the "critical mass" of accessible data necessary for its next step of development. Heretofore, 95% or more of the work done in vexillology has been collecting and organizing facts: very little analytical work has been accomplished because, even today, military historians, museum authorities, social psychologists, political scientists, diplomatic specialists, naval experts, and other interested in the subject have not had a complete and reliable data bank from which to work.

What will CORPUS VEXILLORUM MUNDI look like when completed? What stages are necessary in reaching that goal? First, a list of all independent countries past and present is required. A more difficult task than might be imagined, such a compilation (with over 1,200 listings) has already been established. Next, the "template" of significant questions covering all characteristics for any imaginable flag must be compiled in an orderly format, so that all potentional inquiries to the CORPUS will have been answered in advance, both through the inclusion of data and through accessibility to that data. A preliminary draft of the type of data and its organization has been prepared. The next stage is to extract the necessary flag information from the laws, constitutions, decrees, books, charts, articles, and other sources where they have been recorded. In one sense, that task can never be complete; inevitably, there will always be at least a few points about a few countries that escape us. Nevertheless, I already possess the single largest repository of information on the subject in the world — over 8,000 books and 100,000 documents. I feel confident this will provide — with review and input by scholars elsewhere — the necessary material.

Presentation of the material will be in several forms. Verbal data will be computerized; its published form, including illustrations in color, will also be made available on color microfiche. If the resolution available on video-disk improves in the future, that medium also can be employed for storing and circulating the toal collection. Nevertheless, I recognize that we can miss fulfilling current needs by waiting for improvements in technology. Therefore, the initial presentation of CORPUS VEXILLORUM MUNDI will be a printed version to which archives, libraries, museums, and individuals can refer without special equipment. The authentic full-color reproductions absolutely essential to transmitting flag data suggest that the optimal presentation will be serigraphy.

The last problems to consider are the pace and extent of distribution. The number of flags included (3,000 or more) and the cost involved have so far suggested that the unsubsidized issuance of material would have to be done over a period of years, possibly even decades. It would also mean a distribution limited to those who can afford to pay for production costs — the expenses in researching and editing could never be passed on to subscribers — while clearly, in terms of usefulness, CORPUS VEXILLORUM MUNDI should be in the hands of many individuals and institutions unable to pay for those. This would be especially the case with scholars and research centers in Third World and Socialist countries.

Imagine 3,000 plates of the kind shown here, plus accompanying text pages, covering all countries great and small, past and present: this is CORPUS VEXILLORUM MUNDI as it can be. This program needs be done only once, but it must be done right. My past record of achievements makes it clear that I am capable of the undertaking, given proper resources. It is a new and original concept, a significant step forward in our knowledge of the collective manifestations of human society, and one which is at once feasible and certain of completion.

The 1984 Greenland – Ellesmere Island Traverse

Nancy Van Deren
52 West 27th Street, New York City, N.Y. 10001, U.S.A.

American, born July 9, 1944. Painter of landscape and abstract art.
Educated in U.S.A.; B.F.A. from San Francisco Art Institute in 1966.

The expedition is to travel from Etah in Greenland to Grise Fiord on Ellesmere Island, a traditional route used by Polar Eskimos centuries ago. We plan to climb and explore the east coast of Ellesmere; to do stress studies on the ice; to collect plankton samples in the open leads; and to do metabolic studies investigating our adaptation to cold. The total distance to be travelled is 265 miles.

While on the trip, I will do a series of studies for paintings to be executed in New York City. Over the years, the northern landscape has become important and central to my artistic vision. Nature as a source is louder than the voice of the artist. To be a painter is to be an explorer going a way one doesn't know, using the landscape as a guide.

Background
The Inuit, as the Polar Eskimos call themselves, have like us become estranged from their environment, except in those most northern communities where the traditional huntings skills of observation and patience have been maintained and adapted to the modern world. Grise Fiord on Ellesmere Island and Etah on Greenland are two of the most intact communities.

The earliest explorer, Pytheas (c. 310 B.C.) gave the name Ultima Thule to the lands north of Britain where "there was no night in summer and where neither earth, water, nor air exist separately, but a sort of concretion of all these". He describes meeting with "sluggish and congealed sea, which could neither be travelled over nor sailed through."

It is this sea that we will travel over, and these two tiny communities that will bracket our travel.

Part I
Our group consists of myself, Brian Carey (an experienced Arctic mountaineer), William Mayo-Smith (a student at Cornell University Medical School, with experience in the Arctic, an expert skier, and interested in stress in cold temperatures), Eric Rosenfeld (a veteran of many Arctic expeditions), and Diane Stoecker (an ecologist at Woods Hole Oceanographic Institute, with a special interest in cold water micro environment). We will gather in New York

in March, and fly to Thule Air Force Base in Greenland. From there we will go by charter flight, in a Twin Otter equipped with skies, to Etah in Inglefield Land.

March begins the end of the long polar night, and by April the sun is out all of the time, moving through a circular orbit overhead and dipping only slightly to the north. The polar summer day of 24 hours introduces one to the limitless space and time that is the polar Arctic, and to the infinite shapes and colors of the ice.

Etah, the most northern Greenland settlement, is in Hartstene Bay, 150 miles north of Thule. It is north of Anoritq, where the wind blows continuously, and Pitoravik, a spring camp for hunting walrus. It has a rich history of polar travel, both of long migrations by the native peoples, and as a base for early western explorations. It lies at the end of a fiord dominated by Brother John Glacier.

In the old days, the polar Eskimos travelled on the ice between Greenland and Ellesmere, by crossing Smith Sound at 78°N, a distance of 35 miles. We will cross the sound in this area, keeping to the north of a large body of water that, due to currents, never freezes. Our first stop will be Littleton Island, from which we can check the conditions of the ice to the West. Littleton Island is where the survivors of the famous ship Polaris took refuge. We will travel on foot or on Nordic skis, hauling our gear and provisions on small sleds.

As hunters have done for a thousand years, we will travel dependent upon the weather, the temperature, the wind and the condition of the ice. The ocean surface is broken, pressure ridges form and reform, and leads develop. The ice and weather will dictate our route.

Part II

The second part of the expedition will explore the northeast coast of Ellesmere Island, as we will travel from Cape Isabella to Boger Point with projected camps at Cape Dunsterville, Easter Island and Cape Mouat. From Cape Isabella, we will head south and west, skirting the coast. We will attempt ascents of peaks on the north side of Cadogan Inlet and Talbot Inlet, and also of the offshore peak that is Easter Island. South of Easter Island is a large glacier extending into the sea ice and 30 miles wide at the foot. We will pass this off shore, and then turn west in Mackinson Inlet.

At this point, we will be returning to the area of the Arctic where we have been before. Our 1980 Ellesmere Island expedition took us to the Inglefield Mountains on the north side of Mackinson Inlet and the Thorndike Peaks on the south side. West of Boger Point, at the entrance to the inlet, is Swinnerton Peninsula, a massive headwall and prominent landmark. Here we will find a cache of food and supplies left for us, including a snowmobile and a komatik. The komatik (an Inuit sledge) was built by us for a previous expedition, and was left in the Arctic for the use of the Inuit in Grise Fiord, and by ourselves when we returned.

Part III

We will then proceed to Fish Lake, a traditional spring fishing spot for actic char, to meet with Inuit friends from Grise Fiord, to rest and fish, and then to travel with them the last 70 miles of this 265 mile expedition.

Solar Seismological Observations from a Tethered Balloon Above the South Pole's Clouds

Eric Georges Fossat

Villa Framboise, 44 Avenue Bellevue, 06230 St-Jean-Cap-Ferrat, France

French, born May 29, 1944. Astronomy assistant at Observatoire de Nice. Educated in France; Doctorate in Physics from University of Nice in 1975.

Understanding the evolution of the Universe calls for a precise knowledge of the physical processes taking place within the stars. A major step to this knowledge would be the ability to observe the vibrations of our sun over a period of two months. This project proposes the use of a tethered balloon in place (about one kilometer high, which is said to be technically feasible) above the low-altitude clouds that prevail at the South Pole, as a platform for making observations over the austral summer season.

Finding, Analyzing and Documenting India's Folkloric Medicinal Plants

S. Sundara Rajan

137 (MIG) KHB Colony, Koramangala, Bangalore 560 034, India

Indian, born June 3, 1944. Head of Department of Botany, St. Joseph's College, Bangalore. Educated in India; M.Sc. (Botany) from Mysore University.

The ancient Indian system of medicine, the Ayurveda, traces back perhaps as far as 2,000 B.C., and had made significant progress as early as 1,000 B.C. In modern times it has been relegated to the background, due chiefly, perhaps, to the lack of proper plant identification from the ancient Sanskrit texts. This project involves the detailed study of these ancient texts, along with the field work necessary to find them, and then study them under laboratory conditions. A glossary of the Sanskrit and current botanical names will be made, to help identify these potentially helpful medicines.

The UNICEF Sky-Dive Project

Jack Wheeler

P.O. Box 4174, Malibu, California 90265, U.S.A.

American, born November 9, 1943. Writer, lecturer, company president. Educated in U.S.A.; Ph.D. (Philosophy) from University of Southern California in 1976.

An accomplished adventurer/sky-diver (record for most northerly jump, at the North Pole), this candidate's project seeks to raise money for the needy children of the world through UNICEF, by free-fall jumping within the national borders of UNICEF countries around the world. With a partner, he will hold the flags of the respective countries, and be photographed "in flight" against a prominent national background. Photographs of each "flag jump" would be given to the National Tourist Board of each country to promote tourism, with part of the sales revenues (post cards, posters, etc.) going to UNICEF.

THE ENVIRONMENT

The projects appearing in this section were submitted in competition under the "Environment" category, which was defined in the Official Application Form as follows:

Projects in this category will be concerned primarily with our environment and should seek to protect and preserve, or to improve, the world around us.

The ancient Egyptian art of making papyrus, exemplified in this modern version, should not be lost. The creation of an International Papyrus Institute will help to record, preserve, repair, translate and publish the priceless texts.

Establishment of an International Papyrus Restoration Centre

♛ Hassan Ragab
Honourable Mention — The Rolex Awards for Enterprise — 1984
78 Nile Street, P.O. Box 45 — Orman, Cairo, Egypt

Egyptian, born May 14, 1911. President of the Papyrus Institute. Educated in Egypt and France; degrees in Electrical Engineering, Military Science and Applied Biology.

Whilst all the patrimonial oeuvres d'art, such as paintings, sculpture, etching, pottery, etc., have well established technologies based on recognised scientific principles for preserving and restoring them, papyrus manuscripts and books so far have not enjoyed the privilege of such care. Their restoration, in most cases, is left to the care of personnel with no, or very little, experience in this field. The number of experts gaining practical experience in papyrus restoration hardly exceeds the fingers of one hand. From the alarming rate of destruction to which papyrus has been exposed during the past 20 years, mainly due to air pollution in big cities, stems the need to establish an International Papyrus Conservation and Restoration Centre, to put a stop to the danger menacing the existing papyri with complete obliteration during the coming 50 years.

The objectives of the proposed International Papyrus Restoration Centre will be the following:

1.   To establish a well developed technique based on solid scientific principles and accepted by all important experts working in this field.

2.   To evaluate the method proposed by the author, Hassan Ragab, of using fresh Cyperus papyrus pulp, injected in the deteriorated papyrus by his newly invented machine, details of which are given herewith.

3.   To recruit the most experienced personnel to supervise the restoration works to be carried on in the Centre.

4.   To recruit papyrologists with different languages who would help in collating fragments belonging to the same papyri, together with the aim of revealing the knowledge they contain.

5.   To train a new cadre of papyrus restorers.

6.   To establish a papyrus documentation centre where a micro film, microfiche, or photocopy of all the existing papyri are recorded, and made available to all who need them.

7.   To make exact facsimile replicas of the badly deteriorating papyri, and to expose them instead of the original ones, which are to be preserved in special stores out of the reach of polluted air and other destructive elements.

8. Through having copies of these papyri under one roof, create a special Museum that will attract many visitors, using entrance fees to cover part of the expenses of the Centre.

Method of Papyrus Restoration

In the past, many methods have been tried for papyrus restoration, with varying degrees of success. In all these methods, different types of acid free and Japanese paper were used in the restoration to patch holes, and to stick the papyrus fragments together on such paper.

The method I propose is to use fresh papyrus pulp and fresh papyrus strips to carry on all restoration work. This method will ensure fresh material of the same characteristics and constituency as the papyrus to be restored. In this way, we ensure the best possible repair work, which is certain to last longer than any foreign materials used so far.

The Machine for the Repair Work

After many years of experimentation, I have produced a machine which helps apply the papyrus pulp to the damaged parts of a papyrus. The machine (patent pending) consists of a box containing an electric bulb, the illumination of which is controlled by a variable resistance dimmer. A vacuum is created inside the box by means of an electric aspirator similar to that of a household vacuum cleaner. The vacuum box is covered with a plate of plexiglass having a hole in its centre, and covered with a fine wire gauze. A black rubber blanket is positioned to cover the plexiglass, having a hole in it slightly wider than that of the plexiglass cover. The papyrus manuscript with the recto (written) side is laid on the plexiglass cover. As the vacuum is developed, the manuscript firmly sticks to the plexiglass and the rubber blanket is laid on the papyrus with its hole positioned above the hole in the plexiglass, thus exposing the part to be restored.

The light of the inside bulb is turned on, and its illumination brought to the most suitable degree. With a tweezer, a small piece of previously prepared papyrus pulp is carefully applied to the hole to be filled, and with a sharp tool it is delicately spread on the hole to be patched. The air sucked through the hole helps this operation, and helps to rid the pulp of any excess water. The light against which the hole is seen permits the restorer to gauge the right amount of patching pulp to be applied. Use of a magnifying glass allows this operation to be done very accurately.

Having finished the patching of holes, the manuscript is put between two pieces of felt blankets, and left to dry in a hand screw press. After drying, the patched parts are further reinforced by sticking very thin papyrus strips on their surfaces with any suitable gum.

Location of the Centre

It is proposed that Cairo, Egypt be the site of the International Papyrus Restoration Centre, for the following reasons:

1. Papyrus was invented in Egypt more than 4,000 years ago, and ever since that time Egypt has been the only producer and supplier of papyrus to all the other countries of the world.

2. The Egyptian climate, with yearly average temperatures below 30°C, and average relative humidity less than 60%, which has helped to preserve

papyrus in its soil for thousands of years, would be the most suitable climate in the world for any preservation and restoration work.

3. The existing papyrus plantations around Cairo provide the necessary raw materials for preservation and restoration work, besides patching and reinforcing damaged papyri, and to make exact facsimiles of the manuscripts for exhibition.

4. Hundreds of students graduating annually from the Faculties of Archaeology, Greek and other languages would provide an ample, skilled labour pool for the Centre.

Facilities

The Centre will require a building with an area of nearly 600 m² of usable floor space, which would be leasable in Cairo for a nominal fee of U.S.$2,000 per annum. Major equipment needed includes: a fumigation chamber, a pettifogger or humidifying cabinet, a drum dryer, laminator, vacuum table, standing press, fume hood, pressure spraying equipment, stainless steel sink, drying racks, microscope reading equipment, laboratory benches, work tables, cabinets and sundry equipment. This equipment is expected to cost approximately U.S.$26,000. Staffing of the Centre is expected to cost approximately U.S.$31,000 per annum. Estimated annual operating costs, including staff and miscellany are placed at approximately U.S.$38,500.

Current estimates are that about 150 papyri per year could be processed with the above facilities, at a unit cost of $250 (restored, microfilmed and 2 facsimiles made), for total revenues annually to the Centre of approximately U.S.$37,500, or roughly enough to cover operating expenses.

It is our intention to run the International Papyrus Restoration Centre as a non-profit institution.

The rarely seen "Aye aye" (Daubentonia madagascariensis) one of the least known and most endangered of man's close relatives in the animals world.

Save the Aye aye!

♛ Friderun Annursel Ankel-Simons

Honourable Mention — The Rolex Awards for Enterprise — 1984
2621 West Cornwallis Road, Durham, North Carolina 27705, U.S.A.

*West German, born October 23, 1933. Visiting Scholar, Department of
Anthropology, and Research Associate, Anatomy Department, Duke
University. Educated in West Germany and Switzerland; Dr.rer.nat. (magna
cum laude) from University of Giessen in 1960.*

The objective of this enterprise is aimed at preservation and propagation of the
rarest of primates, the lemur *Daubentonia madagascariensis*, commonly known
as the "Aye aye". A census of living Aye aye will be conducted and a strategy will
have to be developed, to save the Aye aye from its imminent extinction.

Most curious and peculiar of small mammals, and a distant relative of man, is
the Aye aye, the strangest of all lemurs. This creature is small as a cat, with large
eyes that reflect light in the night. It has large, mobile ears, rodent-like teeth,
two wire-shaped, elongated third fingers, dense and coarse fur, and a very long
and bushy tail. This nocturnal, solitary animal has always been rare, which
explains why little is known about its way of life.

Aye ayes build large, leafy nests high in the forks of tall trees, and thus betray
their presence to the searching primatologist. The natural habit of the Aye aye
in Madagascar is rapidly being encroached upon by the expanding needs of
humans for farming land. Steadily, the eastern rain forest of Madagascar is
being burned and cut down.

Ayes ayes were already very rare and endangered when, in 1966, Dr. Jean
Jacques Petter of Paris decided to catch and transfer Aye ayes to the small island
of Nosy Mangabé, at the far northeast of Madagascar in the Bay of Antongil. It
was presumed they would be much safer there than on the mailand. Nosy
Mangabe is a small (5 square kilometer), protected island, far enough away
from the mainland so that it is rarely visited by small boats. It is also taboo to
many of the native people, for it is a sacred burying ground for ancient kings.
The island is owned by the International Union for the Conservation of Nature
as a reserve for the Aye aye. It is covered by secondary rain forest, and is
crowned by a hilltop that is steep and rugged in the southeastern corner, where
the Aye ayes are presumed to be living. There is a one-room laboratory on the
island, built as a shelter for visiting scientists.

In the fall of 1975, an English photographer, Liz Bomford, visited the island
to search for the Aye ayes, and to find out what had become of them during
what was then a nearly ten year period since nine animals were released. She

stayed but one week, and was not sufficiently well prepared and equipped for her venture, seeing only one animal, more by chance than by planning. In 1976, Bomford again returned to the island, and on this occasion, she saw a nest on the eastern side of the island. Since then, newly built nests of the Aye aye, that are placed high up in tall trees and are composed of branches, vines and leaves, have been reported. Also, one female Aye aye, accompanied by an offspring, was seen in 1982 on the island. Nevertheless, no serious, long-term attempt has ever been made to census the Aye aye on Nosy Mangabé, following their transfer there seventeen years ago.

In the meantime, awareness of the plight of the Aye aye has arisen among concerned Malagasy biologists, as well as among the world's primatologists. All are aware of the fact that something will have to be done to save these primates from extinction. Also, techniques for studying nocturnal primates in the wild have greatly improved in recent years, through the development of new equipment, such as the Night Scope, that makes it possible to observe nocturnal animals at a distance in the dark. Much new experience has been accumulated by scholars who have studied nocturnal primates in the wild, and this is also available to researchers.

In February 1983, an agreement was drawn and signed between representatives of the government of Madagascar and a board of internationally accredited scientists (working with lemurs), as well as representatives of the World Wildlife Fund, that will make it possible for concerned conservationists to work in Madagascar with the incentive of preserving such rare animals as the Aye aye. Simultaneously with this application, a request to clear the necessary permits to do this work is being forwarded to the government of Madagascar along the lines laid down by the aforementioned agreement, signed at the Island of Jersey in January 1983. I have been assured of the support, advice and cooperation of such experts as Dr. Patricia C. Wright, who has successfully studied the nocturnal Night Monkey, *Aotus trivirgatus*, with the help of the Night Scope, in the wild in 1976, 1979, 1980 and 1982 in Peru and Paraguay. Also help has been offered by Dr. Jon I. Pollock, who studied lemurs in their natural habitat on Madagascar during 1975, 1976 and 1978. Additionally, cooperation from several Malagasy biologists has been assured.

A survey and census of the Aye aye population on Nosy Mangabé is now mandatory, seventeen years after the original nine animals were released. This survey will serve as a basis for the development of a detailed program to preserve and propagate for the future this rare, endangered and most peculiar relative of human beings, *Daubentonia madagascariensis*.

In the course of this field project, the Aye aye would also be searched for in certain coastal forests, such as that near the villages of Andrianakoditra, south of Tamatave, where an Aye aye was seen in 1980, and near Mohambo, where the nine animals were caught in 1966.

Plans for the expedition call for the acquisition, in early summer of 1984, of the required equipment and its shipping to Madagascar. Items to be included will be the Night Scope, an outboard motor, camping equipment, and climbing equipment needed to inspect nests high in the trees. The preparations for the expedition will be made in Madagascar in September 1984. Actual field work on Nosy Mangabé is planned for October/November 1984, the months of least rainfall in this area. Heavy rains during the rest of the year make a study on the island virtually impossible. To complete the census and survey work, a follow-up study in October/November 1985 is anticipated. The objective is to be able to prepare a conclusive report, with recommendations for future action, at the end of 1985. If the results on Nosy Mangabé are unencouraging, certain mainland sites will be explored and studied.

Water Extractor for Desert Air (WEFDA)

Roman Gammad S. Barba

Barba Research Laboratories, Cataggaman Nuevo, Tuguegarao, Cagayan, Philippines.

Filipino, born February 23, 1941. Inventor-researcher, owner of Barba Research Laboratories. Educated in Philippines; B.Sc. (Physics) from FEATI University, Manila in 1964.

The Water Extractor for Desert Air (WEFDA) is a machine designed to provide potable water for human use, agriculture and industry. It can range in size from that of an ordinary refrigerator to massive superstructures covering several acres of desert land.

In essence, it is a modern version of the ancient "air wells" (used since Biblical times), structures made of stones or rocks, erected in the desert to extract water from the air during night time, or, occasionally, during favorable day time periods.

The main aim of the WEFDA is to make the deserts of the world autonomous, habitable and productive regions.

There are two types of WEFDA machines; the domestic and the agri-industrial complex type. The domestic version is about the size of a refrigerator, and is intended primarily for the production of drinking water. Its output is about 200 liters per 24 hours, and it is powered by solar energy or desert winds, or electricity if necessary. It has an intake port backed by a suction fan, in front of which is a filtering system to remove airborne dirt, dust and organisms. Air brought in is then passed by the condensing unit of the machine, basically a modified refrigerator evaporator, which reduces the air temperature to the dewpoint. This condensed water trickles into a receiving tank, from which it can be dispensed by a faucet for drinking. In this type of domestic WEFDA, a by-product is cooled, dehumidified desert air, which can be used to air-condition a small room, thus allowing the unit to double as an air-conditioner. In this way, the domestic WEFDA is really being used to modify a small portion of the desert (the house itself), into a tiny, comfortable and habitable location.

In the agri-industrial complex type of WEFDA machine, the procedure is the same, except that the machine itself covers a very large area of desert land. In this complex, the roof of the WEFDA is actually the condensing unit, serving multiple purposes; as the roof of the structure, as the condensing area for cooled desert air, and as a collector for the condensed water moisture.

The key element in the agri-industrial complex WEFDA is the condensing system; this is actually the modified evaporator of a refrigerating system, which is integrated into the roofing system of the WEFDA superstructure. Under transparent glass or plastic panels, vegetation is allowed to grow, and carry on the process of photosynthesis. The refrigerants of the WEFDA are recycled within the system under the panels, reducing the air temperatures to dewpoints, and collecting the moisture trickling down the transparent panels.

The refrigerating system is composed of a compressor and its drive mechanism, the condenser, the receiver, the piping, and other components for the reclaiming and control of the refrigerant.

In operation, there are two by-products from every WEFDA machine; waste heat from the desert air, and cooled, dehumidified, filtered air from which the desired moisture has been extracted. These by-products can be used for cooking or processing food for the WEFDA inhabitants, and for air-conditioning the living quarters, shops or offices within the WEFDA structure.

The principle of extracting water from desert air is ancient and well known; the practice of doing it on an efficient scale has not been adequately explored. Using modern materials, the science of refrigeration, the vast amounts of solar energy available in typical desert environments, and the concept of self-contained 'processing units', the WEFDA approach offers a way of combating desertification that deserves further investigation.

The current aim of this project is to construct a working model with a capacity of producing approximately 1,000 liters of water per 24 hour period, in order to evaluate the present designs for the operating system, and to gather data for larger, agri-industrial scale WEFDA's.

In principle, the eventual efficient extraction of water from desert air could lead to very large complexes, housing people, agricultural units, and commercial areas. Once begun, such units could be significant factors in the fight against desertification, and the raising of living standards in desert areas.

Protecting the Flamingo in its Venezuelan Habitat

Miguelina Lentino

c/o Sociedad Conservacionista AUDUBON de Venezuela, Apartado No. 80450, Caracas 1080-A, Venezuela

Venezuelan, born May 5, 1955. Associate investigator for the La Salle Foundation. Educated in Venezuela; Degree (Biology) from Central University.

The only population of American Flamingos, *Phoenicopterus r. ruber,* in the southern Caribbean nests on the Island of Bonaire, but feeds in the marshes and wetlands of the coast of Venezuela. Due to the rapid "development" of these feeding areas and the extreme sensitivity of this species to human disturbance, pesticides and chemical effluents, it is feared that the southern Caribbean population of Flamingos is endangered. The goal of this project is to initiate a monitoring program of the Flamingos and their feeding areas to determine any fluctuation in the population, and to assess the quality of these feeding areas for the period of one year. Monthly counts of the Flamingos at the various feeding areas are to be made; analysis of the age of the birds; a study of their migratory routes; determination of what foods are utilized in Venezuela by laboratory analysis, and an in-depth study of the quality of habitat at the various feeding areas.

In view of the fact that the only reproductive colony of American Flamingos nests on the Island of Bonaire, and that the island does not offer enough food to maintain the population, the birds migrate daily to the coast of Venezuela in search of food. The species can be considered as highly vulnerable, due to its extreme sensitivity to disturbance, its patchy distribution and colonial nature. Most important is the fact that its feeding areas are under immediate threat. These areas are rapidly being developed as resort and/or agricultural areas, with the subsequent destruction of surrounding forests and mangroves, drainage of wetlands, construction of highways and other activities incompatible with the protection of the Flamingos.

Uncleaned effluents from innumerable industries along the Venezuelan coast are dumped untreated directly into the sea or into rivers that drain into the Caribbean. Untreated human sewage from most of the urban population along the northern coast is also washed directly into the Caribbean. Certainly these pollutants play some role in affecting the coastal wading bird populations that rely on the abundance of aquatic organisms in the shallow waters of the coast for their survival. Given the surrounding agricultural areas, the effects of persistent non-directional pesticides on the breeding of the flamingos is at present of unknown dimensions, but may have a substantially adverse effect. DDT is widely used in Venezuela. An analysis of eggshells may indicate that eggshell thinning has occurred in this species.

In the extremely dry years of the recent past (1978 and 1980) when their feeding areas dried up, the birds did not reproduce. If their feeding areas are drastically altered by development, the population will cease to exist.

Among the resolutions of the XV General Assembly of the I.U.C.N. (Christchurch, New Zealand, 1981), was an Action Point "To urge the Venezuelan Government to take all necessary steps to protect the coastal marshes that are critical for the survival of the population of Flamingos on the Island of Bonaire". Letters from concerned societies and government officials to the Venezuelan government about the problem have, to date, produced no national action to protect the birds. The Venezuelan Audubon Society therefore proposes to make an in-depth study of these birds and their habitat, in order to develop a scientific base of proof of the precarious position of this species.

An effort is to be made to discover all feeding sites along the coasts of Venezuela. Site counts will be made both on foot and by air, with a search of the entire coast. Aerial photographs will be taken, and compared with land counts to determine the exact number of individuals in any one area.

Ages of birds are to be determined by plumage color to evaluate the state of the populations.

Counts are to be effected on a monthly basis to observe fluctuations and possible migratory routes, as it has been suggested that when a food source is limited along the coasts of western Venezuela, the birds migrate to the east in search of more productive areas.

The types of food taken by the Flamingos at the various feeding areas are to be determined. At Bonaire, their main diet is the brine fly, *Ephidra cinerea,* and the plankton *Artemia salina,* but of late they have been consuming more mollusks. Additionallly, organic mud/ooze is considered an important food sources for these birds, when it is obtained in sufficient quantity. In the event that food items in Venezuela are the same as on Bonaire, a similar methodology as used on Bonaire will be developed. If different foods are taken in Venezuela, a further and deeper investigation will have to be effected to identify the food items, and determine whether different foods are utilized in different areas.

Quality of habitat will be analyzed on the bases of water analysis (concentration of chlorine, temperature, salinity, concentration of dissolved oxygen, and presence of organochlorides and heavy metals in water and sediments) and human impact (deforestation, land fills, uncleaned effluents, hunting, etc.).

The duration of the project is anticipated to be thirteen months; one year of field and laboratory work, and one month for the preparation and submission of the final report.

Ixias marianne, *Dry season form; Northern Province, Ceylon, Sri Lanka.*
Insert: An example of the lengths to which one must go to get the right picture...

Butterflies of the World: Multi-volume Masterwork

♕ Bernard Laurance D'Abrera

Honourable Mention – The Rolex Awards for Enterprise – 1984
Hill House, Highview Road, Ferny Creek, Victoria 3786, Australia

*Australian, born August 28, 1940. Author/naturalist. Educated in Australia;
B.A. from University of New South Wales in 1965. Fellow of The Royal
Entomological Society of London.*

This project is an attempt to have available, in a series of synoptic volumes, an illustrated systematic guide to all the known species of Rhopalocera (true butterflies) of the world. In total, the work should comprise eight volumes, containing some 20,000 species. Each butterfly is presented in detailed photography, showing its colour and the shape of its wings, and noting various data, such as the food it consumes, the names by which it may be known, and other information. The purpose is to provide workers in more erudite fields, such as genetics, animal behaviour, bio-systematics and conservation with a modern, reliable reference work as a foundation for their own labours.

Not since 1906-1920 has there been a single reliable foundation work available to biologists for reference purposes. That work, entitled *The Mac-rolepidoptera of the World,* was the work of many authors and countless illustrators under the editorship of Dr. Adalbert Seitz. It comprised over a dozen volumes, and was largely out of date by the time it was completed. It also dealt with the moths and skippers *(Hesperiidae)* as well as with the butterflies.

My work deals only with butterflies, because the limitations of time and money are such as to preclude an involvement with the other two groups. Significantly also, the butterflies themselves are by far the most worked and interesting group amongst scientists as well as amateurs.

To accomplish this work, numerous visits have made to the Natural History Museum in London, where the collection of some 3,000,000 specimens is a treasure for the serious scholar of butterflies. Over the years, at least one of each of the species here have been photographed. This work, done with a Pentax 6 x 7cm SLR on a copying stands is much different from that which has been done in the field.

On occasions, unusual tactics are necessary to capture the desired specimens on film. For example, along the rivers of Malaysia, it was found that half a whiskey bottle of urine poured out before breakfast and allowed to 'bake' in the sun was enough to attract some 200 species of butterfly before noon. It seems that the hotter the sun, the faster the ammonia and urea are evaporated, and the more noticeable this becomes to the butterflies. If all such searches were this

Collecting and Painting Endangered Flowers in the Amazon

Margaret Ursula Mee
Rua Julio Otoni 495, Santa Teresa, Rio de Janeiro 20241, Brazil

British, born May 22, 1909. Botanical artist; works internationally exhibited, published, held in collections. Educated in England.

I plan to make a further series of river journeys through the Brazilian Amazon, to discover, collect and paint plants in those remote areas that I have not yet explored, before the destruction which threatens them takes place. I intend to publish a third volume of paintings of some of the disappearing species of Amazon flora.

In view of the rapidly disappearing flora of the Brazilian Amazon, where large areas of forest are being cleared for colossal development projects, it is necessary to record as many as possible of the species of plants which are in danger of extinction due to the destruction of these regions. As a number of species are endemic to limited areas, it is essential to record them before their habitats are destroyed.

For twenty years, I have been engaged in recording, by means of painting, the very rich flora of the Brazilian Amazon. This has involved twelve arduous journeys (average duration; three months), mainly by river. On these journeys, I have usually been alone, engaging Indians and forest dwellers as guides and helpers. I have proved this to be a convenient and successful way of working. To continue my research, I plan at least two more journeys; one to the upper Amazon and one to the lower Amazon.

The Upper Amazon: Rio Japura to Rio Apaporis
This journey of four months duration will be undertaken during March-June, when the waters of the rivers are high, to facilitate collecting in the igapo and igarapes, as epiphytes on the canopies of many trees are then within easy reach. It will also be possible to navigate rapids which would otherwise form impassable barriers.

I would leave Manaus by river boat for Tefe, or Alvaraes – a small town a few kilometers from Tefe – on Rio Solimoes, and proceed up a series of natural canals and waterways into the Rio Japura, having hired a boat in Tefe with a capacity for three to four (the crew and me). We will tow a small canoe, for paddling into the igapo and up the natural canals and small rivers, where much of the collecting is done, and ambient drawings are made.

I would explore tributaries and lakes of the Rio Japura, which are numerous — Lake Parica, Lago de Maracai, Rio Boa-Boa and others — and eventually reach the Rio Apaporis on the Colombian frontier, reputedly a river with a very rich flora.

At Rio Apaporis, I hope to find plants belonging to the families of *Orchidaceae, Bromeliacea, Musaceae, Lecithydaceae, Leguminoseae*, etc., and specifically a species which I have tried to collect for years, and failed—*Qualea suprema (Vochysiaceae)*. I had traced this tree to Rio Cauhy on an earlier expedition. Arriving near the source of that river, however, I found that it had dried up due to devastation, and along with that, all signs of the *Qualea* had disappeared.

After collecting the plants, I will make detailed colour sketches and take notes of those in flower. The living plants will be cared for in open baskets, and protected from excess sun and rain, preparatory to their arrival in Manaus. From there, they will be flown to Rio for care and cultivation in the gardens and greenhouse of Roberto Burle Marx, as well as some being sent to the Royal Botanic Gardens in Kew, England.

In this way, the living plants are preserved, the danger of complete extinction reduced, and new species are propagated. As notes and a diary about the ambient where the plant is found are kept, the origin of each species can be traced later. Great importance must be attached to the fact that all work is done in the field, thus eliminating the chance of errors sometimes made by botanical illustrators, through working solely from herbarium specimens.

To achieve these ends, the following practical steps have to be taken. In Manaus, I make purchases of food for the crew and myself, less some of the basic, bulky ones which can be bought at river settlements — farinha, fruit, etc. A small stove has to be considered, though most guides have their own cooking facilities, however primitive. A hammock with a mosquito net is essential; whenever possible I sleep on board for security and health reasons. Basic medical supplies are bought in Manaus, and innoculation against yellow fever obtained in Rio de Janeiro. Collecting equipment comprises bush-knife, cutters, leather gloves, rope, etc. Fuel for the last part of the journey would be bought in Tefe, as supplies further along are doubtful.

The Lower Amazon: Rio Jatapu and Rio Uatuma

This journey of approximately three months will be undertaken during August-November. On this one, I shall also leave from Manaus, taking a river boat to Parintins, and there hire a boat with a crew of two.

The river level starts to fall in July and reaches a low level in January, so sandy beaches would start to appear in August along the river margins, making entry into the forest quite easy. As both the Rio Jatapu and Rio Uatuma have numerous small tributaries and igarapes, it should be profitable to explore them for plants, using the small canoe in tow.

Purchases of food and fuel will be made in the small town of Parintins, and I will be using equipment similar to that on the Upper Amazon journey.

On each journey, I will be bringing all my sketch books and painting materials from Rio de Janeiro.

Studying and Utilizing the East African Ambatch 'Plant'

Innocent Mbuguje Bisangwa

P.O. Box 3530, Kampala, Uganda

Ugandan, born December 12, 1957. Law-enforcement Park Warden. Educated in Uganda; B.Sc. (Forestry) from Makerere University in 1980.

The ambatch plant, *Aeschynomene elaphyroylon*, not clearly classified as a tree or shrub, grows up to a height of 20-30 feet, traditionally in the marginal waters of Uganda's lakes or slow flowing rivers and streams, and to some extent in permanently soaked swamps. The stem of the plant is short, swollen and quickly tapers almost to a point. The wood from these stems is extremely porous and very light–both in weight and color. There appears to be no distinguishable sapwood. That the tree grows in water, has very light and porous wood and floats buoyantly on water may help to explain its present plight and distribution.

"The Banyankole fishermen fix a torpedo shape of ambatch wood to the shafts of their fish spears. This does not materially impede the spear passage through the water, but causes it to bob up to the surface again in the event of a miss." (Worthington, Inland Waters of Africa.)

Presently, the tree occupies very limited stretches on the shores of Lake Edward — namely between Lake Katwa Village and River Nyamugasani estuary, a stretch of almost two kilometers between River Rwempunu and Rwenshama village, and a very localised area short of Kisenyi Village. On Lake George, the tree is almost extinct, in spite of the apparently wide area where it could survive. It is limited to only one island. In the above places, the plant stock, though healthy, appears to be diminishing.

Its utility, as noted above, has over the recent years been expanded, to be used as a buoy on all fishing gear on Lakes Edward and George and Kazinga Channel. Previously, the fisheries department was obliged to import artificial petroleum-based buoys to be sold to the fishermen on the above lakes. However, since the early seventies, the lack of those buoys forced almost all fishermen to turn to the ambatch tree to provide crude but efficient buoys, and the trees were nowhere else but within the park confines. Since then, it seems the rate at which they were cut greatly exceeded natural replacement. The parks then deemed it necessary to issue permits to specific people for the purpose of cutting and selling these floats. However, there is no identification whatsoever for any of the floats cut without authority. In the late 70's, there was greatly increased illegal fishing on both the lakes and the channel, and this meant very heavy demand on the ambatch trees. To compound this, these pieces of ambatch wood, at maximum, satisfy the fisherman for one week or so, so that weekly or bi-weekly supply of stock is kept up.

The tree used to be a commonplace; it is now disappearing fast.

It may be argued that it is nature at work, but the fisherman may be the major factor directing the whole trend downwards. If no plausible measure is taken, this good plant may sooner or later disappear from our shores.

The most straightforward remedy is to stop the fishermen from using these floats. This would need extensive marine patrols to enforce the law. But what substitute is offered to the fishermen? Fishing is an important part of the life here. The ambatch wood has the livelihood of many people wrapped around it, who do not foresee its uncertain future. The parks cannot sit idle to see the plant decimated, but a balance between the two should be sought.

The most obvious remedy would be to prolong the use of the ambatch floats for, say, a month and a half, by treating them with water repellants (internal or external coatings), marking them as so treated and selling them legitimately. Or, wrapping them in aluminium foils for watertightness, marking and selling them. What is needed now is some scientific investigation to prove the practicability and viability of either of these methods.

This project seeks, therefore, to determine the growth conditions necessary for the plant; correlate these to present distribution and condition; establish growth curves for sites with varying conditions; establish correct harvesting age and means of harvesting that least disturb the ecosystem; treat the wood both physically and chemically to prolong its usage and hence reduce the present high demand; and to mark the wood thus marketed so as to eliminate illegal and destructive acquisition of the wood.

Wood harvested for the study will be measured and assessed with and without bark, and green and dry, with comparisons made of service life. Measured pieces will be floated, and the different sizes correlated to the fishnets they must support while in actual use, to minimize waste.

The measured pieces will be dried and soaked, to assess decrease in density and relate it to loss of water and to establish its water holding capacity. Further tests will be made with various waterproof covering materials, as well as with chemically inactive (in water) water repellants like resins and other similarly oily products to determine buoyancy longevity. Internal coatings will be done at varying pressures and the best internal coat at a give pressure will be discerned. The direction of water absorption (radial or longitudinal) will be studied, and efforts to minimize it by blocking passages with both physical and chemical means investigated.

Finally, depending upon the results, the technique that gives the most durable float will be recommended. This will reduce the demand on ambatch trees, and it is hoped in the long-run, will allow the species to live and serve longer.

Measuring up to 2,7 meters from wingtip, the awesome Griffon vulture is coming back to its old home in the Cévennes, France.

Re-Introduction of Griffon Vultures in the Cévennes – France

♛ Michel Terrasse

Rolex Laureate – The Rolex Awards for Enterprise – 1984
42 Rue Mérédic, 92250 La Garenne, France

*French, born August 11, 1938. Chemist, specializing in clinical biology.
Educated in France; certificates in Bacteriology, Hematology, Immunology
and Parasitology.*

The aim of this project is to rebuild a population of Griffon vultures, *Gyps
fulvus,* in a part of the Massif Central in France, where these great birds used to
live in large numbers until 1920. Beginning at about that time, the toll of
shooting and poisoning of the birds outpaced their ability to reproduce or
survive in this natural habitat; by 1940 they were completely gone from the
area.

I am the Director of a group, Fonds d'Intervention pour Les Rapaces, that
has chosen to bring about the restoration of these magnificent creatures to a
land that was their traditional home. To accomplish this work, it has been
necessary to involve the aid and cooperation of both the authorities and the
public, not only in France but in other countries as well.

Fortunately, today, there exists a large and well administered protected zone
in the area, the Parc National des Cévennes. Following the establishment of our
program, it was possible to obtain the help of the park officials, who are now
locally responsible for the execution of our plan, at the site wherein the
program for reintroducing the Griffon vultures is located. The area is ideal for
our objectives, as it is undisturbed by agricultural and industrial development,
and the local pastoralism is still going well — an important factor in both the
original destruction of the vultures and our restoration program of today.

By the beginning of 1983, we have succeeded in establishing a wild
population of 17 vultures, all of which have come to us from captivity. Over the
next ten years, it will be necessary to release about 10 Griffon vultures per year
in order to establish a self-sustaining population.

The originality of this experience is complete: there is not a similar example
of such a program anywhere in the world. As a result, many countries have
evidenced interest in our results and and our approach to the task.

The overall view of our efforts includes the following elements:

Obtaining the Vultures
Though the birds had long ago disappeared from the Cévennes, they had not
become extinct. Through dint of patient work and searching, we were able

to establish contacts with two sources for the birds; institutions and animal treatment centers.

The institutional sources were those private or official centers and zoos that have captive breeding programs involving Griffon vultures. Examples of these are the Zoological Gardens of Paris, La Garenne in Le Vaud (Switzerland), and Villars-les-Dombes in Ain (France). Each of these institutions have decided to help this project by giving us all the vultures they succeed in rearing in captivity.

The second source is the network of animal rehabilitation centers we have located. In Spain and France, in the Western Pyrénées, many vultures are brought to these centers for care and attention after having been found in the wild; poached from nests, poisoned, injured, etc. After treatment and recovery, these birds are now being sent to us.

Added to these two sources of new birds are our own efforts at building up the stock of the Griffon vultures. From our own aviaries in the Cévennes, we have been able to match reproductive pairs from the injured birds sent to us. In this way, we have obtained seven new young vultures.

Where to put the Vultures?

A necessary part of our program in the beginning was the need to convince the zoos and rehabilitation centers that we were equipped to handle the birds in proper fashion. As part of the initial work, we selected a site just above a wild canyon in the Cévennes where the birds uses to breed in 1920. We built three aviaries there, as acclimatization quarters. After a period of captivity, the vultures are released near the aviaries and then maintain a continuing contact with the captive birds. Two people watch the aviaries to make sure the birds are undisturbed, and to give them food.

Education

This is perhaps as important as the actual work with the birds. We have developed a large and on-going program to educate the public to the value of these large birds. There was a time, many years ago, when the size and appearance of the Griffon vultures, along with their feeding habits, were looked upon as forbidding. The sight of a soaring Griffon vulture, with its wing-span of up to 2.8 meters, was once a fearsome view for many. As carrion eaters, the Griffons would be seen at the carcasses of the local sheepherders sheep, and it was assumed that they had done the killing. Little was understood on the part of the local people in those days of the ecological value and role of the vulture in nature's own system. The killing commenced, and, as noted above, by 1940, the birds had disappeared from the area.

With the passage of so much time, our task of re-introduction has perhaps been made easier, as we were able to identify the potential enemies of the birds before we began the program, and to go about an education program. For this purpose, we organized conferences with films and slides for presentation in the villages and farm areas of the region, to prepare the farmers, shepherds, sportsmen and tourists for the coming back of the birds. These presentations showed the value of the vultures to the local people, how they contribute directly to the economy by efficiently disposing of those animal carcasses that farmers or sheepherders would otherwise need to bury in order to prevent spread of disease. It was also possible, through publicity given by newspapers, radio and television, to make the return of the vultures a major event for local pride. An important part of our education program regarding the Griffon vultures is directed to children, who have come to look at the great birds as sources of wonder and fascination. As an indication of the effectiveness of this part of our program, we have to date a perfect safety record; no vulture released in the Cévennes has been shot.

Preparation of the Vulture "Restaurants"
In our original program, we foresaw the need to provide artificial feeding areas for the newly released Griffons. Though we were prepared to do this, we soon learned that the local sheep-farming in this region was adequate to provide sufficient carcasses in natural conditions to support the vultures without any large extra effort.

Observing the Released Birds
We currently have a full time researcher devoting one year to surveying the freed birds. With the help of the Cévennes National Park guards, he is attempting to stay in continuous contact with the released birds. This effort is carried out through radio-telemetry, or radio-tracking.

Each vulture, prior to being released, is fitted with a very tiny radiotransmitter, located in its tail feathers. Each bird has its own special frequency, allowing it to be identified easily. Distance and direction of the bird's location is found by the position and the intensity of the "bip-bip" signal, which allows us to track them up to 20 kilometers away.

Thanks to these transmitters, we are able to carry out significant continuing work. First, they enable us to help avoid serious problems with recently released birds by being able to locate them quickly if something appears to have gone wrong (injury, illness, etc.). Secondly, we are obtaining a great amount of data concerning foraging areas, social behaviour, courtship display, reproductive behaviour, food searching, etc. Further, we are especially interested in observing the behaviour of the vultures as they meet other scavenger birds, such as kites, ravens or eagles, and to learn of the natural interaction between them. Thanks to the presence of the vultures, we hope to make feasible the comeback of other disappeared species, such as the Egyptian vulture.

Future of the Project
We have been encouraged by many aspects of our work, but perhaps most of all by the birth, in the spring of 1982, of a young vulture born in the wild, the first such wild fledgling in 50 years. At the beginning of 1983, there are 17 vultures flying in complete freedom; within this group, there are three nesting pairs currently incubating eggs.

This success is, for us, very interesting and encouraging. But to be fully successful in our program, we must continue in the same way during many years ahead, in order to constitute a viable population of free birds, and to continue to learn more about the needs for the safe survival of these grand creatures.

A Network of "Space-Bridge" Terminals for Group Consciousness Transformation

Joseph Goldin
Ujinsky Lane 9 apt. 3, 103104 Moscow, U.S.S.R.

Russian, born November 21, 1939. Free-lance author. Educated in U.S.S.R.; graduate work at U.S.S.R. Academy of Sciences, Institute of Bio-physics (1965-69).

"Humanity", according to Norbert Wiener, the founder of cybernetics, is too wide a term to adequately represent the sphere of activity of most types of social information, because any community is always limited by the extent to which its information can be transmitted. From this it follows that "humanity" — in the true sense of the word — does not yet exist. The myth of "humanity" appears in the form of political declarations, humanistic images, religious beliefs — but "humanity" has yet to emerge as a real, living community.

Our proposed conception of a Network of Space-Bridge Terminals as a new channel of direct, live, multi-lateral communication between large groups of people around the world could turn out to be the informational resource with which man is at last able to transform the myth of "humanity" into reality.

The President of France, François Mitterrand, recently announced an International Contest in which architects from around the World were to submit their designs for an International Communications Center to built at La Défense in Paris. 893 entries from 45 different countries eventually were accepted by the judges. One of the Soviet groups chose to endorse my idea of creating a so-called Space-Bridge Terminal as a key element of the Communication Center. Such a terminal would enable large groups of people from around the world to communicate with each other using a new kind of video technology.

The decision by the Soviet group of architects to endorse my proposal was taken only in light of a direct, live, two-way satellite hook-up between California and Moscow that I helped to accomplish on September 5, 1982. American Robert A. Freling, Research Assistant on the Space-Bridge Project, described it well. An abridged summary of his description follows.

"On that day, using the remarkable "Diamond Vision" videoscreen made by Mitsubishi Corporation, a landmark communications event took place. A very large "Diamond Vision" screen was set up in San Bernardino, some 60 miles from Los Angeles. It was used to receive a feed, via satellite, from Moscow and relay this back to the originating Moscow studio. When the several hundred Moscovites assembled in front of the TV studio videoscreen saw the massive "Diamond Screen" appear and then saw themselves on that same screen in

California, and realized that the hundreds of Californians were waving their arms to the beat of music coming from Moscow, they were affected by a new emotion of "distant proximity" — and they were happy. A true "Space-Bridge" had been created for people-to-people communication on Space Ship Earth!"

"I have spoken with Soviet students who participated in this hook-up. Their accounts of what happened definitely fall into the "peak experience" category. The artificial boundaries of "me versus them" suddenly and dramatically gave way to a new feeling of unity as the Soviets were able to see themselves reflected in the eyes and smiles of their American counterparts."

"A prophet or seer I am not, but something tells me that this new form of human contact, if used wisely, could give birth to a planetary consciousness that until now has been realized only by an enlightened few. A worldwide network of satellite-linked Communication Centers could act as a cosmic mirror with which we, as citizens of the world, could look back at ourselves and grasp our essential unity as never before."

In a telegram sent to California on the day after the California-Moscow hook-up, I wrote: "The world community has become the real object of natural science investigation, thanks to direct, live communication with video-feedback between large groups of people in Moscow and California... Such gigantic videoscreens used for bilateral communication between thousands of people will soon become a traditional element of the environment, just like the public squares in the Greek City-States, or the Forums of the Roman Empire... The new feelings of "distant proximity" experienced by millions of people all over the world will create a new self-awareness and inevitably lead to a radical transformation in the way we deal with global problems... The scientists who created the Bomb, and the Artists who conjured up Anti-Utopia, will be challenged to find new ways of using their creative energy."

Imagine a small city connected to the Network of Space-Bridge Terminals (in the Soviet Union, 'small cities' are those with under 100,000 people, and include almost 90% of our cities), with the entire population able to participate in multi-lateral communication with other cities around the world. Immediately, the temptation will emerge for truly creative scientists and artists to join forces to help ordinary inhabitants be transformed into citizens of the world. From learning languages to communicate better, to staging live theatrical events with a 'global stage', ordinary people will be initiated into the zone of expanded awareness.

It can be done. Our project seeks to develop the resources and interest needed to bring such a Network of Space-Bridge Terminals into being; "humanity" in its truest sense is what we are seeking to achieve.

Cultivating Medicinal Herbs in Bangladesh

S. Sikander Ahmed
Bhuiyan Manzil, 23/1, M.M. Ali Road, G.P.O. Box No. 245, Chittagong, Bangladesh

Bangladeshi, born December 1, 1933. Sole Proprietor of an importing company and a multi-purpose farm. Educated in India; B.A. from St. Xavier's College, Calcutta in 1952.

Bangladesh is one of the poorest countries in the world. The approximately 5% of the literate population either congregates in the main cities and towns, working for government or private business, or pursues petro-dollars in the Middle East. The vast majority of people (90%, or over 80 million) reside in villages, left to their fate and to ignorance, poverty, squalor, despair and the worst forms of exploitation by their fellow man.

As a city-raised businessman and importer, I became aware some time ago of two things that concerned me and about which I decided to do something. First, I knew that Bangladesh, in spite of having a wide variety of flora possessing considerable medicinal value, imports medicinal products from elsewhere in the world with currency the country can ill afford to spend. Second, I knew next to nothing regarding the requirements for attempting to produce valuable plants on an agriculturally commercial scale, which I could see as being one hope for a happier future for Bangladesh.

Consider two small examples. Bangladesh currently imports refined or semi-refined castor oil worth over Taka 10 million (U.S.$400,000), yet the castor plant grows in profusion, wild, along the thousands of canal and river banks all over Bangladesh. The papaya is one of our most common fruits, yet large amounts are spent to import Papain and Pepsin for our pharmaceutical industries from countries where these do not grow.

This made no sense to me, and I set out to prove that a managed approach to the commercial culturing of medicinal plants could prove feasible, and that it could also be the basis of improving living standards in what are now terribly impoverished villages.

Primarily, my project is a pioneering one in Bangladesh, the main aim of which is to create awareness, both in government circles and by the general public, about our country's unique flora, and the rapidity with which it is being depleted by indiscriminate, unceasing assault by an ignorant, greedy and burgeoning population and an apathetic bureaucracy, on the few virgin tracts of forest still left in some parts of Bangladesh.

Generally, my aim is to bring to the fore our rich heritage in plants having medicinal and/or commercial possibilities; I intend to a) collect, b) preserve, c) propagate, and d) exploit as many of these plants as possible.

Specifically, I have undertaken, as a first step, the large-scale cultivation of *Rauwolfia serpentina*, the plant used in the making of the drug reserpine, as this has the potential of making the venture financially independent and self-generating.

The story of the attempt to lease land for the project is long, and filled with frustrations, but, finally, in 1979, I was allotted possession of approximately 50 acres of hilly land, 16 miles from Chittagong. The land consisted of low sandy hills, interspersed with low valleys. Overgrazing, the total denuding of the land from the constant search for firewood by local villagers, strong winds and the annual monsoon rains (over 4 meters between June and October) had laid the land virtually waste. My initial task was to repair the land; armed with book-learning, voraciously devouring all information I could find, I have succeeded in turning this property into a viable, multi-species farm. Through the use of contouring, proper irrigation techniques, the planting of fast growing legumes, and fruit and shade trees on the hilltops, there has been a marked improvement in the land.

In mid-1980, I was able to undertake the 10-day journey (by country boat) necessary to procure a few surviving plants of Rauwolfia Serpentina from a nursery set up in a remote part of Bandarban. These were carefully nurtured, and despite adverse climate and soil conditions, were slowly propagated to 211 in early 1981 and nearly 4,000 by early 1982. By May of this year I expect to have over 35,000 saplings from the 49,247 seeds collected during June-December 1982 and already sown in seedbeds. I expect to come close to my target of 3 to 4 million plants by the spring of 1985, when the first token exports of roots are anticipated. Optimum production of 40/50 tons of roots is expected by 1989.

The three species of Rauwolfia Serpentina being grown commercially at Faireen Farms (the name of the property) are, however, only the beginning. As it would be patently impossible to cultivate even a fraction of the herbs native to this area, it is my hope to convert Faireen Farms to a "mother farm", where as many plants as possible would be collected, preserved and propagated. Already more than 20 species of plants and herbs have been collected.

An important part of the future lies in the prospect of teaching the local hill tribe peoples, who have for years practiced the 'slash-and-burn' type of agriculture, to turn to this more productive kind of cultivation. Government resettlement programmes, aimed at creating cooperative farms and villages, will be able to act as "satellite farms", where plants and herbs having commercial values can be cultivated easily by the settlers, with handsome returns. With such 'contract growers', we can envisage the future production of our own basic materials from processing and/or alkaloid extraction plants.

There is no expectation on my part of completion of this project; to the very end, I shall go on seeking something new that can contribute to this country.

Reed boats are one of man's earliest sea-going accomplishments. The makers of these miniatures are among the last surviving craftsmen who know the ancient skills.

Documenting the 'Paper Boat'; Its Construction and Use

Cornelis L. van der Spek

49 Mercia Avenue, The Basin, Melbourne, Victoria 3154, Australia

Dutch, born July 14, 1952. Container controller in shipping company. Educated in Holland; archaeology studies at Leiden University (1973-76).

Observations made during a survey of reed bundle craft in South America undeniably suggest that their construction is a dying art. The situation in other parts of the world, where these craft have survived since prehistoric times, is not much different. For reasons that are not quite clear, but possibly due to a persistent lack of confidence by maritime historians in the qualities of reed as a ship building material, very little has been documented on these craft. Following the success in recent years of experimental, off-shore voyages with reed ships, these qualities have become better recognized, yet no specialized study on these craft has been done. New interpretations of archaeological and anthropological evidence have recently attributed a far greater importance to the reed boat in the evolution of the ship than was hitherto assumed, but to further formulate this evolution process it is essential that these craft be properly documented.

It is the purpose of this research, then, to fully record the construction and use of reed boats before extinction makes this impossible. Furthermore, the disappearance of the reed boat may have an immediate bearing on the economic existence of the reed boat builder. This project, therefore, also wishes to establish cause and effect of the decline in reed boat building, and what measures can be adopted to protect this industry from extinction.

During a preliminary survey of surviving reed bundle craft in Ecuador, Peru and Bolivia in 1976, I noted that the Bolivian 'balsa de totora' ('balsa' is from the Spanish, meaning not only the tree, but something that floats, like a raft), surely the most sophisticated of all still existing reed boats anywhere in the world, is increasingly being replaced by cedar-wood 'botes'. Surviving reed-boat-making skills find their major application in the construction of models sold to tourists, generating a cash flow that enables the Aymara Indians to invest in longer lasting, and therefore increasingly favored, wooden boats; this, of course, results in the further loss of skills as far as the working of the reeds is concerned. The present generation of reed boat builders on the Bolivian side of Lake Titicaca could well be the last one, and whilst life-size reed boats are still being built, the prolific production of miniatures reminds me of the last reed boat — also a model — I saw on Easter Island that same year: the silent reminder of a maritime past, and yet the all-telling story of another closed chapter in the history of ship building.

The totora reed boats built by the Uru Indians on the Peruvian side of the lake will probably continue to exist as long as the population of the floating reed islands is capable of maintaining its identity. With the Uru themselves dwindling away — their manner of life reportedly being continued by groups of Aymara and Quechua Indians who flee to the reed beds as a refuge from the tax collector and the conscription officer — this seems only a matter of time.

The situation in other parts of the world where reed bundle craft are still in use is not much different. In Ethiopia, the papyrus boats made by the Laqi (Zay) and Weyto tribes are likely to give way to 'progress'. With the disappearance of the Sardinian Fassoni, this type of vessel has only recently become extinct in the Mediterranean. The papyrus boats built by the Buduma tribe on Lake Chad will suffer from the warfare surrounding them.

What concerns me most in view of all this is the fact that very little has been documented on this type of craft. References to reed boats classify these craft in a generally denigrating way, as floats, rafts or a type of canoe. The recent off-shore experiments with reed craft conducted by Gene Savoy ('Cuvique', 1969) and Thor Heyerdahl ('Ra', 1969; 'Ra II', 1970; and 'Tigris', 1977-78) have resulted in a greater acceptance of the maritime qualities of reed boats, but no specialized study has as yet been done on this type of vessel.

Publications sometimes stress the short-lived character of the reed as a boat-building material, and conclude that as a result of necessary reproduction, this type of vessel is unlikely to show signs of dying out.

Is it possible that marine historians and anthropologists alike consider these craft to be a record in their own right, their continuously repeated construction ensuring an unending and immediate availability whenever required for review or discussion?

The objective of my project is to research the distribution, construction and use of surviving reed bundle craft, and to record the many different aspects of their existence before extinction prevents any further study. The eventual aim of this research is the publication of a complete study on surviving reed bundle craft, the need for, and the importance of which stem from and have a bearing on, in my opinion, 4 different areas:

1.   New interpretations of archaeological and anthropological evidence have recently attributed a far greater importance to the reed boat in the evolution of the ship than was hitherto assumed. There is reason to believe that the dug-out canoe was not the only progenitor of the modern ship's hull, as was thought previously, but that the concept of water displacement by air in a watertight hull was preceded by the wash-through, self-buoyant reed bundle craft. The ancient civilizations of Egypt, Mesopotamia and the Indus Valley all used reed bundle craft as the common form of water transport. The earliest wooden ship known to man and only recently discovered in Egypt, the superbly restored Cheops Boat, not only has the shape of a papyrus boat, but more significantly maybe, is built up of planks which are sewn together. I firmly believe surviving reed bundle craft should be properly documented to trace this evolution process, where the binding ropes of reed bundles and the stitched planks of wooden boats offer a common denominator.

2.   From an ecological and environmental perspective, we have much to learn from the harmonious relationship with nature achieved by the reed boat builders, whose technological sophistication and ingenuity resulted in beauty

rather than destruction.

3. If the reed boat disappears, of course, it will also mean the economic extinction of the reed boat builders; I hope that my research will result in a pilot study which aims to establish cause and effect of the decline in reed boat construction, and what measures can be adopted to protect this industry from disappearing.

4. A greater appreciation and a better understanding of the construction and used of reed bundle craft could lead, within the disciplines of archaeology and anthropology to a change in our interpretation of the entire culture within which the reed boat existed. As far as South America is concerned, it seems paradoxical that the reed boat is one of the very few surviving exponents of the Pre-Columbian and Inca civilizations. We generally identify with these civilizations through, and know them by, their legacy in stone. Yet, the skills of the stone-masons have been lost, for the enduring quality of their material required no further maintenance or re-construction, while the reed boat, as a result of the extreme opposite, has survived. Remains in stone and pottery impress us, yet the ultimate appreciation of these civilizations would require us to witness the use of materials to which they owed their daily functioning. With the disappearance of one of the last great technical achievements of this perishable world, the Keshwa Chaka, or 'straw bridge' of the Andes, only the reed boat survives to remind us of the true character of civilizations that have vanished. Documenting and preserving these craft is significant for a proper understanding of the past, and of invaluable help in recognizing the correct perspective within which archaeological and anthropological data should be viewed.

There are three main areas of effort in my project:

— A survey of the available literature pertaining to reed boats, which falls into two categories; a) Scientific and popular scientific publications dealing with the craft and the (pre) history of navigation, and b) Travelogues written by people who have visited areas where the craft are still used.

— Experiments with the construction of reed boats on a model scale, in order to learn and record the methods of construction, techniques, size/weight/cargo ratios, etc.

— Further field-work in the surviving centers of reed boat construction, where first hand observations can be made and recorded as the necessary summary for the previous two phases.

I wish to ensure that the intricate art of the reed boat builder is not lost; such a study is needed now, before it is too late.

taught to how to eat on its own again. In stages, more and more daylight will be allowed into the brooder, until it regains its natural aggressiveness.

8.   *Training*: When ready, the animal will be moved to an unheated mews (a large closed closet) outdoors and trained to perch, to become more aggressive and re-learn to fly. Progressively more and more sunlight will be let into the mews until finally the doors will be left open and the bird can escape out into a large flight cage. Flight exercise continues, usually in the presence of other patients of the same species, where they will be forced to fly around and around for exercise, as well as flight training by a family member wearing a cover-up poncho to minimize human contact.

9.   *Live Food Lessons*: During training, if an animals is judged likely to be able to make it on its own in the wild, live mice, rats, chicks or rabbits are substituted for the whole dead food it has been given. A Raptor must be able to catch, kill and tear apart live food prey in order to make it on its own in the wild. After our new construction phases are completed, feral Starlings and pigeons will be added to the food list. Now we buy rats and mice from University labs, are given some mice by college labs, are given rabbits by the Rabbit Breeders Association, and are given chicks from a hatchery. We hope to raise some of our own rats, mice, rabbits and pigeons and trap Starlings to reduce costs.

10.   *Disposition*: After record photographs have been taken (ventral, dorsal and natural poses), we notify Ken Nieland (Zoo), Buck Del Nero (Warden) and David Yee (Audubon) that a bird is ready for release, and a place, day and time is determined for release to occur. After the release, copies of the documentation on the particular bird are transferred to the Zoo office. A Raptor that can't make it on its own can either be kept for a captive breeding program, or used for educational purposes by a school or museum.

11.   *Final documentation*: The records are updated and computerized by family members.

Our Raptor Center has been a labor of love from the beginning, started by our boys, then 7 and 9, and has grown through the last eight years, involving the whole family in "The Five Mile Creek Raptor Center". It now requires more than $2,000 per year for food, housing, etc. No financial assistance has ever been offered or requested. On hearing of the "Rolex Awards for Enterprise", we considered for the first time the time-saving implications of an outside financial award. Such support would be an opportunity to speed up our research and our usefulness on the Raptor project by about eight or nine years, compared with continuing on our own more limited funds.

Our long-term program is based on a very complete study of materials needed, their costs and installation, as will be seen by the attached detailed documents.

The Creation of a Multi-Volume Nature Guide to the Indian Ocean

♛ Thierry Victor Marie Robyns de Schneidauer
Honourable Mention – The Rolex Awards for Enterprise – 1984
189 Avenue Molière, B-1060 Brussels, Belgium

Belgian, born November 9, 1929. Free-lance naturalist and painter. Educated in Belgium and Holland; B.Sc. (Engineering) in 1974.

The objective of this project is to establish a collection of nature-guides, focused on the broad area of the Indian Ocean, especially made in order to make people more interested in conservation, in nature, and to aid development of a more intelligent tourism oriented to a concern with the environment.

I write and illustrate these books, under the series title of "Nature Guide to the Indian Ocean"; five volumes will make up the complete series, of which one is already published, and submitted with this application as an example of the work I am doing. The five books in the series are divided by geographical areas as follows:

1. *Western.*

This includes Madagascar, Comoros, the Seychelles, Mauritius and La Réunion (already published in French).

2. *North-Western.*

This includes the South Arabian coast between the Gulf and the Red Sea, Yemen, South Yemen, Oman and the United Arab Emirates (now in progress).

3. *Central.*

This will cover the Maldives and Ceylon.

4. *Eastern.*

This will cover Sumatra, Java and Bali.

5. *Southern.*

Presently planned to include the islands of New Amsterdam, Crozet, Kerguelen, and Prince Edwards, though this last area may be exchanged for the western coast of Australia.

As can be seen from the copy of the first book, I aim at making these volumes an approach for the scientist, a handbook for the teacher, and a guide for people interested in nature and travel.

The project generally divides into four phases: a) Exploration in order to gather the needed elements and obtain sufficient experience and knowledge, b) Documentation research in scientific institutions and libraries, c) Creation of the text and illustrations, and d) Establishing contacts for publishing and distribution of the books. The present program calls for publishing a new volume every 2 or 3 years.

A vanishing art form? The intricate patterns and creativity of the early manhole covers are giving way to today's "efficiency".

Preserving the Art Under Our Feet: Manhole Covers

Hertha Bauer

145 East 37th Street, Apt 2 F New York, New York 10016, U.S.A.

Austrian, born December 28, 1919. Photographer, and Rare Books Librarian. Educated in Austria and Czechoslovakia; Ph.D. from University of Prague in 1945.

My project is to record and document the history and geographical range of street hole covers. I would accomplish this by photographing covers in the streets of town and cities in Western Europe, North America, and possibly the Near and Far East, and by researching local history materials. The result would be a book illustrating and explaining the 150 year history of the covers.

Cast iron covers for the holes permitting access to sub-street systems (sewer, water, power, transportation, and communication) have appeared in the streets of our cities and towns since the early middle nineteenth century. Long after the original purpose for decorating the lids, to assure safe footing for horses, had vanished, the lids continued to be decorated with patterns of stylized flowers, stars, foliage and geometric designs. The many and varied patterns of the lids suggest an innate human urge to decorate even objects unlikely to be noticed. As more components of the infrastructure are housed beneath the streets, more covers appear, but most of the new ones are standardized grids, while simple rectangular designs replace some of the older operative covers; others completely disappear as coal holes are cemented over, and electrical handhole covers designs are filled in with cement. Thus, progress eradicates wonderful expressions of the human urge to decorate even unlikely objects. Still, the variously designed manhole and handhole covers can be found and saved in photographs.

During the last five years, I have photographed more than 800 street hole covers in the United States, Canada and Western Europe, and have exhibited both my photographs and photo-etchings of them in New York and California. My original interest in the designs of the covers led to an interest in their history, the history of the companies and foundries producing them, and the histories of the systems beneath the streets. Friends have sent me photographs and information about covers in England, Israel, Malaysia, China and Japan. As a result of my exhibits, individuals and representatives of utility and transportation companies have contacted me both to seek and to offer information.

Streethole covers, such as manhole, coal shute, handhole and tree hole can be characterized by their designs, age, usage, the companies of public institutions

Preventing Ecological Imbalance Due to Bamboo Flowering in Mizoram

Tawnenga Pachuau

Mission Compound, Aizawl, Mizoram 796 001, India

Indian, born March 1, 1956. Lecturer in Botany. Educated in India; M.Sc. (Botany) from North Eastern Hill University, Shillong in 1977.

The project aims to study the physiology of bamboo flowering in Mizoram, which is believed to be the cause of outbreaks in the rodent population.

Mizoram is one of the Union Territories of India, situated in the extreme northeastern corner of the country. With an area of 21,087 square kilometers, it had a population of 487,774 in 1981. The people in the Mizoram area depend mostly on agricultural cultivation, in which rice is the chief cereal crop grown. Rats, when they increase sharply in population, have caused great damage to the rice fields, which has brought about famine and misery.

This project intends to survey and investigate the ecological imbalance that appears to be brought about by the flowering of bamboo in Mizoram, with the objective of checking or controlling the famine, misery and fear brought about by these imbalances.

A number of different species of bamboo grow in Mizoram, the dominant ones being *Bambusa tulda, Dendrocalamus longispathus,* and *Melocanna bambusoides.* These bamboos traditionally flower in a cycle of about 48 years, plus or minus 2 years. There are two cycles, each running independently of the other. In the first, *Bambusa tulda* and *Dendrocalamus longispathus* are associated. The second cycle is that of the flowering of the *Melocanna bambusoides.* Their flowering is always followed by their death. With nearly a century of data, we see the following picture:

Bamboos in Flower	Flowering Years	Flowering Cycle
B. tulda; D. longispathus	1880-1884	48 (+/− 2)
Meloccana bambusoides	1910-1912	48 (+/− 2)
B. tulda; D. longispathus	1928-1929	48 (+/− 2)
Meloccana bambusoides	1958-1959	48 (+/− 2)
B. tulda; D. longispathus	1976-1977	48 (+/− 2)
Meloccana bambusoides	2007 *(expected)*	

Every time the bamboos have flowered, there has been an unbelievable increase in the rodent population. The common, widespread belief is that at the time of the bamboo flowering, rats increase at exponential rate after eating the bamboo seeds, and then go about eating grains from jhums (rice cultivated lands) and homes. These rats are identified as *Rattus niviventor* and *Rattus nitidus*.

About 75% of the half-million population of Mizoram depends on jhuming. The flowering of bamboos, accompanied by an enormous increase in the rat population, is nowhere so marked with apprehension and fear as in Mizoram, where the bamboo flowerings bring tidings of famine and misery.

Apart from this, the flowering of bamboo in 1958 led to a very bad political result. Some segments of the Mizo people felt that the Government of India did not take good enough care of the people during the famine of 1958 that coincided with the flowering of *Melocanna bambusoides*. This feeling led to rebellion against India. The rebellious movement is still going on in Mizoram, and because of this, tragic events are taking place; brutal killings, burning down of villages, imprisonment of innocents, curfews, etc.

In my project, I wish to accomplish the following:

1. Make a thorough investigation and survey of the difference species of bamboo that flower.

2. Conduct a comparative study of these species of bamboo with the same species in other countries. This will aid in understanding the effect of environmental factors in the flowering of bamboos.

3. Study the physiology of bamboo flowering.

4. Test the application of chemicals as a means of delaying, or changing, the flowering cycle.

5. Assess bamboo population control. The Mizoram bamboos are second generation successors that came into existence following the clearing of fields for cultivation. Application of chemicals, or environmental agents, at early growth stages may be a solution to population control.

6. Study the inter-relationship between bamboo seed and the rat. This will require chemical analysis of the seed, the necessary collection of rats for experimental purposes, and the testing of the effect of the seeds on the reproduction pattern in rats.

7. Assessment of other possible factors that may have enhanced the rat populations, such as possible diminishment of natural predator populations (hawks, owls, snakes, etc.).

8. Control of the rodent population. This is the core of the problem, and will include evaluations of means to increase predator populations, application of poisonous chemicals, or possible means to employ labour to kill rats.

9. Testing and evaluating the possible use of chemicals to protect the rice from the rats.

Siberian crane (Grus leucogeranus) in the marsh of Keoladeo Ghana, near Bharatpour, North India, where 30 to 50 of these species come back punctually to spend winter.

Development of A Chinese Crane Center at Zha Lung National Reserve

George Archibald
Shady Lane Road, Baraboo, Wisconsin 53913, U.S.A.

Canadian, born July 13, 1946. Director and Co-Founder of the International Crane Foundation, international conservationist. Educated in Canada and U.S.A.; Ph. D. in Ecology and Systematics from Cornell University; Ithaca, New York, U.S.A.

The headquarters of the International Crane Foundation (ICF) near Baraboo, Wisconsin consists of a mixture of administrative, educational and propagation facilities to meet our objective of saving the earth's endangered cranes. ICF has achieved global recognition in the conservation community for its success in conserving the wild cranes and their habitats in Asia and Africa, in public education (particularly through mass media), and in the captive breeding of rare cranes in captivity. One of ICF's primary objectives is to catalyze events where similar conservation centers are built in the nations to which the cranes are native. As a consequence of our work in Japan, a crane center was built in Hokkaido in 1974, and through our influence and example, the USSR established their crane center in the Oka State Nature Reserve in 1979. Currently, we are working with the Chinese to establish the China Crane Center at the Zha Lung Natural Reserve.

Zha Lung Natural Reserve encompasses 560,000 acres of shallow wetland in northeast China, only 25 kilometers from the city of Qiqihar. The area was protected because of its importance as a nesting area of the rare Red-crowned Crane — that beautiful black and white crane that is frequently featured in Oriental art. Four other endangered species, the White-naped, Siberian and Hooded Cranes, and the Oriental Stork, also depend on the wetland.

A small administrative center has been built on a knoll of ground near the northwest corner of the marsh, and motel-like facilities have built for tourists who come from distant points to see the rare birds and their magnificent habitat. But the area is extremely flat, the vegetation tall, and it is extremely difficult to see the birds without unacceptable disturbance.

The Chinese want to build an ICF-like China Crane Center (CCC) at the headquarters with public education, research, and captive propagation facilities, and they have asked ICF to help them build the CCC. In 1982, ICF was officially made the sister center to Zha Lung, and later that year the Governor of Heilongjiang Province visited ICF.

Unfortunately, ICF does not have the funds to help the Chinese build CCC, but we do have the ability to instruct as to how the center should be established

A Comparative Study of Alphabets in Sculptural Form

Peter Haden

29 Chemin Moïse-Duboule, Petit-Saconnex, 1209 Geneva, Switzerland

South African, born April 30, 1939. Sculptor. Educated in South Africa; BA in Psychology from University of Witwatersrand, Johannesburg; and U.K., Physics at Oxford University.

This project has several objectives. I wish —

First; to make a large outdoor sculpture, symbolizing the importance of communication, with small replica sculptures as annual awards to those fostering understanding through communication.

Second; to consider the design of a universal alphabet for a future auxiliary language in conjunction with the expertise of professional linguists.

Third; to make a comparative study of the shapes of letters of various cultures.

Fourth; to study and analyze alphabets through the artistic media of graphics, sculpture, books and film, and,

Fifth; to increase awareness of the importance of alphabet symbols in the history and culture of mankind.

Elementary Knowledge is a project which, in part, challenges the question of language communication. Notwithstanding the Biblical story of the Tower of Babel, the tongues of man have been confused from their inception, and that confusion has been increased by the multiplying of language and dialects in use on the planet. If we believe in a future that will include plenty for all, we will eventually be confronted with the necessity of having ease of communication with an extended number of humans in every country of the world.

It is debatable today, as it was in the past, that a totally new set of alphabet symbols would be too difficult for the majority of people to learn with ease. The detractors of the concept of a universal auxiliary language frequently stress that Anglo-American ought to hold this place; however they have not taken into account the fact that Anglo-American is not a politically neutral language. In this project, I am tracing man's first written communication through its evolution to the present, and looking at some needs for the future. My intial interest lay in the forms of the letters, and to this end I sculpted the Arabic, Greek, Hebrew and Roman alphabets after researching their history and mythology. The intention is to eventually treat this subject in graphic and book form. A large measure of research toward this objective has been carried out, and I am currently in the process of writing a book on the subject.

Alphabet letters in sculptural form allow for the possibility of viewing these symbols of communication from many different angles. In so doing, various properties of the letters emerge showing common features between cultures which suggest a distant shared ancestry. In general esthetic terms, the letters have been refined and, over the centuries, a purity of shape has evolved in them which is artistically satisfying. In a world of enormous artistic diversity, the letters alone are both "abstract" and "representational", and remain "contemporary" at all times.

The idea which has attracted me to the Rolex Awards concerns honoring those who continue to believe that mankind has a destiny based on communication. Practically, this would take the form of an outdoor sculpture dealing with the subject of communication through the generations and toward the future, based on alphabet symbols. Thereafter, the idea includes making small replica sculptures of the large piece which, in themselves, would constitute prizes given to perhaps twenty deserving individuals on an annual basis. The recipients would come from all walks of life, people who are not seeking prizes for their efforts, but undertaking difficult communication problems from their personal convictions. The Pugwash delegations are the first group of people I would envisage honoring in this way.

I am enclosing reference materials on the completed sculptures of the four alphabets, as well as other works and macquettes dealing with the project, including a macquette for a large sculpture on communication through the generations.

Elementary Knowledge began in 1975. It is more than half way to completion. It includes, to date, over 100 sculptures, a large amount of graphic work (including a portfolio of 23 lithographs on the Hebrew alphabet, two published books, and one large book in final preparation). It might be completed in two years.

Studying, Treating and Protecting Tropical Freshwater Fish

Colette Santerre

Château de Mauvoisin, 35650 Le Rheu, France

French, born October 1, 1945. Manager of tropical freshwater fish research center. Educated in France; State nursing qualifications in 1965, plus further medical training.

Siluriform fishes, and especially the Loricariidae species, are among the most often imported ones, for aquarium hobbyists. Due to this trade, some these species are, or have been, endangered. They are also among the least understood hobby fishes, and many of them die because importers and aquarium owners have no detailed information about the requirements of these fish. Because of this, I have created a center for their acclimatization, breeding, and disease treatment; there is a special focus on the genus Farlowella, which is the most difficult to acclimatize. After two years of study, I have developed a correct procedure of care for Farlowella, which keeps them in good health and results in their successful breeding of offspring that survive until adulthood.

As the result of this experience and knowledge, I am now preparing a book on these fish, which considers their anatomy, physiology, biology, ethology, taxonomy, pathology and behaviour in both wild and captive states. My particular concern is with ichthyopathology, where I am promoting a veterinary service well equipped to handle tropical fresh water fish diseases with appropriate treatments.

My tropical fresh water fish center resembles a human hospital, with only one difference; fishes in good health are more numerous than diseased ones. I handle all the duties involved with acclimatization, aquariae management and cleaning, feeding, fish observation, breeding, pathology (diagnosis and treatments), ichthyological research by personal observations, and documentation (anatomy, physiology, etc.), and all the secretarial duties. I am aided in biological research (analysis, anatomopathology and pathological germ identification and culturing) by the local "Departmental Veterinary Research Service"; in radiological research by a specialized doctor from a human hospital; and in surgical interventions by one or two veterinary doctors.

A major part of the center's activities has to do with acclimatization, wherein arriving fish are kept in biologically balanced quarantine aquariae for variable terms (2-3 months, sometimes longer). Once the arrivals species are determined, the aquariae are adjusted to specific requirements (water temperature, etc.), and the fish are observed regularly and data on them recorded. The quarantine aquariae permit; the immediate establishment of the arriving fish in sanitary conditions, where they may be observed for any serious diseases; optimum adaptation possibilities due to specifically tailored environment, and the opportunity for making detailed observations regarding the social, individual, territorial, feeding and other habits of a given species. These steps allow for sharply improved aquaria management when the fish move on to their ultimate destinations.

An important part of the center's performance is based on the keeping of very complete records. Each aquarium has its own number and its own register. In these registers are recorded the following data; complete listing of materials present, lighting (source direction, natural or artificial, intensity, length of time), management data (monthly plant growth, routines, etc.), modifications (nature, date, time), water (physical and chemical values, renewal dates, etc.), additives (chemicals, reasons for use, dates, etc.), cleaning (dates and nature of products, etc.), and other information pertaining to the physical habitat of the fishes.

For the fish themselves, specific registers are maintained for each specie. Divided in three parts, these registers first record information on taxonomy, paleontology, and such faunistic data as geographic origin, habitat description, water requirements, inter specie relationships, social organization, feeding, laying, breeding, survival of young, longevity, and pathology. A second section is devoted to the anatomy, physiology and biology of the species. The third section is devoted to observation on behaviour and adaptation of the species in captivity.

While the center handles many species of fish, there is a particular interest in the Loricariidae, and especially the Farlowella, which are often purchased for their capabilities as algae scrapers in hobbyist aquaria. They cannot be kept in good health without eating natural algae, and when there is not a sufficient amount, these fish will generally die. Those that do not die generally suffer metabolism problems. Work done at the center has resulted in the development of a diet that suits these fish that do not know how to eat moving food, as well as a very large body of information about this particular species and its requirements.

I have shown that it is possible to study tropical fresh water fish on a scientific basis and make the results of that information known to the hobbyist and truly interested fish owner. Ichthyopathology is not a specific study in veterinary schools, but there is no reason why it should not be available to interested owners. A better knowledge of fish life must surely be an aid in the cause of attempting to prevent extinction of species, as part of our overall concern for our global patrimony.

Corals: things of beauty, delicacy and sensitivity. Cultivating them may help ward off their destruction.

Establishment of World's First Underwater Gardens of Cultivated Corals

Adham Ahmed Safwat

632 Cordova, Davis, California 95616, U. S. A.

Egyptian, born October 16, 1935. Organizer of diving tours and exporter of corals. Educated in Egypt, studying law and economics at the University of Cairo.

The project I wish to undertake is the establishment of our world's first cultivated coral gardens — underwater — just like botanical gardens on land.

In 1974, I started cultivating corals underwater in the Red Sea. The experiment was successful, and the corals are growing in a lagoon in the Red Sea located near the island of Giftun Al-Saghir. During May 1982, I began cultivating more corals in a lagoon in the Pacific Ocean near the island of Bora-Bora in French Polynesia.

As these experiments have been successful, I would like to establish these lagoons as underwater parks or gardens. The protection of beauty is very important to man and, if we can overcome hunger and pollution, then beauty will be man's urgent concern.

Underwater coral gardens are protected in a few places as national parks. However, to my knowledge, nobody has tried to cultivate corals underwater; and when the corals die in these parks, nobody tries to replace them.

The technique of planting and growing corals is very simple. A small branch is cut from a living coral colony (a tree), and then fixed to a stable, heavy piece of concrete, or tied to a strong nylon line tied between two concrete pillars. The larger the transplanted branches, the more rapid growth to a larger colony.

The critical aspect of transplantation is choice of a site where there is a current constantly bringing nourishment to the corals. Also, a barrier reef is needed to protect the site from heavy surf. Planting takes place at a depth of about six feet below the low tide mark; if the planting is too shallow, hot surface water may kill then; if too deep, there may be inadequate sunlight. Of course, there are certain strong corals growing near the surface and exposed to heavy surf and hot surface water, as well as others that grow deeper than 30 feet. I am concentrating, however, on cultivating the beautiful and delicate coral varieties that cannot withstand either the hot, turbulent water, nor the great depths.

Marine biologists and divers around the world are reporting that coral reefs are dying, and that more dead than live coral is being seen. Some have blamed the animal called the "Crown-of-Thorns", though I do not believe it is primarily to blame. I have seen with my own eyes large areas of coral reefs die at once in less than five days. Everytime this phenomenon happened, there had been a week of dead calm weather, with no winds to stir the surface waters and mix them with deeper, relatively cooler water. In the summer heat, the temperature of the surface water rises and the delicate corals die. Their polyps, the living part of the coral colony, die off and leave the hard skeleton, floating on the calm surface of the sea, forming a brown, sticky substance all over the reef. A few days later, looking at the corals, you note that the ones that died are white as snow. About a week after that, you look again and all these clean white skeletons are covered with greenish algae.

How can some areas of the reef survive the periods of intense heat in the summer? Some corals, like so-called Brain Coral, are tough enough to survive. Delicate corals that live deep enough to escape the hot surface water will survive. Some corals of the most beautiful and delicate varieties will survive because they are exposed to strong currents that keep the water temperature around them ideal for their growth. It is these delicate corals that I want to propagate, increase and protect.

There are no serious obstacles to overcome before establishing a cultivated coral garden underwater, like a botanical garden on land.

The first requirement is a suitable location. The two I have mentioned above are highly satisfactory. In the lagoon surrounding Bora-Bora, there is an area given over to being a protected reserve, and the local police do their best to keep coral collectors and spear fishermen out of this area. In the Red Sea, lagoons are protected by horse-shoe reefs, and the government is trying to enforce laws prohibiting the collection of corals and spear fishing in this area.

Secondly, seed money is needed to plant or cultivate the corals on a large scale. These funds will enable me to proceed, and to grow thousands of coral colonies, or trees, of many different species. I expect more publicity for these underwater gardens through publications such as the National Geographic Magazine, and, after this, more funds will be available to keep the project going.

The project can be completed in less than a year. Three months of work will be enough to plant each coral garden with trails marked for swimmers. A strong swimmer will need about two hours to follow all the trails, the length of which will be approximately four kilometers.

Money is needed for the establishment of the parks, not for salaries. I believe marine biologists will be glad to monitor and study the corals growing in these gardens, and the local authorities and police will be capable of protecting the coral gardens.

If this sounds interesting, I can furnish you with more detailed information and pictures of my experiments — underwater movies showing how it is actually done, and charts of reefs showing the exact locations where I propose to carry out this project. I am now 47 years old and in good health. I can devote all my time and energy to this project.

Timshel — Power Bank

Buddie Gordon Miller

c/o 'Millerbros', 41 Tudor Street, Bridgetown, Barbados, West Indies

Barbadian, born May 8, 1945. Company Director/General manager; manufacture and marketing of cosmetics and toiletries. Educated in Barbados; engineering studies at University of West Indies.

The "Timshel" — Power Bank project proposes the installation, in "Power Banks", of sets of people-powered pedal/rowing type machines. These machines, under state control, would feed electrical energy into the national power grid.

This system would provide direct and immediate benefits in:
— Reduction of consumption of fossil fuels.
— Reduction in emissions generated by burning of fossil fuels.
— Substantial reduction of unemployment.
— Creation of an unlimited employment bank.
— Creation of significant spin-off and downstream industry.

The machines would be fitted with individual generators/alternators, and would be metered. The output from each machine would be fed into the National Power Grid, and would replace its equivalent power now being generated by the burning of fossil fuels.

Each unemployed person — as identified by the State — would be issued an identification/meter card, and would have unlimited 24-hour access to any of the "Power Banks" located at strategic sites in and around his community. He could then work at the times and for the periods relevant to his needs. His "Power Credits" would be encoded on his identification/meter card, and he would cash these credits as he desired at various offices licensed by the State.

It is important to note that the fact of the average human's output of measurable energy being relatively small is scarcely relevant to this project. Once a basic daily output level is determined — by the State — anyone achieving this output should earn at the bare minimum that "unemployment benefit" that he or she now earns for no output at all.

In terms of fuel substitution, we may estimate that the average fit human being, working conscientiously for 230 days per year, would replace some 200 lbs of coal, or its equivalent in another fossil fuel. This would indicate that each 1 million persons, now unemployed and utilizing the "Power Bank", would replace some 200 million pounds, or 100,000 short tons of coal (or its equivalent; 300,000 barrels of oil) per year. At the time of this writing, the unemployed of North America and Western Europe exceed 25 million workers. With 50% of these people utilizing the "Power Banks", we may envisage a fossil fuel substitution on the order of 1.2 million tons of coal, or 3.6 million barrels of oil per year.

The "Timshel" — Power Bank programme would:

— Reduce; fossil fuel consumption and importation; emissions generated by burning fossil fuels, the extent of acid rain, neutralization/repair costs due to acid rain damage, government subsidies on fossil fuels, social and political stresses of unemployment.

— Reduce and redistribute peak traffic loads on roads.

— Eliminate unemployment — direct and indirect costs.

— Eliminate "Special Works" projects.

— Generate down-stream and spin-off employment.

— Stimulate light industrial sector in manufacturing and service of machines and meters.

— Assure fair and equal wages for equal efforts.

— Enhance physical fitness on a national level.

— Reduce costs of National Health Services.

— Allow the state to determine the compensation rate for optimum social benefit, with possible reduction in direct taxation.

— Utilize currently idle commercial/industrial floor space.

— Fit easily into multi-storied layouts.

— Redeploy administrative staff currently engaged in unemployment services.

— Allow government to influence population distribution in a direction away from urban congestion.

Funding for the "Power Banks" is available from a variety of current sources, such as:

— Current cost of electricity from Municipal Plants, via fossil fuels directly replaced, and government subsidies on cost of KWHs.

— Costs of unemployment programs — direct and indirect.

— Costs of "Special Works" programs.

— Emission Controls — Capital and Operating costs, and government subsidies.

— Costs of neutralization and repair of emission damage.

— Taxes generated from associated manufacturing and services industries.

— Reduced costs of National Health Service programs.

On an individal basis, benefits accruing from this system include:

— Choice of times and periods of work most convenient to oneself.

— Wider choice of work locations.

— Elimination of indignity of accepting unemployment handouts.

— Elimination of the social and personal stresses of unemployment.

— One earns in direct proportion to one's own effort.

— Precludes discrimination of sex, race, creed, education, political persuasion, etc., in the workplace.

— Other personal activities (e.g., reading) practical while working.

— Access to emergency funds.

— Direct health benefit.

It may be argued that this program does not provide the emotional gratification so necessary in a regular occupation, and that it offers "sweat shop" dehumanized labour. So it does, in the context of an unconventional solution for an unconventional world circumstance. In an effort to improve in this area, we would propose:

1)   The quickly obvious pay-off of a rate and accumulator indicator built onto each individual meter.

2)   An educational/entertainment/audio/visual console attached to each work station. A relay would allow access to this system only when a basic minimum of power output has been attained and maintained. For specific, approved educational programs, the power absorbed by the console would be credited back to the individual's own meter card.

Another anticipation is that a number of people would permanently abandon the conventional workforce, in favour of the freedom and flexibility of the "Power Bank". This problem is minimized by two factors:

1)   Any such migration should leave corporate positions open to the 'corporate' types, and hopefully generate a well-motivated work force.

2)   The government retains final control, in its ability to restrict and/or restructure the program so that it best serves the needs of the period.

A second phase of the programme may be envisioned when individual work stations could be installed in homes, offices, factories, etc. These would allow the population at large to supplement incomes while generating the national benefits associated with substitution for fossil fuels. A number of technical problems would need to be resolved before this phase would become feasible, and this phase is not being considered for the purposes of this presentation.

The project is divided into three sub-sections. The first is Completion of Economic Analysis (Dec '83), the second is Completion of Design of Electrical and Mechanical Elements (Jul '84), and the third is Publication of the Detailed Program (Sep'84).

The Small Reliefs of Venice

Hansjörg Gisiger
28 Chemin de la Colline, 1093 La Conversion, Switzerland

Swiss, born December 26, 1919. Sculptor, copper-plate engraver, lithographer, lino cutter, art critic, former Professor of Sculpture. Educated in Switzerland.

Venice is the only city in the world where a considerable number of houses are adorned with small reliefs and sculptures, sometimes being just ornamental, sometimes being very elaborate. Considering their obvious unpretentiousness, there is little chance than anyone would venture upon saving these works, which are all the more doomed to disappear through the combined influence of the climate and the pollution.

This project envisages taking as complete an inventory as possible of these works, including: location of the site, historical information (if available), and comments of aesthetic nature.

Venice, like Amsterdam, is a city of painting. The atmospheric humidity, which erases the outline of objects, thereby plunges them into a pictoral atmosphere that has probably contributed very much to turning these two cities into the great Western painting centers.

This assertion also acquires value through its negation: Venice is a city without monumental sculptures, except the works of Riccio and some monuments of the 19th Century. There has not been, and there is not, a famed Venetian sculptor.

In matters of sculpture, however, Venice shows a peculiarity it doesn't share with any other city in the world: its houses — the palaces as well as the dwellings of the middle-classes and even the buildings of the common quarters — are often adorned with small reliefs (the dimensions of which rarely exceed 80-100 cm) or embossings. They include simple decorations, geometrical ones, fabulous animals (often, of course, the winged lion), or religious subjects, having a predilection for the Madonna with the coat, or free subjects, sometimes referring to the crafts, and, finally, heads rising above vaults, but also fitted into the façades.

If a great many of these reliefs look as if they had been created for the given place they hold, an almost equal number seem to have been added to the walls after having been intended originally for another site, from which they were removed to their new destinations. This is above all the case with the houses in the common quarters.

Hence, it is a genuine manifestation of folk art, in which all classes of the society obviously participated, from the aristocrat who adorned his palace with elements of sculpture to the modest craftsman who wanted to give the façade of his house an artistic touch and who, to do this, executed himself a naïve adornment or, during a demolition, "took away" what had served already. But these elements of ornament are always of modest size and discreet appearance; so discreet indeed that until today apparently nobody has thought of taking their inventory.

If one studies the styles of these works, one discovers that they range from Romanesque to the post-Baroque, and even to the realism of the 19th century. Therefore, one may consider that it is a custom that has maintained itself for almost a thousand years.

Of course, it is reflected in the state of preservation as it is dependent on the quality of the stone and the place where the work is situated. Some of these reliefs are hardly recognizable, the aspect of many others is altered by the ruining caused by the rainwater, while others did not withstand the modern pollution brought on by the exhaust gas of the "motoscafi".

One can't see very well how these works might be preserved, restored or saved. There are two reasons for this:

a.   For the great works of art, the city Venice has a renovation plan which is so laden that the city will never be able to devote its aid to the restoration of these modest works.

b.   Technically, the problem would be almost unsolvable, since every time the stone ought to be removed from the facade for making a copy. One can't re-cut a sculpture without reducing the volume, and consequently, disfiguring it.

Therefore, the only way to keep a remembrance of these reliefs is to take as complete a photographic inventory as possible, which afterwards would placed in the "L'assessorato alla Cultura", or in the archives of the city.

Of course, later on, a selection of these photographs might be published in a book — which might induce others to take an interest in the problem — but that is not the main object of the present project.

Program. There are three stages in the project:

1.   The systematic search for these reliefs in the six "sestieri" of the city, photographing, notation of the place, of the dimensions, of the nature of the stone, etc.

2.   Filing of index cards, including a photograph, and possible historical or aesthetic remarks, and.

3.   Possible summary according to subject matter, topography, style, etc.

Timing. The project is anticipated to take 4 years in total, allowing for one-month visits over a three-year period to cover Stage One, and another year for Stages 2 and 3.

Conclusion. Considering the enormous environmental problems which arise from present day life, this salvage by photograph of the small reliefs of Venice might seen without interest, almost futile. But the writer believes that the artistic patrimony forms a whole, and that there are no minor problems. Certainly, there is no choice if it means to save a Titian, but a small anonymous work may have contributed to the creation of the environment from which arose the chefs-d'oeuvre. Culture is much talked about these days. Maybe culture is this: a modest craftsman who can't conceive of his house without an artistic touch...

By using algae to create works of art, Mrs. Takezawa is heightening awareness of
the multiple uses and values of these ubiquitous plants.

Algae, the Sea, and Human Life

Mieko Takezawa
Yumigahama Beach 1600, Minamiizu-machi, Kamo-gun, Shizuoka-ken,
415-01 Japan

Japanese, born January 19, 1949. Businesswoman, conservationist, algae researcher. Educated in Japan; graduate of marine biology course at Marine Research Center of University of Tsukuba in 1979.

I propose to travel to eight localities throughout the industrialized and developing world, to draw attention to the importance of algae and the sea to human life, through public exhibitions at which I would carry out the following activities: 1) Harvest algae from the sea nearest the locale and demonstrate how the local algae can be used artistically, 2) Show how the algae can be used as food, and prepare samples for exhibition visitors to eat, 3) Exhibit my Japanese algal art, and 4) Promote understanding of the importance of algae and the sea to human life through the use of informative posters and charts.

Many people are concerned about the ecology of terrestrial plants, but very few give much thought to the plants in the sea. My project is specifically aimed at heightening worldwide understanding of the importance of seaweeds and other algae to the sea and to human life.

The Problem

Most people who are reached by mass media are aware of the environmental destruction caused by atmospheric pollution. But fewer people recognize the fact that our oceans, too, are continually being polluted by urban sewage, industrial wastes and other cultural garbage. At present, however, marine pollution is advancing on a global scale. Moreover, it is not only a problem for people who live off the sea, but is a life-death issue for more terrestrial beings.

The ocean covers 70 percent of the globe, but the parts of the ocean that are most susceptible to pollution, namely the coastal regions, are also the places where practically all marine animals and plants exist. If we look carefully at the things that live in the ocean, we notice the existence of plant planktons and seaweeds.

Seaweeds are the simplest kinds of water plants. They are most generally classified as green, brown and red. Coloration varies greatly within each of the three major classifications; e.g., red seaweeds may be pink, red, purple or dark orange. The green seaweeds are found in relatively shallow water, the red seaweeds in deeper water. Throughout the world there are over 3,000 major kinds of seaweeds. In Japan alone, there are over 300, of which about 60, or one-fifth, have been eaten as food from antiquity. On the basis of this ratio, it is estimated that about 600 of the world's seaweeds are edible.

These simple forms of marine life play an extremely important role in the ecology of the sea, and they have an inseparably close relationship to the propagation of fish and other marine animals. Moreover, the seaweeds and vegetable planktons that have proliferated in the oceans from time immemorable are photosynthetic like terrestrial plants. This means that they absorb carbon dioxide and produce oxygen, and thus share the vital role which makes it possible for us to breathe.

Another contribution of sea plants is that they are becoming increasingly important as food. Most seaweeds contain surprisingly large amounts of protein, calcium, iodine, iron and a variety of essential vitamins. Every year, millions of tons of seaweeds, which have come to be called sea vegetables, are naturally produced along seacoasts throughout the world. Practically all of this spontaneous production is going unused, however, at a time when much of the world's population is malnourished, if not underfed. Even in Japan, where nutritious seaweeds have been eaten from prehistoric times, there is still considerable potential for their utilization by humans. Most terrestrial plants are limited by their need for fertilization and arable land. Algae, on the other hand, thrive freely in the fertile environment of the sea, and moreover they are easily cultivated.

Seaweeds also have many untapped uses in industry and medicine. Thus, seaweeds and other algae, which are basic links in the marine food chain, promise to play an umbilical role as a potential resource for humankind, and as a factor in marine conservation.

My Involvement

I live on the southern tip of the Izu peninsula, about 200 km from Tokyo. The area is part of Izu National Park, one of Japan's best known natural splendors.

The park boasts a beautiful coastline and many fine beaches, and its coastal waters are said to yield the greatest variety of algae — or seaweeds — in the world. Although I was raised in Izu only since my early teens, for much of my formative life its beaches were my playgrounds, and I came to love the sea as the mother it truly is. But over the years, as I witnessed the coastal waters near my home become increasingly polluted, I recognized the need to study the problem academically. Even when armed with formal knowledge, however, my basic concern remained emotional: I continued to be upset by what seemed to be a widespread lack of appreciation for the beauty of the sea and its life forms. It was at this point in my life, about six years ago, that I endeavored to create an art form using algae, through which I could make the kind of aesthetic appeal that would help open the hearts of more people to the scientific facts that aid a deeper understanding.

After studying marine biology at the Marine Research Center in Shimoda, on Izu peninsula, I proceeded to conduct my own research into methods of preparing seaweeds as food. I also succeeded in developing an art form using seaweeds, which unlike terrestrial plants are rarely seen and little understood by people not familiar with the sea.

Through exhibitions of my seaweed or algal art, mainly in Japan but also in Australia and soon in the United States, I have appealed to my audiences to witness the beauty of marine life as I have. From this essentially aesthetic starting point, I have endeavored through lectures and graphic materials to impress upon my audiences the importance of the sea as a natural resource, and to lead them to an intellectual awareness of the fact that the sea—the global commons which has been the mother of all life, including mankind—is being threatened by marine pollution caused by all of us, who in turn suffer.

The Project

I am applying for a Rolex Award to enable me to extend my activities internationally. I propose to travel to locations in Western Europe, Eastern Europe, West Africa, East Africa, East Asia, South Asia, North America and South America to draw attention to the importance of algae and the sea to human life by demonstrating how algae can be used as a food and as an art medium. As far as local conditions permit, at each of the eight locales, I wish to carry out the following plan:

1) Reserve a low-cost facility where I can both demonstrate algal cooking and exhibit algal art.

2) Publicize the time and place of the demonstration and exhibition.

3) Harvest algae from the sea nearest the locale.

4) At the publicized time and place:

a. Display the kinds of algae that are locally available.

b. Show how this algae can be used as food, and prepare samples for the visitors to eat.

c. Demonstrate how the local algae can be used in art.

d. Exhibit my algal art, in which algae harvested from seas around Japan have been used in designs based on the spatial concepts of *ikebana*.

e. Promote understanding of the importance of algae and the sea to human life and culture through posters and charts that show; the kinds, numbers, colors and shapes of seaweeds, the role of algae in the food chain of higher marine life, the nutrients and food values in seaweeds, the damaging effects of pollution on marine ecology, and the effects of marine pollution on human life.

While scientific solutions to our environmental problems are needed, I believe they are insufficient without some spiritual approach, which involves affective more than cognitive interaction with nature. The visual arts have a universal appeal; even peoples not familiar with the sea can appreciate the beauty of algae when it is presented in an aesthetically creative manner. Thus, algal art may be thought of as an emotional key that opens the door to a wider technical understanding of marine life in general. By adapting the spatial concepts of *ikebana*, the traditional Japanese art of flower arranging, to two-dimensional pressed-flower form, I have sought to depict life in the sea in a condensed and integrated form.

I have considerable experience travelling overseas. I also feel that I have sufficient diving experience and background in food and nutrition, as well as the physical strength and language ability to successfully carry out this project. With the aid on one assistant, I wish to exhibit at the eight localities over a two-year period, each locality taking about one month. Spacing the exhibitions over a two-year span allows the flexibility necessary to schedule each locale for a season when the local algae can be harvested.

Simple Solar Heating / Cooling System for Sub-tropical Regions

Milan Pospisil
Villa Delfina, Alessandro Curmi Street, Tal Virtu, Rabat, Malta

Czechoslovakian, born April 10, 1936. Physics teacher. Educated in Czechoslovakia; Ph. D. from Technical University, Prague, in 1975.

My project involves the design and construction of a system of air and water cooled solar collectors, connected to an accumulator which will provide for a three-day storage of energy. Used to provide hot water, this system will provide space heating in the winter. In summer, air collectors are used for ventilation, water collectors for the production of hot water. The basic orientation of the collectors is optimized for winter time.

During winter time, students in schools and people in homes in subtropical regions shiver with cold, even though the solar radiation that falls on the flat roofs and vertical walls of local buildings would be sufficient to heat the interior rooms.

In schools, the heat energy is required in the morning, whereas for homes, the heating requirement is mainly for afternoon hours. For maintenance of temperatures, the air cooled collectors are considered for the morning peak hours, and late in the day requirements. For periods when there is insufficient sunlight, the system of water collectors and the three-day energy storage facility is used.

Three stages of the project are considered, as follows:

1. A one room experiment. This should deliver data for evaluation and optimization of the equipment requirements.

2. A one floor experiment. This effort, to be conducted within a school building, should deliver practical experience, as well as reactions of the students concerning adequacy of the system.

3. Application of the system to an entire school. The successful operation of this phase will provide a model to be used by other schools facing similar heating needs and possessed of similar solar energy resources.

Assumptions

In our work, we are using data that are valid for Malta, where the project is to be carried out. These include:

1. The daily temperature in winter averages 10°C.

2. Three or more successive rainy or cloudy days in Malta is an extremely rare phenomenon; it is on this basis that the accumulator design elements are configured to store energy over this period.

3.   The lowest average daily amount of direct sunshine is five hours, during the month of January.

4.   The average power density of solar radiation during time of direct sunshine *in winter* is 800 W/m².

5.   The 'climate control' requirements are that heating should be provided from October to March, and cooling from June to September.

6.   If the average solar energy available is combined with the normal bodily heat of average classrooms of students, an energy factor of 20 kWh/day per class is required.

7.   Power delivery should be designed to peak at 8 a.m., as classes begin at 8:30 a.m., and therefore needs to be delivered from storage.

8.   Temperature constancy during the day is maintained by air collectors; only on cloudy or 'non-solar' days does energy have to be taken from storage.

9.   Energy storage tanks are recharged during the hours of afternoon sunlight, and on Saturdays and Sundays.

10.   As noted above, three days of energy storage is considered sufficient.

11.   The energy of diffused radiation has not been taken into account in the design of this system; it remains as an obvious plus to the underlying system design, which is based on direct radiation.

Energy Estimates

Consumption Requirements for one class.	20kWh/day
Energy received by one square meter of collectors during 5 hours direct sunshine.	4kWh/day
Efficiency of the overall system in delivery.	50%
Useful energy (available for use).	2kWh/day
Collector area required for one classroom	
Water cooled = seven square meters	
Air cooled = three square meters	
Size of tank required for temperature change of 40°C, and stored energy of 50 kWh.	1 cubic meter

Construction

The water collectors will be situated on the flat roof of the school building, set at a slope of 55° to the horizontal, facing south. The slope is optimal for energy reception during January. The energy accumulator is placed above the collectors. It operates with a thermosyphon, to activate it in case of power cuts, which are frequent in this area.

The air collectors are vertical, and situated on the building walls, under the floor level of the classrooms. Their non-optimal slope results in a loss of 20% of the potential solar energy during the month of January.

It is our desire to be able to record all data on a personal computer, in order to register, compute and analyze the pertinent measurements (temperatures, radiation levels, heat/cooling output efficiencies, etc.) as we proceed.

Beaver Transplantation from North America and/or Eurasia to North Africa

Gerard Johannes Timmer

Route 1, Box 135, Goldendale, Washington 98620, U.S.A.

American, born January 12, 1913. Retired abdominal surgeon. Educated in Indonesia and the Netherlands; Ph. D. in 1942 from Marburg/Lahn.

This project concerns the translocation of beavers from North American and/or Eurasia to North Africa in order to counteract the rapidly progressing deforestation and desertification of the African continent. Wherever there exists even a small but continuing source of water, beavers will be able to start and create ponds. The only materials required are small trees, shrubs, mud and rocks to build the water-retaining dams and their own lodges. The airlift and subsequent transport of the beavers should be arranged by the United Nations under the motto: DROP BEAVERS — NOT BOMBS.

Coordinating many steps by many people, I wish to arrange what I believe will be a successful translocation of beavers to familiar climates, above the 30th parallel in North Africa. Live-trapping of 3-4 year old beavers in 'donor countries', followed by immediate air transport to pre-selected recipient areas, offers a good chance of success. Personal observation of beavers has convinced the author that these animals are wise, persevering, tenacious, and even intelligent. They are an excellent symbol for this totally peaceful project with many beneficial side-effects. Beavers do not harm other animals; their only defense when attacked is the flight under water to their lodges, or dens. When 'donor' countries and 'recipient' countries understand each other, actual transplantation can begin.

The beauty of this totally peaceful project lies in the fact that after the transport and delivery, there is very little else to do: the beavers will do the work. Recipient countries must handle potential poachers by declaring the beaver a protected animal, explaining its local purpose, and understanding that the results are long term. Beavers first create ponds; these ponds, after the beavers have left for "greener pastures", become meadows, and finally forests will appear: a long process.

Physically, beavers are more impressive than most people realize. They average about 4 feet long, and weigh between 40-60 pounds (the largest on record was 110 lbs). The jaws are very strong, with 2 large upper and 2 large lower incisivi that grow constantly, making it mandatory for the beavers to use them always. With webbed hindlegs for swimming, clawed or toenailed front legs for manipulation, the beaver is able to stand, and walk like a human. Large lungs and liver allow the beaver to stay underwater for up to 15 minutes, during which time they can travel up to 1/2 mile.

They live in freshwater, can move on land by walking on 4 feet or just the two hindlegs, and will use their broad tail as a tripod point when standing up to cut down a tree. Monogamous, the females are generally the builders, while the males collect the building materials, working from dusk to dawn. While water is a requirement, beavers will work to provide it, given an initial small source, such as a stream or a brook, or even a spring.

In general, they construct two types of housing; dens and burrows (tunnels) along streams, or beaver lodges (houses, mounds) on river banks, midstream or in the ponds. The size of the ponds can be in the tens of acres. During the winter, temperature in the beaver house is around the freezing point. The warmth emitted by the beavers prevents the plunge-hole from freezing up.

Transport pathways (beaver slides) can be up to 6 feet wide. When constructed between bends of a stream, canals are produced: a shortcut for fast transport of logs and water.

The well-known dams are marvels of construction: woodsticks, twigs, tree-trunks, stones (up to 8″ in diameter), mud, and finally, waterplants. The soundness and integrity of these dams is well and constantly guarded. (Trappers take advantage of the beaver's quest for maintenance by making holes in the dams for the placement of traps.) This need for water is essential; no water — no beaver. If for any reason the water supply is cut off, the beavers will move on to an area where food and water is available (this 'trek' being the most dangerous aspect of their existence).

The steps in the transplantation process are as follows:

1. Live-trapping, with Bailey 'live-traps' in donor countries.
2. Preparation for transport (treatment for parasites).
3. Airlift to recipient country.
4. Resettlement in selected test-areas: overland transport and release of beaver pairs, or, in inaccessible countryside, helicopter transport and air-drop in wooden crates. Choice of location involves valley width (the wider the better), and valley grade (0-6% excellent, 7-12% good), and preference for slower rates of water flow (reduced erosion rate, higher sediment deposition rate and water infiltration).
5. Management follow-up, to include seasonal timing of beaver colony counts by counting beaver food caches, and preventing cave over-population by trapping.
6. Economic benefits, to include trapping for pelts, castoreum (the beaver 'musk') for the perfume industry, and recreation (trout fishing, ranching, etc.).

It seems logical to approach the problem of actual arrangements for beaver transplantation via the good services of the United Nations. The recent discoveries, with the help of radar, of ancient rivers in North Africa and Asia Minor will help the peoples of these lands, who are troubled by increasing drought, to look for new ways to combat drought's devastation. Just as the dolphins are now helping mankind, so the beavers could via vegetation improvement.

Top: Speciosum Lily — *Lilium speciosum* — Family Liliaceae: Sepal (29), Petal (22), Pistil (23), Stamen (30), Flower Bud (3).

Bottom: Pitcher Plant — *Sarracenia purpurea L.* — Family Sarraceniaceae (in longitudinal section, with 1 sepal removed, petals and stamen fallen.): Superior (33), Ovary (17), Ovules (18), Placenta (24), Locule (14), Stigmatic surface (31), Style (32).

A Botanical Glossary of Vascular Plants, Illustrated by Photos

Virginia Vernon Anderson Crowl

248 West Lincoln Avenue, Delaware, Ohio 43015, U.S.A.

American, born June 20, 1909. Naturalist and retired Science Librarian. Educated in U.S.A.; B.A. in Biology and English from College of Wooster, Wooster, Ohio, plus numerous courses elsewhere.

I wish to produce and publish an illustrated glossary of botanical terminology, manageable in size and price, which would be extremely useful to both high school and college students of botany, to taxonomists, ecologists, naturalists, agriculturalists, gardeners, and interested lay people.

There are many plant glossaries in print, but none of the sort envisioned here. The most common are either abbreviated lists included in local flower books or pamphlets generally used by casual naturalists. At the other extreme are large, detailed hardcover books devoted specifically to botanical terms, but having restricted use, mostly to research botanists. There are good lists in regional taxonomies, but few with more than line drawn illustrations of a small portion of the terms that are defined.

The diagrams in most glossaries are helpful, but they are just that — diagrams; usually small, and unable to show actual plants. The advantage of photographs over diagrams lies in making use of actual plants and plant parts labeled to show the structures defined. A single term used to describe a variety of similar structures will be labeled in several different photographs to show the range of variation encompassed by that term. The glossary will indicate the location of each photograph where the term is illustrated.

My literature search has revealed no published materials like that which is proposed here.

I have worked for years with professional botanists, college students on field trips, high school students and on a variety of field trips. I have found a plateau of knowledge beyond which the nonprofessional generally does not care to go because of the disproportionate amount of effort needed to tackle the separation of species by means of the taxonomic keys. What I am proposing should make that effort easier.

It would seem appropriate to make this glossary available to the public in National and State Parks, Nature Preserves, etc., as well as to the memberships of organizations such as the Audubon Society, The Nature Conservancy, Sierra Club, and other similar groups.

In over 30 years of traveling and photographing plants, I have amassed a collection of 15,000 slides from which this book will be formed.

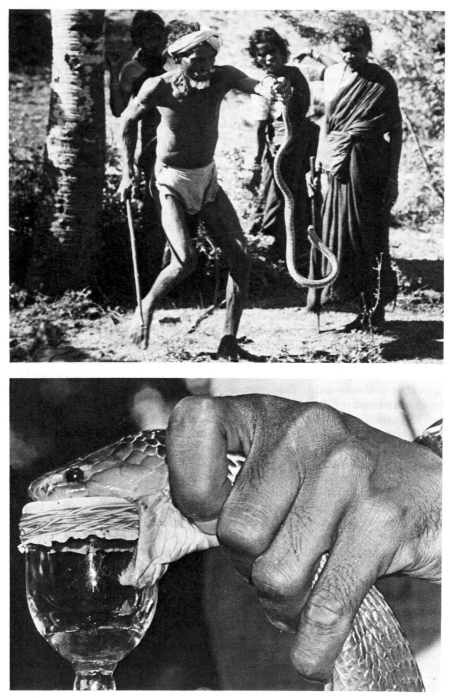

Top: The art of defying death. An Irula with a freshly caught cobra.

Bottom: Venom extraction from a cobra. Making the most for the tribals from a sustained yield resource.

Establishment of a Snake Venom Production Industry for Irula Tribals

W Romulus Whitaker

Honourable Mention — The Rolex Awards for Enterprise — 1984
Madras Snake Park, Raj Bhavan P.O., Madras 600 022, India.

Indian, born May 23rd, 1943. President, Irula Snake Catchers Industrial Cooperative Society Ltd., Head of the Madras Snake Park, and The Crocodile Bank. Educated in India and U.S.A.

In conjunction with the Madras Snake Park, which I head, I have formed the Irula Snake Catchers Industrial Cooperative, as an organization designed to enable the Irulas to continue to utilize their skills and experience, and at the same time to improve their precarious economic situation. By tradition and heredity, the Irulas of Chingleput District, Tamil Nadu, are a hunter/gatherer tribe of people who for generations have specialized in the art of catching snakes, rats, termites and collecting herbs for medicinal use. Because of diminishing natural resources, they have found themselves without a livelihood, forced to subsist at a hopelessly inadequate level with no land for cultivation and little chance to compete on the job market. Previously, many were engaged in catching snakes for the snake skin industry. For ecological reasons, this became illegal with India's imposition of the Wildlife Protection Act of 1972.

The aim of the Cooperative is to enroll at least 1,000 Irula members and their families, employing their skills in an economically productive way within the bounds of the law. Such essential services as health, education, and savings/loan facilities will also be provided. Since 95% of the Irulas are illiterate, these services will be designed in the specialized manner required by the Irulas. Self help will always be emphasized, and technical guidance only given where needed.

In December 1982, with the permission of the Tamil Nadu Government, the Cooperative began its first tribal self help scheme, a venom extraction project. This project began with an initial membership of 26 Irulas, and has since grown to 52 members. It involves four species of poisonous snakes being caught by the members, who then sell them to the Cooperative.

When bought, the habitat and district in which the snakes are caught are noted, and they are weighed, measured, sexed and clipped with a code number. Venom is extracted several times from these snakes, over a period of a month, before they are then released in reserved forests outside Madras.

The venom that is collected is vacuum dried and sold to recognized institutions in India for the production of antivenom serum. This serum, which is a vital medical product for a country like India, is scarce. It is the only effective treatment for venomous snake bite. In India there are at least 10,000 deaths annually reported, in part due to the shortage of antivenom serum. There is also a potential foreign market for snake venom; it is used in the production of several allopathic medicines, including coagulants and pain relievers, and is being experimented with in the treatment of such diseases as cancer.

It will take several years until the Cooperative can operate successfully and supply India's venom needs. Meanwhile, a considerable investment of time and money is necessary to make the venom project lucrative for the Irula members. Eventually the Cooperative plans to start catching and extracting from the different venomous snakes that vary with each region of India. The procedure will be to operate a mobile venom van that will travel to different parts of India to catch, extract from, and release the venomous snakes indigenous to that region.

This venom extraction project has a number of unique characteristics. It is the first Indian project that is legally catching wild snakes for venom extraction and has made venom production a self help cottage industry for tribals. It is the world's only venom production scheme which liberates the snakes after extraction. The venom industry in India has been operated in the past by entrepreneurs who illegally buy snakes from the tribals, with no interest in the fate of the tribal or the snake. The Cooperative, however, ensures the maximum returns for the tribals. As a member of the Cooperative, the Irula receives money for both the snake and also its venom, effectively eliminating the middleman. Ecological effects on the snake population are minimized, as the snakes are released in healthy condition after extraction. The code number that is clipped on each snake ensures that if the snake is caught again, it will be re-released and not used for extraction. The scientific data that is collected on each snake will add considerably to our knowledge of the distribution and population dynamics of local snakes and guarantee that the population is not over-exploited.

The conservation of Indian snakes, however, cannot be attained until the illegal snake-skin industry is brought under control. As long as European, Japanese and American markets exist for snake skins, this industry will continue to compete with the Cooperative for the snakes caught by the tribals. Only by expanding the venom project and providing other alternate employment can the Cooperative wean the Irulas away from illegal and ecologically destructive exploitation of snakes.

Other activities planned by the Cooperative for Irula employment, including crocodile, deer, butterfly and medicinal herb farming, will ostensibly result from the profits and infrastructure created by the venom project.

Research in Biology of Malaysian Rajah Brook Birdwing Butterfly and Establishing its Farming.

Angus Finlay Hutton

Garaina Farms, Scrubby Creek Road, Gympie, Queensland 4570, Australia

Australian, born April 8th, 1928. Retired Tea Planter and Wildlife Officer, presently farming. Educated in England and India.

The National Butterfly of Malaysia, The Rajah Brook Birdwing, *Trogonoptera brookiana albescens* (Rothschild — 1895), is rapidly becoming endangered due to excessive collecting for the trade and destruction of its habitat for 'development' of plantations, logging, etc. Incredibly, though the Butterfly has been around for nearly 100 years, since it was first described for Science, its early stages and life history are still unknown and even its larval food plant is not known (though it is suspected to be a species of *Aristolochia*, on which other related species feed exclusively).

My project is, firstly, to make an exhaustive study of the Butterfly in the jungles of Malaysia; by intensive and patient observation of the adults, to determine its larval food plant and then to observe and study its early stages (with full photographic coverage), from Ova to Imago.

Secondly, using the information obtained, it is intended, by means of habitat enrichment and education of the Aboriginal people who are most closely involved with collecting, to establish "Butterfly Farms".

Thirdly, I intend to train Malaysian Wildlife Officers in the special techniques of butterfly farming.

The Rajah Brook Birdwing is the largest Butterfly in Malaysia, and, being particularly beautiful, it is greatly sought after by collectors and manufacturers of souvenirs alike. For the past 20 years or more, collection of wild specimens has increased at an alarming rate to satisfy this demand, and it is clear that butterfly numbers (not only of this species, but of many others that are used in the 'trade') have been greatly reduced and are rapidly reaching the endangered situation.

Habitat destruction, both for logging and development, for new plantations and other forms of agriculture as well as human settlement, has also been a major factor contributing to the decline.

The Butterfly Trade in Malaysia. To understand the problem, it is necessary to briefly describe the organisation of its operations.

At 'jungle level', Aboriginals, mostly of the Senoi and Jakun tribes (commonly known, incorrectly, as Sakai), have put aside their blowpipes in favour of Butterfly nets, and scour the jungles for Butterflies and other insects. Everything is 'taken', and the percentage of unusable and badly damaged specimens is considerable. The collectors then take their catch to buying points, usually Chinese controlled, where they are paid according to species, size and quality. Rarely does anything exceed ten cents per specimen. The middleman carries out more sorting and some rejection of obviously badly damaged specimens, adds his 300-400% mark-up, and forwards the specimens to the processors in Kuala Lumpur, Penang, Singapore, Hong Kong and elsewhere. At those points, more wastage takes place and specimens are set into cabinets, embedded into plastic, or used in many other ways for souvenirs for the tourist trade. At this point, the mark-up in price is at least 500%. Few, if any, of the specimens are sold as 'papered' for serious collectors, with any data on them, and it is clear that identifications and nomenclature leave much to be desired.

Butterfly Behaviour. Years of observation in India, East Africa and Papua New Guinea have given me an intimate knowledge of the habits and lives of tropical butterflies. For instance, by observing the flight pattern, I can tell whether a female is seeking a mate, feeding, seeking a place to lay her eggs, or just travelling between two points.

Observing a male can similarly indicate his occupation and often, if there are pupae of the same species in the area, the male can recognise a potential mate and will settle on the pupa and stroke it, though ignoring the male pupa. Where male and female butterflies congregate in the same area, often it is only the males that come to ground level, the females preferring to fly high.

In the case of the Rajah Brook, this difference in habit results in about 100 males being captured for every female. The females are found at a higher elevation in the forest, usually in the canopy, but they do descend in the late evening to ground level, where they will still not accept the baits that attract the males at any time.

Study Methods. Taking advantage of past experience and using very powerful binoculars, plus infinite patience, together with a stoic attitude toward mosquitoes and other discomforts, I shall determine the prime habitats of the females. Under constant observation, they must sooner or later lead me to their unknown food plant. As this is certain to have its foliage high in the jungle canopy (up to 200 feet above ground), access can be a problem. However, the jungle Aboriginals are skilled climbers, and I'm not too bad myself.

The first priority is to get specimens of the leaves, and then carefully trace the stems to ground level; this, too, can be difficult in thick jungle. Sometimes young plants can be found growing, or even seeds or fruit. Cuttings can be taken and planted in small baskets, or plants raised from seed, or even from transplantings. These small baskets can then be hauled up into the canopy by a strong cord fired over a branch with bow and arrow. Passing female butterflies, preferring to lay on fresh lush growth, become attracted and oviposit on the leaves. It is then relatively simple to lower the baskets down to ground level and breed out the butterfly in a cage without more ado. It is a proven method I have used for many years to deal with these high flyers in Papua New Guinea, and I foresee no difficulty using the method in Malaysia. Sometimes from a ridge or hilltop or above a cliff, it is possible to get an eye level view of the jungle canopy, which makes the task of observation much easier. This part of the project could take several weeks, and in the case of plant propagation, would necessitate return visits to the locality for some years, for short periods.

Capture of gravid females is another method of studying the life history, particularly if the caterpillars will eat and survive on an alternate, commoner larval food plant of the same Genus. This method was also used successfully in Papua New Guinea to determine the life history of *Ornithoptera goliath,* a closely related Birdwing, using fertile eggs, expressed from a wild-caught female. It is almost impossible to get them to lay in captivity. Sometimes one has the good fortune to stumble on feeding larvae, or find pupae hanging up; in such cases they can be bred out and all stages recorded. I have succeeded in hand pairing an ex-pupa virgin female Birdwing with a wild caught male of the same species (and, for an experiment once, with another sub-species, thereby producing a hybrid), and from this obtained fertile ova.

Farming Methods. Basically, environmental enrichment is the key issue and it is remarkable how effective this is in increasing numbers in the wild. It is environmentally sound, and is often best put to full advantage in secondary forests, where logging has been completed. Even the high flyers will change their habits and oviposit on such planted areas once they have located them. This makes the task of collecting up to 50% of the pupae much easier, ensures against overstocking, and, ultimately, it produces an on-going renewable resource of top quality, thus making the capture of wild specimens with baits and nets unnecessary.

Full details of farming methods, used successfully in Papua New Guinea for the past 8 years, are in the attached literature. I would mention that the system of Butterfly Farming that was pioneered by me in Papua New Guinea, starting back in 1974, has proven successful, both as to the conservation of the species and their environments, and as a valuable source of income for the farmers. It has stabilized prices, and by cutting out the middleman to a great extent, has improved the lot of the farmers. It has also proven to be a self-policing influence in the villages; the former rapacious plunderers and smugglers of PNG fauna are a thing of the past. Thanks to the actions of villagers protecting their butterfly interests, most of those people were deported from Papua New Guinea by 1978.

The principle of combining Conservation with Development, as enshrined in the World Conservation Strategy, is very relevant. The American National Academy of Science Panel, which visited Papua New Guinea, in 1982 expressed the view that my pioneering methods could be used for the benefit of many other developing countries in South East Asia and Latin America.

That my work did not go unnoticed by the Papua New Guinea government was amply demonstrated when, in 1977, the Prime Minister, Mr. Michael Somare, presented me with the country's highest honor — The Papua New Guinea Independence Medal — of which I am very proud.

A Simulated Natural Environment for the Study and Conservation of Marmosets and Tamarins

Jürgen Wolters

Department of Ethology, University of Bielefeld, Morgenbreede 45, 48 Bielefeld, West Germany

West German, born February 26, 1951. Doctoral candidate in biology. Educated in West Germany; M.Sc. (Biology) from University of Bielefeld in 1978.

Marmosets and tamarins, the *Callitrichids*, are the smallest true primates, barely larger than a squirrel. There are twenty species and about forty sub-species distributed throughout the tropical forests of South America. Today, these small monkeys have become an index of the destruction of the South American rainforests; six of the twenty species are directly threatened with extinction, and it is feared that each year more species will become endangered.

The solution to this problem is naturally difficult, and involves the culture and economy of an entire continent. However, even if one concentrates solely on designing conservation measures directly for the marmosets and tamarins, one is faced with inordinately difficult problems. The small size and arboreal natures of these animals hinder the behavioral and ecological investigations upon which such conservation must be based. Therefore, research on captive populations of marmosets and tamarins in carefully planned environments takes on a special meaning. It is the design and innovation of such captive animal research studies that is the focus of this application.

I have set up the Bielefeld Callitrichid Station as a research facility at my university in order to carry out studies on captive populations of Callitrichids, with the ultimate aim of designing appropriate conservation measures to help the threatened populations in South America.

The station, which is now eight years old, houses four species of marmosets and tamarins (currently 150 animals in total), all of which are immediately threatened with extinction in the wild.

We are carrying out research on the social structure, social dynamics, communication, and, in particular, the development of the young and their integration into the group as fully competent members. We have also done, and are continuing to do, studies on improved methods for maintaining and breeding these animals in captivity, including investigations on dietary physiology, optimal structure for the artificial environment, and rearing techniques, all with the aim of simulating in important respects the conditions under which these monkeys live in the field.

We have developed an artificial rearing system that is adapted, as far as possible, to the natural rearing conditions of Callitrichids and that avoids the normal pitfalls of classical hand-rearing methods. We have already used this facility for cotton-top tamarins, *Saguinus oedipus,* and are now preparing the system for use, if necessary, on the pie-faced tamarin, *Saguinus bicolor bicolor,* one of the rarest primates in the world. At present, there are not more than a dozen of these tamarins in captivity. The Bielefeld Station recently obtained four of them, three of which may have problems in rearing young (one female is quite old, and two animals may lack early social experience valuable in parenting). Our special artificial rearing system can help not only to save the life of expected infants, but can also provide a solid basis for normal development of the young (which should contribute to their own later success in breeding).

By using maintenance techniques developed at the Station, we have already built up self-sustaining breeding colonies of two endangered species; the cotton-top tamarins, and the white-faced marmoset. With the latter, we have recently organized an international breeding program to better insure proper genetic management of the captive world population.

We regard our applied work and basic ethological research, however, only as a first step in providing information useful to conservation efforts in the field. Conservation measures must certainly be founded on more than the biology of individuals, or even of isolated social groups of individuals, the normal basis for behavioral research in captivity. Such measures must be based on populational studies, specifically the interactions between groups.

To this end, we have designed a research facility that is entirely new in concept, and which will allow the study not only of groups and group members, but, for the first time with Callitrichids, the interactions between groups. The primary innovative feature of our new facility will be that the cage-systems which house the individual monkey groups will be able to be connected with each other by a network of enclosed gangways (up to several hundred meters long) that can be connected in any desired combination. This means that a social group will be living in an environment, albeit artificially created, that can be constructed to simulate the spatial and temporal characteristics of its natural habitat. Thus, the qualities of different territories can be altered by varying not only the physical parameters (tactile, visual, acoustic, olfactory, etc.), but also by varying the amount and kinds of food available, density levels (virtually isolate to actual crowding), and other factors.

Our ability to measure the effect of these factors via, e.g., hormone levels sampled from urine, could not be duplicated in the field. By making the facility available to other primate researchers, particularly those from countries with surviving Callitrichid populations (i.e., Brazil and Colombia), we will be developing conservation measures that can actually be put into effect.

Top: High in the mountains of Papua New Guinea, the Agaun Cattle Farmers built and use this water supply drain from a local stream to provide electricity.

Insert: At this river, 5 km from the village and having a 50-meter drop, the group will be building a new turbine to bring 15 KVA to the village for improved facilities.

The Agaun, Papua New Guinea, High-Mountain Village Hydroelectric Project

Alwyn Mamuni Taigwarin

Via Alotau, Agaun, Milne Bay Province, Papua New Guinea

Papua New Guinean, born August 14, 1955. Manager, Agaun Cattle Farmers. Educated in Papua New Guinea.

The Agaun mountain area is one of high rainfall, and until recent years, one in which our people have had little contact with the 'outside world'. Beginning about a decade ago, we commenced various undertakings by ourselves that would broaden our contacts with our country, and improve our own quality of life. I was leader of a local, unpaid team of volunteers who constructed the first, and only, access road to our isolated village. Made by hand and digging sticks, this road is now completed and able to take bullock and horse traffic. We have started to widen and smooth this road, to enable pack horses to bring in stores, and, in the future, to allow tractor transport. Part of the reason for originally building this road was to allow us to introduce cattle to our small local economy. I introduced cattle for the first time to the Agaun Mountain Area in 1972; over the last ten years, we have developed six associated cattle projects, as members of the Agaun Farmers Group.

The current project is called the "Agaun (Papua New Guinea) Mountain-Village, High-Rainfall Area Hydro-Electric Project". It is based on our solid success in building, on a small local stream, two small hydro-electric plants that supply a minimum amount of power for our Farm Building lights and our small deep-freeze, in which we preserve our locally grown meat for later sale, as well as the butter we produce. Our local people have a protein deficient diet, and this supply of meat and milk products is important for our health.

The plan described in this project, for a larger home-made Hydro-electric Turbine, to be built (funds allowing) on the larger local river, will greatly benefit the environment, and provide power that will enable us to run hospital and school lights, as well as power for the small sawmill, plus power for more meat freezing.

Our mountainous homeland has only since 1952 begun to emerge from the Stone Age, with the establishment of our first school (the last in the province) and Church. The tools used by our fathers include the stone-axe and the wooden crow-bar for digging the ground; and these are still in use here. The first Government patrol officer arrived in 1963, and a local Government Council was established in 1966. A medical team estimated at that time that the average local life-span was 41 years, due especially to a deficiency of protein in our diet. As one answer to this problem, our young farmers determined to introduce to these mountains (1,750 meters) a herd of beef and dairy cattle.

All the young men in my group of the Agaun Cattle Farmers had, by 1972, completed grade 6 at school, but none of us was among the very few who have been able to go to Secondary education. We had no prospects for employment in the towns, so we have decided to remain in our home area, to marry here, and to stay here for our lifetimes. We do not wish to swell the numbers of unemployed in the towns. Here, high in our mountain home, we have spent ten years making our own cattle projects. We are quite cut off from the coast; our place is too small to warrant the huge expense of a machine-made road from the sea up to 5,000 feet, through rugged, near vertical slopes in places, and deeply cut ravines.

The airfreight to our small airstrip is now 0.5 kina/kg, and the cost of diesel fuel, for powering our lights, freezing equipment and small sawmill, is prohibitive. We need cheaper power for these, and also for the local hospital, church and school. To this end, we have already completed and use two home-made turbines on a small local stream. Our project now is to enlarge this concept by harnessing the larger river nearby.

The novelty of our project is that we will use the rear-wheel section of discarded tractors as turbines. We have no funds to buy factory-made and much more complicated turbines. We recommend this idea to any other friends in non-developed countries, with the good fortune to have high mountains and sufficient rainfall; using the natural, ever-repeating cycle of rainfall, it will provide electric power, cheaply and easily, for villages far from the cities and towns.

Our existing hydro-electric plant serves as our prototype and is explained as follows. Made at negligible cost and considerable effort, it is built with scrap-metal machinery parts, sprockets and chains, a discarded belt-drive wheel, half-inch galvanized water pipe spokes on an eleven-foot overhead water wheel, a shaft from a disused boat, and a 3KVA alternator salvaged from beneath the debris of a volcanic eruption. The water supply drain is hand made, with no expensive drain, running off from the stream at normal level, with overflow and flood gates to minimize flood level changes. It has 35 feet of water pressure pipe dug into the mountainside, set in one square-yard of river stones with a six-inch diameter hollow concrete core going beneath the water table to prevent subsidence. The water to the overshot wheel is delivered by bent corrugated roof iron helped by open-ended 20-litre drums.

The second alternative power unit at the same site is a home-made turbine driven by water from the pressure pipe. It is of very simple construction; a 1.5 inch shaft with two circular plates welded to the center of the shaft, and curved plates welded to the discs to take the water under pressure. The casing is cut from a 200 litre drum and shaped and welded at the seams, with a removable cover for the maintenance of the turbine blades. A flywheel is attached to the end of the shaft, with rubber "v-belts" driving the alternator, seated above the turbine for easy variation in the belt tension. The two bearings are protected from water intrusion by small discs, which in spinning throw off any water leak from the turbine case. The voltage is regulated by the pressure pipe taps, and is used at a constant fixed load.

The current project involves building a similar turbine for the large river nearby. About 5 kilometers away, the river descends over several waterfalls, totalling a drop of some 50 meters, with a dry-season potential of 15KVA. The turbine is simply the two rear wheels of a discarded tractor, with the differential removed and the two splined shafts cut slightly shorter and welded in the center, to provide a direct connection from one wheel to the other. The wheel hub on one side has a series of flat steel plates easily welded between the flanges, and the other wheel hub requires no alteration to provide an excellent flywheel, from which the rubber "v-belts" can run directly to the alternator. In this case, the alternator (400 volts) needs to be at double the required voltage to compensate for the voltage loss from the turbine site to the village. At the village, the voltage can be stabilized without expensive equipment, by simply having a fixed load, and a voltage adjustment load placed on resistance wiring (such as electric cooker elements) which provides incidental, but useful, heat.

There is minimal finance available for this project, which is basically very cheap. Some items will have to be bought; the 15KVA 400 volt alternator, the 4-inch galvanized water piping for power line pylons from the turbine to the village, the copper/aluminium power cables, the one-foot steel flanged water pressure pipes, and the 5KVA electric motor for driving the small sawmill.

In this vicinity, there are three rivers of similar size, all of which may be used repeatedly at successive levels as a source of future power for our Agaun area.

This project is put in the category "Environment", because we wish to preserve the naturalness of our inherited environment, without unnecessary noise and pollution, and at the same time improve the life of our people, in this very remote and technically undeveloped area.

Our enterprise does not compare with that of those who engineered the micro-chip, or the control of nuclear power, but we present our small project for what it is worth in human terms.

I make this application to The Rolex Awards for Enterprise on behalf of my friends in this and other similarly undeveloped countries, to encourage the inexpensive use of the natural gift of water. It is a means for our countries to enter more fully in the modern world of technology, while at the same time maintaining our heritage of closeness to the natural elements and unpolluted surroundings that our ancient cultures have bequeathed to us.

Desert Irrigation from a Simple Solar Desalting System

John Albert Stephen

P.O. Box 931, Soldotna, Alaska 99669-0931, U.S.A.

American, born May 17, 1935. Commercial artist and wilderness canoe guide. Educated in U.S.A.

I call the work I am currently doing "Project Desert Life". It seeks, as the name implies, to bring agricultural capabilities to certain arid portions of the world.

The project centers on a system designed to desalt sea water and to function as an irrigation system, all in one continuous stream operation. The desalting portion of the operation is based on a solar distiller composed of easily prepared materials that are available, for the most part, from commercial plumbing markets. In addition to providing fresh water, which is flowed into an irrigation component, the system produces salt and other minerals as natural and economically useful by-products.

In certain arid regions, the lack, or scarcity, of fresh water prevents optimum utilization of the land, even though large quantitites of salt water may be readily available nearby. The massive and expensive desalination plants built in some regions, notably the oil-rich Middle Eastern countries, attest to the urgent need for better land utilization for agriculture to provide for expanding populations. Such systems are available to a few lands, but are beyond the means of many localities whose needs are equally as pressing. This project seeks to provide fresh water from salt water in those areas where such a system can be used effectively.

I have designed a system, and built a small working model of the design, that proves the distillation of sea water into usable fresh water within the context of an irrigation system is possible, using only low-technology, inexpensive equipment.

The fundamental requirement of providing an expansive surface area of water for evaporation is met easily with an ABS, or similar type plastic material, readily and inexpensively available from a wide variety of sources. This material is formed into long, rectangular aqueduct basins of flat or round bottom construction. Perhaps the simplest manufacturing approach to this requirement is through the splitting in half, lengthwise, of twelve-inch ABS type plastic tubing, to form the half-round aqueduct basin sections.

In either case, flat or round bottomed, the aqueduct is fitted at each upper outside edge with one-inch rigid plastic tubing that has also been split in half lengthwise. The one-inch, half-round rigid (ABS) plastic is welded with an ether-base glue to this outer edge of the aqueduct to form a fresh water collector gutter.

In the next step, a clear glass or plastic cover 'lens' is formed into an Λ-shaped roof. This roof is fabricated so that its lower outside edges will nest easily in the one-inch, half-round fresh water gutters at the outer rim of the aqueduct base.

The ends of this construction (which can be of virtually unlimited length, depending on terrain levels) are then sealed so that no wind or air currents can flow about the inside of the distiller.

Salt water is piped into the distiller base unit tank, where its level is constantly regulated by a float valve. Solar energy, shining through the 'lens roof' of the aqueduct, distills the salt water into fresh water through accelerated evaporation. This water collects on the underneath side of the roof. The fresh water trickles down the side of the roof, and collects in the half-round gutters.

The collector gutters are fitted about every three or four feet with quarter-inch diameter drain hoses (trickle tubes). These feeder hoses carry the fresh water from the distiller to the soil where the vegetation is planted.

Length of the units can be determined by the installer. The materials I have referred to here are normally available in twenty-foot lengths. Apart from the forming of the "lens cover roof", a simple process, manufacturing requirements are minimal. It should be noted that the lens cover must be removable, so that mineral deposits can be retrieved and the units cleaned when necessary.

As protection for the unit against high winds, etc., the lens cover is securely wired into place with non-rusting nichrome wire. In areas where drifting sand is a problem, the lens cover should be sealed to the aqueduct base along the outside edge with readily available (and easily removable) ductal tape to keep sand and dust out.

The environmental philosophy behind this project is as follows: The need for expanded farm lands is evident in most countries. Valuable timber land is often sacrificed to agriculture as a result. Resources being spent to fight such agricultural expansion by environmentalist groups could, at least in part, be better spent on reclaiming desert land for farming, this taking the pressure off the forest acreage.

I have already made a small working model of this system, and know that it works as described. I should now like to install a working pilot project either in the state of Texas, or in the Baja Peninsula, Mexico, in order to gather data on a complete experimental cycle.

The Keyline System of Land and Water Management

Percival Alfred Yeomans

11a Dalley Avenue, Vaucluse, New South Wales 2030, Australia

Australian, born March 8, 1905. Keyline consultant; rural, mining and urban projects.

"Keyline" is a system of land and water management that takes control of all water that falls on any productive area, or can be brought to fall on it, so that it is used to the maximum productive benefit, and never allowed to act destructively; by so doing, it cultivates the land on the basis of a continual enhancement of its fertility and productivity. The Keyline System is applicable to both fresh water and reclaimed water, storm water and effluents.

My project aims at securing world-wide recognition of the principle and practice of the Keyline System, preferably through the establishment of a Keyline Education Foundation.

Not all elements of the Keyline System are novel, but the synthesis of these elements is. Its principles have been applied, and we wish now to demonstrate it to the governments and people of the world.

The name "Keyline System" derived from my recognition that the natural contours of the units forming the consistent pattern of the landscape present the possibility of logical planning. I named three shapes of the land; the primary valley, the primary ridge and the main ridge. The normal break of slope of the primary valley is named the *Keypoint*. Contour-related lines to left or right define the *Keyline of the Valley*.

The system based on this concept was first worked out and put into practice in undulating land west of Sydney, Australia, with spectacular results. It has since been proved appropriate for many and diverse landscapes. Numerous publications have explained and noted results, and Keyline System projects have been carried out in Australia, New Caledonia and the Texas Panhandle in the U.S.A.

Various applications of the Keyline System include:

1.   The original water-and-land development project, in a medium rainfall area having a significant range of elevation. The key principle is the identification of the Keypoint at the break of the slope of the primary valley. Near-contour lines store surplus rainfall and subsequently distribute the stored water, particularly to the larger ridge areas.

2.   For irrigation and/or on flatter or flat land, a sufficient head of water is contrived to compensate for lack of any significant slope. The principles of aeration and minimal flooding contact with the land apply.

3.   Urban development and redevelopment — laying out of a settlement, village or town. As in rural development, the basis is the water lines of nature and those of Keyline control.

4.   Productive use of sewerage effluent and other so-called wastewater. Wastewater renovation through biological processes in the fertile soil of a "City Forest" and prevention of pollution of rivers and lakes from inappropriate enrichment.

For each of the above, there are documented example references within Australia.

There are several basic claims made for the Keyline System management of land and water, as follows:

1.   It identifies and builds on the inherent forms and shapes in any landscape situation and designs its development accordingly.

2.   It controls, uses, purifies and re-uses all available water.

3.   It transforms subsoil into soil, ensuring the permanent enrichment of land and environment.

4.   By controlling water and rapidly enhancing soil fertility, it assures soil conservation with little need for conventional structures.

5.   Its methods are energy conserving, both in the working of the land and in greatly reducing the need for synthetic fertilizers and pesticides.

6.   It provides economical structure plans for water controls and related road systems on which planning, architectural and other skills, and the reticulation of power and water, can best be applied or supplied by their respective expert practitioners.

7.   In urban development, it lends itself both to primitive and to sophisticated implements and methods. People with poor financial resources could do the work themselves with minimum outside aid. Housing could be simple to begin with, but within this basic framework could be progressively upgraded to any heights of excellence.

8.   It avoids salt-waterlogging, which has destroyed irrigation-based civilizations across the world and over the centuries.

9.   It makes comprehensive use of trees for food, shelter, fuel, bush timber and to emphasize the water lines.

10.   Overall, it presents a great simplification of water-and-land usage, with universal applications.

It is affirmed that the matters above are of such potential multi-billion dollar value for the world's people that we are desirous of finding sources of funding that will allow them to be widely explored and implemented as a matter of urgency. The Keyline System has a hopeful message to the world's people. It has much to offer the world in terms of food production and also in energy conservation in a manner and by methods that cannot threaten the environment, but which will positively and increasingly enhance it.

Our project enterprise is the launching of a *Keyline Education Foundation*. Such a foundation is necessary to ensure a new generation of competent Keyline System practitioners.

The "Aquascope" *(top)* is an important part of this project's success in bringing the wonders of exploring the undersea world to children *(below)*.

"Continent 6" — The Children's Village Beneath the Sea

René Eydieux
Sub Aqua de Porticcio, 20166 Les Marines de Porticcio, Corsica, France

French, born July 19, 1947. President of Sub Aqua Club of Porticcio. Educated in France; State Diplomas in diving, skiing, swimming.

"Continent 6" is a semi-submerged resort designed for children of six years upward. The aim of this structure is to stimulate children and to allow them to discover, master and take control of the shallow (less than 3 meters) underwater coastal surroundings.

"Continent 6" favours a gentle approach to the sea by developing two principal ideas:

1. To instigate the reconsideration of underwater penetration by means of creating new machinery, and above all, pedagogic instructions for future sea-classes.

2. To insure, through the children's enthusiasm, a greater recognition in adults of the problems (and their possible solutions) imposed by this "civilization of the sea".

In our general program, "Continent 6" develops an important biological activity. This consists of the following:

a. Demonstrations; cultivation/sea farming, breeding of fish, harvesting, preparation and tasting of certain species (mussels, oysters, fish, etc...).

b. Playing games; making discoveries on information sheets concerning the various underwater species.

c. The creation of an underwater garden, under the responsiblity of the children.

d. The fulfillment of certain operations involving re-stocking of fish (with the participation of a fisherman).

e. The creation of a photographic card index on the Mediterranean fauna and flora.

Technical Objectives
"Continent 6" will be a workshop of the sea, and will contribute toward the perfection of new techniques (mechanical and electrical) and better energy utilization (solar and hydraulic).

Positioning
"Continent 6" should be positioned in a privileged site: shallow, natural protection, near a diving center, access to rescue facilities, etc. Among the

different prospective sites under consideration is a beach at Porticcio, in the Corsican Gulf of Ajaccio, which provides these characteristics. Due to its semi-mobile structure, however, "Continent 6" can be quickly repositioned at any other appropriate marine site. Accessible by sea only, the resort of "Continent 6" creates no problems for the environment whatsoever.

Program of Functioning
The program is designed to function all year round, and every day between June and September (i.e., the school holidays). The resort is, by definition, reserved for children, but on several days during the week, adults will have the opportunity to participate in its running. Additionally, the adults will benefit from:
— an observatory from which they will be able to see the children working.
— a closed-circuit monitoring television.
— an explanatory "model village".
— an aquascope, functioning outside the village, allowing them access to fish cages, a wreck, and an oyster and mussel farm.

During term-time, "Continent 6" will be at the disposition of children and teachers, and will also welcome sea-schools, as well as science research schools (i.e., biology, archaeology, etc.). The program is designed to achieve the mastering of underwater surroundings by children, and their introduction to a wide range of biological activities. Several times a week, excursions at sea will take place to provide the children with a better knowledge and experience of the sea.

Instruction
There will be teams of specialists made up of; a medical team to insure the children's permanent safety, a team of qualified instructors (French and foreign) to insure supervision of the dives, biologists and teachers who hold the responsibility for presenting the different workshops, and hostesses to welcome and accompany the children. The time spent underwater does not exceed thirty minutes per session, and the depth does not exceed three meters.

Aquascope
The Aquascope is a semi-submersible mobile vessel, with a seating capacity of 8 persons, that allows non-divers or the children to observe the shallow underwater surroundings easily through a wide glass screen permitting panoramic vision. This Aquascope, which has already participated in the "Aquabulle" experiment in 1981, will be permanently based at Porticcio, and will permit the program of "Continent 6" to begin, by allowing a large public to participate in the activities of the Sub Aqua Club.

The Aquascope will be used in a variety of ways, including the sinking of old ship hulls to provide shelters for fish, the equipping of a cultivation zone with fish cages and oyster and mussel fields, demonstrations by children from the diving school, demonstrations of old diving techniques, shallow underwater rides and nocturnal dives.

Overall, "Continent 6" is planned to serve as a "university of the sea". We have completed detailed architectural and design plans, produced a scale model of the "Continent 6" resort, and are staffed with highly competent divers who are committed to instructing young people in the ways of our oceans.

Development and Construction of a Multi-functional Solar Observatory

Scott Thomas Grusky

15954 Alcima Avenue, Pacific Palisades, California 90272, U.S.A.

American, born December 12, 1961. Environmental Design student at University of New Mexico School of Architecture. Self-employed importer/ exporter of Mexican handmade clothing, jewelry and art.

With the aim of developing functional sculpture, I propose to design and construct a self-evident, easily accessible solar observatory. This observatory will enable its users to accurately follow the apparent motions of the sun through the day and year by directly indicating several aspects of the relationship of the earth with the sun. It will precisely reveal daily solar time, standard clock time, daily solar declinations, yearly calendar time, and the position of the sun on the ecliptic. I will accomplish this by carefully shaping and strategically locating certain geometric forms. These forms will replicate basic lines, planes and circles, such as the axis of the earth, the plane of the equator, and the path of the ecliptic. The primary massing of these forms will be built from a mixture of adobe and cement, in order to permit precision and durability while maintaining a vernacular style. I expect that this solar observatory will enrich the landscape and promote an awareness of the sun.

Introduction

The solar observatory is a system of points, lines, planes and circles which are placed in such a way so as to clearly and precisely reveal the apparent motion of the sun. For instance, the observatory indicates a line parallel to the axis of the earth, a plane parallel to the plane of the equator, the meridian for the place on which the observatory stands, and a circle concentric with the ecliptic. These geometric elements are represented by carefully shaped adobe and cement walls. The walls forming the observatory can be resolved into four main components. First, there is the gnomon, which is a triangular wall. The gnomon establishes a meridian and its hypotenuse is parallel to the axis of the earth. Second, there are the major arcs, which are concentric with the ecliptic. The walls forming the major arcs are parallel to the plane of the equator and perpendicular to the hypotenuse of the gnomon. Next, there are the minor arcs, which are perpendicular to the major arcs and form walls adjacent to the gnomon. Finally, there is the analemma tracer, which a system of strategically placed indentations in the gnomon wall. The gnomon and the major arcs work together to provide the sundial function, which shows daily solar time. This solar time is determined by the position of the shadow cast by the hypotenuse of the gnomon onto the major arcs, which are marked at equal intervals. The gnomon and the minor arcs work together to provide the sextant function. This function reveals the declination of the sun and the altitude of the sun. Both these angles are determined by the position of the shadow cast by a steel rod (which runs perpendicular to the hypotenuse of the gnomon and contains the center points of the arcs) onto minor arcs, which are marked at intervals of half a degree. The analemma tracer operates independently. It shows the equation of time, the declination of the sun, and the date of the year. Therefore, the four components of the observatory work together to perform three primary functions, which independently indicate important aspects of the apparent motion of the sun and together pinpoint the exact position of the sun on the ecliptic. In order to adequately present the solar observatory, I will now describe the basic construction and placement of its components. For the sake of clarity, I shall assume that the observatory will constructed in Albuquerque, New Mexico, at a longitude of 106°39′W and at a latitude of 35°05′N.

The Gnomon
The gnomon of the observatory is a three dimensional right triangle, of which the hypotenuse is parallel to the axis of the earth. Perpendicular to the plane of the horizon, it runs north-south, with its upper edge sloping downward from north to south. This edge must be parallel to the axis of the earth; in Albuquerque, this means it must intersect the plane of the horizon at an angle equal to its latitude, or 35°05′.

Resting on a concrete foundation, the gnomon will be made of adobe, cement, plaster, stucco and iron. Precisely 11.19 feet from the northern end of the hypotenuse, a steel rod will be set into the upper edge of the gnomon. It will protrude perpendicularly six inches from both upper edges of the gnomon, and will be parallel to the horizon, serving to indicate the center points for the arcs. Engraved plaques on the north face of the gnomon will explain to the observer how to use the various functions of the observatory.

The Major Arcs
The major arcs (east and west) will be constructed of adobe and cement. For smoothness, durability and precision, the final surfaces of the arcs will be coated with plaster, concrete and stucco. Iron numbers and gradations (forged by a blacksmith) to indicate solar time will be bolted onto the arc surfaces. The numbers, which represent hours, will be spaced at intervals of 2.50 feet, starting with the number 12 at the point where the arcs intersect with the gnomon. From there, the numbers will decrease on the west arc from 12 to 6, and increase on the east arc from 12 to 1 to 6. In addition to the numbers, there will be gradations every 2.5 inches, which corresponds to five minutes.

The surfaces of the two major arcs of the observatory are each formed by the edge of a disc which is bounded by two planes parallel to the plane of the equator. Consequently, the surfaces are at right angles with the plane of the equator, while the arcs themselves are parallel with the plane of the equator. The center points for both arcs are where the steel rod on the hypotenuse intersects the respective western and eastern edges of the hypotenuse. Both arcs lie in the same plane. The north-facing sides of the two arcs will intersect the horizon at an angle of 54°55′. The south-facing sides, however, will intersect the plane of the horizon perpendicularly, and thus will be slightly curved.

Both major arcs will be 90°, with radii of 9.55 feet. I selected 9.55 feet because on an arc with such a radius, one inch of shadow movement along the arc corresponds to precisely two minutes of time, or 1/2° of the arc. Consequently, the arcs can be easily marked at equal intervals, readings will be quite accurate, and observers will clearly see the sun's apparent movement.

The Minor Arcs

Constructed in the same fashion as the major arcs, these will each have a strip of steel 8.33' x 5" x 1/4" bent to fit the surfaces, and then securely attached. These strips will be marked at intervals of 1/2°, which corresponds to 1". The steel strip of the east minor arc will be marked to show the altitude of the sun at local noon, while the west minor arcs strips will be marked to show the daily solar declinations. Both arcs will indicate solstices and equinoxes.

Lying in planes parallel to the plane formed by the gnomon, they form walls to the east and west of the gnomon, measure 50°, are six inches thick, and share the same center points and radii as their respective major arc partners, to which they are perpendicular.

The Analemma Tracer

This is a system of indentations into the gnomon wall, the most important of which is a steel plate rising out of the hypotenuse perpendicularly to the horizon and facing due south. From the upper middle part of the plate, a small hole will cast a spot of sunlight onto an adjacent, steel-plate-topped table north of the vertical plate.

This will allow tracing of the analemma. Each day at exactly 12:00 Mountain Standard Time, I will mark the spot with a center punch, and date it. On cloudy days, I will have to interpolate from previous and later spots for accurate locations. At the end of a year, the spot will have returned to its original position, and a figure eight will have been traced on the table.

Social and Environmental Benefits

The solar observatory will provide a wide range of benefits for the community in which it is constructed. Being a rather unusual combination of solar technology and artistic sculpture, it will reach the people of the community through several different channels. Its primary role will be a resource center, to expand people's awareness of the sun and earth, and to enrich the surrounding landscape. Specifically, builders and architects will consult it for help in selecting solar sites, and in designing and positioning solar panels. More generally, it will help everyone who observes it to understand the basic solar rhythms and their relationship with earth.

The observatory will be a durable sculpture that will continue to contribute to the community and the environment for many years, while requiring very little maintenance and minimal supervision.

Screening for Carcinogens Using Fruit Fly Wings

Janos Szabad

Csiz u. 5, H-6726, Szeged, Hungary

Hungarian, born December 31, 1945. Research worker at Institute of Genetics, Hungarian Academy of Sciences. Educated in Hungary, Switzerland and U.S.A.; Ph.D. (Biophysics) from Jozsef Attila University, Szeged, in 1972.

In the course of evolution, many mechanisms have evolved by which organisms deal with the hazards of their environments. However, advances made by modern technology have introduced over a short period of time—too short for a response by evolution — a great variety of physical and chemical agents that are mutagenic. It is estimated that 30,000 new chemicals are produced in the world each year. Many of these are introduced to our environment (insecticides, pesticides, herbicides, medicines, etc.) and the possibility of exposure to chemicals mutagens has increased. Although these chemicals are tested for toxicity, they were formerly not tested for mutagenicity or carcinogenicity. It appears that most, if not all, mutagens are carcinogens and agents that were not known to be carcinogenic were not found to be mutagenic. It is difficult and expensive to determine whether a particular compound is carcinogenic. The mutagenicity of carcinogens, however, affords the opportunity to identify those agents that owe their carcinogenicity to their mutagenicity. Several rapid screens for detecting mutagenic compounds have been developed lately. These test procedures are also of great value because they provide identification of compounds that can induce mutations in all living organisms and it has been well known that the great majority of induced mutations have deleterious effects on the carrier. Two genetic dangers can be attributed to chemicals, namely that they cause mutations in germ cells, which are then inherited, and that they may produce mutations in somatic cells that result in cancer.

The fruit fly, *Drosophila melanogaster,* is particularly useful for mutagenicity screening because it has capacities for drug metabolism which appear to be comparable to those found in mammals. It is a multicellular eukaryotic organism. Its well known genetics and developmental biology provide unique possibilitities for use in mutagenicity screening procedures. There is absolutely no doubt that the genetic events detected in the fruit fly are of the sort which give cause for the greatest concern in man.

We have developed a mosaic test, the underlying principles of which are as follows:

In the course of embryogenesis, groups of precursor cells of adult organs are set aside. These cells form small "imaginal discs", remain diploid and divide through the larval life, for about four days. The number of cells in such a disc increases from about 20 to some 40,000 within four days. Cell division stops shortly after pupariation and imaginal cells produce structures characteristic for adults, like eye colour, hairs, bristles, etc. If a genetic event happens to *one* of the disc cells during larval life, that feature is propagated for subsequent cell generations. Progeny cells stay together and form a mosaic spot, or clone, on the adult body. The well known genetics of the fruit fly provides easy and convenient "genetic labelling" of these cells, and recognition of mosaic spots by the use of the so-called marker mutations.

The principle of using mosaic spot formation to detect mutagenesis rose some ten years ago. However, the early attempts came to a deadlock largely because eyes were screened for mosaic spots, and this system does not allow identification of small clones, is unsecure, insensitive and tedious. These difficulties can be overcome when wing mosaicism is an indicator of mutagenesis.

We make use of two marker mutations: MWH (multiple wing hairs) and FLR (flare) and screen for wing mosaicism. Both mutations are located on the left arm of the 3rd chromosome (that comprises roughly 20% of the fruit fly genome). Larvae are produced in a genetic cross of MWH/MWH and FLR/TM2 flies. (TM2 is a multiple inversion carrying chromosome.) Larvae — with the genotypes MWH/FLR and MWH/TM2 — are exposed to the agent to be tested. This may be added to the food or can be administered as vapour. (Toxicity of the agents to the larvae is determined at this stage.) Adults that develop from the treated larvae are sorted out according to their genotype (MWH/FLR or WMH/TM2), their wings are isolated and screened for the presence of mosaic spots under a compound microscope.

Usually 50-100 wings are analyzed. If the agent possesses no mutagenic activities, wing cells produce one straight hair each. However, as a consequence of mutagenicity mosaic spots form in which cells produce 2-7 hairs each (MWH) and/or woolly hair. Any significant increase in the frequency of clone induction — as compared to control — reflects mutagenicity.

The extent of the mutagenicity is given by evaluating f, the fraction of cells that were exposed to the mutagen and in which clone induction took place. The nature of the mutagenesis — if it is based on chromosome breaks and/or point mutation — can also be determined from the mosaic wings.

The mosaic test has other features worth mentioning: simultaneous mutagenesis detection in somatic and germ line cells, and the prospect of X-chromosome nondisjunction detection. To the best of my knowledge, no test systems provide simultaneous detection of chromosome breaks, point mutation and non-disjunction while analysing both somatic and germ line cells.

WHALE SICK BAY

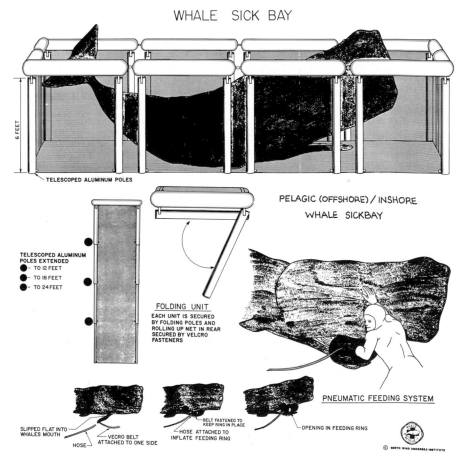

6 FEET

TELESCOPED ALUMINUM POLES

TELESCOPED ALUMINUM
POLES EXTENDED
● — TO 12 FEET
● — TO 18 FEET
● — TO 24 FEET

FOLDING UNIT
EACH UNIT IS SECURED
BY FOLDING POLES AND
ROLLING UP NET IN REAR
SECURED BY VELCRO
FASTENERS

PELAGIC (OFFSHORE) / INSHORE
WHALE SICKBAY

PNEUMATIC FEEDING SYSTEM

SLIPPED FLAT INTO
WHALES MOUTH

VECRO BELT
ATTACHED TO ONE SIDE

HOSE

BELT FASTENED TO
KEEP RING IN PLACE

HOSE ATTACHED TO
INFLATE FEEDING RING

OPENING IN FEEDING RING

© NORTH WIND UNDERSEA INSTITUTE

As part of a complete 'rescue package program' for beached whales, this Whale Sick Bay enables highly qualified teams to perform needed medical treatment before returning the whales to their natural environment.

The Whale Rescue Kit:
A Humane Rescue Technique

♛ Michael Ira Sandlofer

Honourable Mention – The Rolex Awards for Enterprise – 1984
610 City Island Avenue, Bronx, New York 10464, U.S.A.

American, born March 4, 1948. Director and Curator of North Wind Undersea Institute. Educated in U.S.A.; Divers Institute of Technology, Seattle, in 1971.

Based on previous experiences, including the successful healing and return to the sea in 1981 of a sick, stranded sperm whale, North Wind Undersea Institute is developing safe and humane procedures for handling stranded marine mammals. Whale rescue gear — a floating whale breakwater, anti-chaffing whale harness, whale sickbay, pneumatic feeding system, dual fendering system, pneumatic undersea bandage, sea tip and mammal sea sled — and rescue techniques are being tested and perfected by North Wind personnel and an international team of marine experts.

Documentary films and a manual on whale rescue are being produced to aid the rescue team in spreading these procedures world-wide, particularly among existing 'stranding networks'.

The combined "whale rescue kit" and educational efforts around it will help to preserve a magnificent and endangered species for future generations.

There are two areas in which the great whale population has been threatened. The first is in the commercial slaughter of whales. The International Whaling Commission has voted by a margin of 25 to 7 to ban all commercial whaling by 1986. Unfortunately, the opinion is not shared by all nations of the world and the commercially hunted whale is still in great danger of becoming extinct.

The whale's inexplicable predilection for beaching itself is the second great threat to its survival. "Beaching" is the term used when a whale comes into shore and becomes stranded on the beach with no independent ability to return to the deeper waters where he can swim and maneuver. Whales are known to beach when they are too old and feeble to swim against the inshore currents, when their sonar echo systems are faulty in unfamiliar waters, when they are in pain or sick, or when they attempt to rescue a beached member of their herd who is sending out distress signals.

Except in the cases of the sick or old whale who can no longer function, most beached whales are otherwise healthy mammals. Hundreds of whales are known to beach themselves each year. In New Zealand alone, in the period of

1873 to 1975, there were 35 mass strandings reported that involved 2,129 whales.

The fact that twice as many sperm whale strandings were reported in the five year period 1970-1975 than in the previous 96 years suggests that earlier sightings were inaccurately reported. The apparent increase in beaching in the last 20 years may be due to more accurate reporting or to the effects of greater pollution in our waters. But these whales can be saved!

North Wind has developed a twofold rescue program to save many of the marine mammals that are beached each year. It encompasses the creation and perfection of whale rescue gear in conjunction with a world-wide educational campaign to promote the procedures for use of such gear.

The first component of the program is the development of apparatus and procedures to aid a stranded whale. I designed these based on my experience in saving "Physty", believed to be the only whale ever restored to health by human beings from a prolonged stranding and returned to the sea (Fire Island, New York, April 1981). Each apparatus has been reviewed and approved by other marine scientists and experts. They are summarized as follows:

A Floating Whale Breakwater. Designed to provide a calm environment, free from surf and surge, for the rescue attempt to begin. By creating a lee, three things are accomplished; the mammal, freed from the turbulence of the surf, can rest and breathe more easily, a medical team can then safely enter the water and approach the stranded whale, and calm conditions allow thorough observation of the animal to determine the best course of action.

An Anti-Chaffing Whale Harness. Developed to move the animal to deeper waters where it can be further treated or set free. It has four important features; 1) It will not chafe the mammal, being made to be neither painful nor injurious to the whale's skin, 2) It tows the whale in a forward position with its head above water, considerably lessening the stress on the whale caused by a backward tail tow, 3) It is quickly and easily applied and released, and can be put on the whale in either shallow water or on the beach itself, and 4) It includes flotation gear, capable of compensating for two tons of negative buoyancy.

A offshore/inshore whale sickbay. Designed to proved space for a sick whale to rest and recuperate. A weakened whale, towed out to sea, may beach again or become a victim of a shark attack if he has not had the opportunity to recover his strength and health. The whale sickbay is light-weight, easy to handle and allows the whale to rest comfortably and safely. Designed in two-meter sections, it can extend up to 8 meters in depth through its telescoping feature. The sickbay is mobile, accommodates all sizes of whales and has a quick application, quick release feature.

A pneumatic feeding system. Developed to provide safe, comfortable and effective means of feeding a sick whale the medicine and food it needs to recover. A whale must open its mouth to accept food or medicine. Because he may be confused, disoriented and frightened during the rescue ordeal, he may not be willing or able to cooperate. An inflatable device has been designed that will slip in between the whale's jaws. As it is inflated, the jaws will be parted safely and painlessly to enable administering of food and medicine. A channel running just off center through the apparatus allows tubes for the transporting of food and medicine to be inserted in the whale's mouth.

Dual Fendering Apparatus. This protects the whale from barnacles and other abrasive matter. An air-filled buffer will be used when the whale is confined for treatment in an area between a stationary or floating object and support personnel. This procedure is extremely necessary to keep the whale comfortable during treatment. Optional equipment: water can be pumped through the fendering system and sprayed on the whale to cut down heat build up. The fendering apparatus doubles as a stretcher which can be used to transport the stricken mammal. It has an egg crate foam surface for better support and circulation of water.

A Mammal Sea Sled. Designed to transport small stranded marine mammals, it is inflatable, making it mobile on water as well as on land. It allows the mammal to be towed from the stranding area to a treatment center or to deeper waters where it can be treated or set free. The great advantage to using the sled is that the stricken mammal need never be transferred to another conveyance.

A Sea Bandage. This is a pressure bandage that can be made water tight by pumping oxygen into the bandage, causing its water tight seal to be activated by displacing the water. This is an important aid to be used for humans as well as marine mamals, as protection from the contamination of polluted waters.

A Sea Tip. This is a swab that comes in 2 and 3 foot lengths, and offers a water tight seal to protect the specimen from contamination and exposure to the elements.

The second component of the North Wind project involves communication and education. To ensure an effective, world-wide whale rescue program, North Wind Undersea Institute has a three-way approach to teaching correct use of the first aid apparatus. This effort will be spearheaded by an international team of professional men and women.

The first aspect of the educational program is to send the whale rescue team to meet with groups around the world, including existing stranding networks, to stimulate an exchange of ideas and demonstrate these new whale rescue techniques.

A documentary film, showing the use of the breakwater, harness, sickbay, pneumatic feeding and fendering systems, represents a second, broader method to communicate rescue procedures. It will be shown with an existing North Wind film, "Physty — Rescue I", which documents the actual rescue of a great sperm whale, the first to be saved by man from a prolonged stranding. This film will be screened at the International Whaling Commission's "Whales Alive" conference in Boston, in June 1983. Both films will reach audiences world-wide, educating them about the problems and procedures of stranding marine mammals.

The third approach to communicating the whale rescue program is through a printed *Whale Rescue Manual*. It will expand upon the films in explaining and demonstrating in detail the correct techniques for using the first aid apparatus.

Whales are now dying needlessly for lack of a coordinated world-wide effort that would provide uniform standards and procedures. North Wind Undersea Institute wishes to correct this situation through its dual program to produce whale rescue gear and to teach its correct use to the world community.

Saving the Early Masters of Argentine Painting

Cristina Margarita Navarro

Paraguay 2669, 1425 Buenos Aires, Argentina

Argentine, born April 20, 1927. Confidential Secretary in multi-national company. Educated in Argentina and U.S.A.

My project involves the compilation and recording of the works of the early Argentine painters, which are disseminated around the country and abroad, in museums and in private collections. I wish to obtain all available data on these artists and color slides of their works, for purposes of publishing a book that will fill the currently blank area of information on these artists, and preserve their work for future generations.

To accomplish this, I plan to utilize the following means:

— Visits to the cities in which the artists lived and to those other cities in which they sometimes worked.

— Contacts with owners of private collections, in order to obtain information on works produced by these artists and permission to photograph their paintings.

— Visits to museums, art galleries, churches, etc., where the works of these artists can sometimes be traced.

— Visits to public libraries, to consult those books, mostly out of print, with data relating to these artists, their works, and the time and historical circumstances in which they lived.

The work will start with the travels of foreign painters who came to Argentina early in the 18th century, and cover their works, their pupils, the places they visited, etc.

From this, the work will then begin to trace the lives and artistic achievements of the pupils, some of whom formed the so-called "Cuyo School of Painting": Benjamin Franklin Rawson, Gregorio Torres and Procesa Sarmiento, as well as the life and works of Ataliva Lima, a self-taught painter who took after this artistic movement for his own production.

The reason for starting this work with the "Cuyo School" is because this school is the only case in Argentina of a school of art that appeared in the 18th century. Several masters worked independently of one another, in different provinces, with little chance of communication between them. The "Cuyo School" is thus an important part of our patrimony, and should be preserved as an entity. It was formed by the following people:

Benjamin Franklin Rawson, born in San Juan, Argentina in 1820, son of a U.S. citizen who became a permanent residence and his Argentine wife. After studying in Buenos Aires, he returned to San Juan, and then moved to Chile for five years, where he was in contact with the best painters that had come from Europe at that time (Maurice Rugendas, of Bavaria, and Raymond Monvoisin of Bordeaux). When he returned to Argentina, Rawson painted mostly portraits of prominent personalities in San Juan and Buenos Aires, as well as miniatures. He died of yellow fever in 1871.

Gregorio Torres was born in Mendoza, Argentina, in 1814. After being taken at a very young age to Chile by his parents, he returned to Mendoza, where he almost immediately met Raymond Monvoisin, who suggested Torres go back to Chile with him. Torres accepted, and studied portrait painting in Chile with Monvoisin. In 1863, after he had moved to San Juan, Torres married, raised three children, and continued his work there, until late in his life, when he returned to Mendoza, where he died in 1879.

Procesa Sarmiento was the first woman painter in Argentina, devoting most of her skills to portraits of her family. Done in a delicate manner reflecting her intimate feelings, they also benefited from her having been one of Monvoisins's pupils in Chile. Her album is full of beautiful miniatures that register the style and fashion of that period.

Ataliva Lima, a self-taught painter, was very advanced for his time. His selection of colors is almost impressionistic, though that school of painters came much later. He followed the artistic tendency of his period, but put much of his own creative style into his paintings. Born in San Juan in 1833, he went to Chile with his parents as a boy, and started painting at the age of 7, under his own direction, following his own inspiration and devotion. His works have a distinctive touch of personality that make them very attractive. He returned to Argentina, and also lived in Mendoza, where he painted a great deal, though without great economic success. He married Carmen Gonzalez, and had many children. He passed away in 1882.

The lives and works of these artists, as well as their skills, will be extensively described in my book, which will end with a collection of color slides of their paintings. Of these, the paintings of Franklin Rawson are the most remarkable.

To date, in my work on this project, I have gathered over 200 color slides on the work of these artists. I started in Buenos Aires, working with Procesa Sarmiento's album. Since then, I have been to San Juan and Mendoza, and obtained the pertinent permissions from the museums to have the paintings they own photographed by local photographers. I have been working on this project for the last four years, during my vacations (three weeks per year), and assume I need approximately three more years to have the book printed and the color slides ready.

Due to Argentina's peculiar economic situation, private collections are sold, resulting in their dissemination all around the world. No records are kept here for future generations. Tracing the paintings takes time, but the value of having a compilation of the lives and works of the early masters of Argentine painting is well worth the effort I make.

The Nutrition Rehabilitation Centre of Kitale, Kenya

Kulwant Singh Sembhi

P.O. Box 862, Kitale, Kenya

Kenyan, born June 13, 1948. Managing Director of a construction firm. Educated in Kenya; 6th Form in Kagumo High School, Nyeri, Kenya.

Undernourishment in the developing countries is one of the biggest threats to human life. Because of this, we have established our Nutritional Rehabilitation Centre to teach, as well as treat, the patients we have so that they can further spread the knowledge learned in the centre to other adults. In this way, we hope some human life somewhere will have a chance to survive. This institution is charitable, which means that its continuation is not guaranteed; because of this, funds are required to provide this essential service to our needy people.

"It is human's duty to minimise human's suffering; but it is God's will to give reward for the rendered duty or service," as I have written.

A man of good will cares for his community. He is keenly interested in its welfare, in its health standards and in its progress.

He belongs to an 'international family', with an obligation to help and assist his family members, no matter what class, color, creed they may be.

One must realise that without the support of individual citizens, problems can never be solved. One must understand this duty and do all one can to solve any problem.

The Nutrition Rehabilitation Centre in the Transnzoia District of Kitale is one institution started jointly by a doctor from India, a doctor from Holland and a doctor from Kitale, in 1976.

Basically, the Centre acts as a focal point for undernourished children, who come to it with their mothers. The children are treated here for undernourishment, while the mothers are taught how to prepare food which is nourishing for the children. Surroundings in the Centre are very similar to those the mothers have back home. In two week's time, the mothers are taught to prepare nutritious food, how to look after poultry, and how to carry out general gardening to produce green vegetables and other essential foods.

The Centre comprises 5 huts, of which one is used as a classroom. As a charitable organization, the current national and international 'cash flow' problems are threatening the halt of the Centre's work.

Added to this difficulty, the Transnzoia District is experiencing a population explosion, with current growth rate at approximately 4% per annum. At the same time, there is an influx of people from the surrounding areas. This is due to re-settlement schemes, whereby big farms are taken over and sub-divided into smaller farms to provide settlement for more people. Although our district is a farming area, the local population does not seem to benefit from it; malnutrition is on the increase.

The age group at risk are pregnant and lactating mothers, infants either breast or bottle fed, and children in the weaning stages.

What can be done? We could admit these people to our local hospital, treat them and send them back home. The problem is that there is no teaching done here for the mothers; they simply take medicines, feel better and go back home.

The Rehabilitation Centre was started with the purpose of teaching the mothers on the spot how to grow and prepare nutritious food for their children. We explain to them the relationship between the sickness of their children and their way of feeding these children.

We do not, however, have enough teachers, nutritionists or educators, nor do we have adequate transport. We do have a few facilities for training. This allows us to not merely tell mothers what to do, but to make it possible for them to actually do so, using familiar surroundings and their own crops. These mothers, when they return home, are then able to teach their neighbors, and we have had success in this transfer of knowledge. Within the Centre, there is a vegetable garden. Mothers cook the food they raise in this garden on a "Kuni" stove (an arrangement of three stones), similar to that they have at home. The supervision of the cooking is carried out by a nutritionist.

In each of our four thatched roof houses used for living quarters, we are able to house one or two families comfortably. Due to requests for admissions, we have had to hold as many as 3 or 4 families in one house, and even then are turning down applications.

This project is seeking to build at least four more houses, make needed repairs and renovations in the existing houses, add two more latrines and a washing area, build a store house, provide fencing for the Centre, build a shelter to keep poultry and rabbits safe from stray dogs, and add more staff. We have shown that we can make progress against undernourishment in our area through proper rehabilitation and education. Our work needs to be expanded, and for that we need new funds.

The European otter (Lutra lutra) is a relatively large (weight up to 10 kg.) member of the weasel family, Mustelidae. This endangered species is adapted to an amphibious way of life. Otters have large home-ranges, occupying up to 30 km. of river and they also live by lakes and along some sea coasts.

A Conservation Strategy for the Mediterranean Otter

♛ Christopher Frank Mason

Honourable Mention – The Rolex Awards for Enterprise – 1984
7 Mede Way, Wivenhoe, Colchester, Essex, U.K.

British, born December 28, 1945. Lecturer in Biology, University of Essex. Educated in U.K.; D.Phil. from University of Oxford in 1970.

Over the last two decades, the European otter, *Lutra lutra*, has declined sharply in numbers in many parts of its range. The main reasons are pollution, habitat destruction and increasing human disturbance. While most countries have protected the otter, there is a lack of public interest in the survival of the species, particularly in the south, while the present status of the otter in many countries is unknown.

The aims of this project are:

1. To carry out field surveys for otters in the countries of the Mediterranean basin (Phase I).

2. To train local groups in the techniques for monitoring the otter population.

3. To initiate the establishment of long-term otter conservation strategies within these countries (Phase II).

Background to the project

The otter is a relatively large (weight up to 10 kg) member of the weasel family, Mustelidae. The species is adapted to an amphibious way of life, with webbed feet and a strong, muscular tail. Otters have large home-ranges, occupying up to 30 km of river and they also live by lakes and along some sea coasts. They are specialist feeders on fish, preferring the slow-moving coarse fish and eels.

Recent reviews of the status of the otter have indicated that the species has declined throughout much of its European range. However, this information derives largely from Northern Europe; in many of the Mediterranean countries, the present distribution of the species is unknown.

The otter is now protected in most European countries, although in southern Europe the present lack of public awareness or concern, the supposed competition of otters with fishermen, and the difficulties of law enforcement result in some illegal exploitation of the species.

In Britain, the main period of the decline of the otter (late 1950's) has been associated with the widespread introduction of persistent hydrocarbon pesticides into the environment. These compounds are now banned in Britain, and the fish food of otters is but little contaminated, no longer posing a threat. However, the otter is still declining. The presence of heavy metals, the condition of bankside vegetation, particularly sycamore and ash trees whose extensive root systems provide holts (dens) and secure lying-up sites, and other habitat factors have been well documented, and provide considerable insight into the ecology of otters in Britain. Prior to the commencement of our project (my partner is Dr. Sheila Macdonald), no data existed for the Mediterranean basin countries.

The Resolutions of the Second Working Meeting of the IUCN/SSC Otter Specialist Group recognized that the otter was declining through most of Europe and recommended further studies of the species and its habitat. Subsequent meetings gave top priority to the need for surveys and the establishment of conservation strategies, particularly in the Mediterranean countries.

We had carried out a pioneer survey into the status of otters in Norfolk, England in 1974-75. Largely using the techniques developed in that work, national surveys were carried out in Wales (1979), England (1981), Scotland (1981) and Ireland (1982). This also led to the Vincent Wildlife Trust's setting up the Otter Haven Project, of which S. Macdonald is national co-ordinator; more than 200 havens are now managed for otters in the U.K.

Much of the available information on otters in other parts of Europe has been based on questionnaires sent to hunters and game wardens. This is a superficial and unreliable method of collecting data, with little comparative value. For one thing, hunters tend to be ignorant of basic otter biology, and reticent to share their views on protected species. We have found otters to be common in districts where both hunters and local biologists claimed that otters had long been absent. Questionnaires provide little useful information concerning habitat or its utilization by otters. The advantages of rapid surveys by experienced otter field biologists will be demonstrated below. The techniques developed in Norfolk can be applied to other countries.

Phase I.
This part of our work is progressing on schedule and is financed by the Vincent Wildlife Trust, with a generous donation by World Wildlife Fund Italia for the Italian Survey. No additional funds are sought for Phase I.

The aim of these surveys is to cover as large an area as possible within the time available. Sites are selected for ease of access; for example, bridges or places where a river runs close to a road. At sites, a minimum of 200 meters is searched for signs of otters (i.e., spraints and footprints), while notes are made on habitat at all sites. If a search of 600 meters of river proves fruitless, otters are considered to be absent. The results of these surveys provide:
— an overall picture of distribution of the otter.
— identification of specific regions holding good otter populations.
— identification of regions where fragmentation of otter populations has occurred.
— relationships between habitat types and density of otter signs.
— identification of habitat features which are of importance to otter survival and which should be catered for in a conservation strategy.
— information on otter feeding habits.
A summary of the survey results obtained so far:

Country	Year	Survey Days	Sites Surveyed	Per Cent Positive
Portugal	1980	12	90	70.0
Greece	1981	19	200	62.0
Spain	1981	32	176	40.0
Italy	1982	20	188	8.5
Tunisia	1982	14	75	40.0
Yugoslavia	1982	30	131	43.0

In addition to these summary figures some of the key features of the surveys are given below:

Portugal. A healthy otter population, many animals living in very disturbed conditions close to man. The underdeveloped agriculture, with comparatively small use of pesticides is probably the reason for this situation, which could change dramatically if Portugal joins the E.E.C. Many Portugese streams dry up to fetid pools in the summer, and these were heavily marked by otters, probably defending a dwindling resource. Otters were eating a substantial number of snakes.

Greece. Some evidence of decline in places, but some very healthy populations. The management of drainage ditches in the heavily agricultural areas is such that otters survive in good numbers, but a change in management practice could have very damaging effects.

Spain. Evidence of decline and fragmentation, with the best populations in the West, toward the Portuguese border. Most animals were found in the upland areas of the country.

Italy. The otter is on the verge of extinction, with probably only one viable population remaining. The habitat along rivers is considerably managed and degraded, many rivers are polluted and hunting pressure is intense.

Tunisia. The central and southern areas of the country are too arid for otters and many of our sites were outside its range. North of the main R. Medjerda, 71% of the sites were positive. The evergreen shrub *Nerium oleander* is particularly important in providing cover.

Yugoslavia. The distribution is similar to that in Spain, with fragmentation occuring, the south holding the better populations. The middle and upper reaches of rivers hold the larger numbers.

Surveys are planned for Morocco (April, 1983), Algeria and Turkey, while the feasibility of carrying out studies in Romania and Bulgaria is being assessed.

The results of Phase I provide valuable and essential information to be utilized in Phase II.

Phase II.

Building on the knowledge of otter distribution and some of the factors important to their survival, we are now concerned with the formulation of detailed conservation strategies. At this stage, it is becoming necessary, indeed essential, to involve local personnel who will put the strategy into practice and monitor its effectiveness. A good understanding of local watercourse management and agricultural methods is required, and considerable lobbying of government organisations may be needed. To this end we have, as a starting point, produced a bibliography of publications on otters which will enable people new to the field to familiarize themselves with the relevant literature.

A Grass-Roots Approach to Preserving the World's Endangered Cultural Heritage

Meredith H. Sykes

132 Rue de Rennes, 75006 Paris, France

American, born April 4, 1939. Architectural historian and preservationist. Educated in U.S.A. and U.K.; M.A. (Art History) from Columbia University in 1964.

The built environment and its artifacts embody cultural identity. Efficient curatorial management of this heritage on the local level is the principal means for its protection. Identification and description, using cultural property inventories, is the first step in the preservation process.

This project will put in the hands of curators in the developing world, and at the grass-roots local level, standardized training materials on inventories and information handling that will give them the same facility for directing preservation activities that heretofore has been the domain of large, centralized systems.

The Need for Cultural Resource Inventories

A sense of the past possesses such cohesive and stabilizing force that nations devote important resources to enshrining their symbols of community and nationhood in museums, and in the restoration of sites and monuments. Beneath this iceberg tip of protected cultural properties lies a massive quantity of unprotected structures, archaeological and historical sites, and related artifacts. Deterioration, demolition and new construction take their toll of this patrimony. Often it is not appreciated until it is gone.

All cannot be saved. What is needed is a rational policy for choosing what will be preserved, and insuring its integration into the fabric of the present. This policy requires accurate information about existing cultural properties, their relative merits, states of conservation and potential costs of preservation; in short, it requires the information collected in cultural resource inventories and surveys.

While information collection and handling is essential to curatorial management, the methodology and equipment skills to carry it out are seldom part of the training of the curator and officials charged with preservation tasks.

Project Objectives

The purpose of this project is to prepare a two-part training package containing; 1) inventory methodology, and 2) advice for choosing and using information handling equipment. Both methodology and equipment will focus on the use of microcomputers and general purpose data management software now available in many parts of the world. A supplementary objective

418

of the project will be to identify potential users of the training package throughout the world so they, and the cultural properties they manage, can benefit from its availability.

Background

Cultural resource inventories created during the last two decades were generally centralized to guarantee compatibility of results through a region or nation, and, if computerized, to have access to adequate data processing capacity. I designed two such systems, for the historic buildings of Canada, and for the urban cultural resources of New York City. Both systems were large, computerized and employed many specialists, from the separate worlds of liberal arts and data processing. This approach was virtually mandatory up to the recent arrival of inexpensive, powerful microcomputers. Now curators anywhere in the world can have essentially the same inventory management possibilities as the earlier systems without the need for a costly data processing superstructure.

The essential goal of this project is to transfer existing survey methodologies to the new technologies, and to help cultural resource managers establish working systems.

Inventory Methodology and the Search for Standardization

Every survey of cultural properties, on rock art, Greek vases, or Gothic cathedrals, interrogates its materials to bring out certain information. The following questions could be asked of all properties:

— Identification and physical location?
— Degree of significance as a cultural property?
— Date, historical associations and authorship?
— Physical description?
— State of conservation, restoration and preservation?
— Existing documentation, bibliography, photos, plans?
— Accession and recording data about the item?

These questions fall into two categories; those describing any property (date, location, etc.) and those specific to one type (e.g., vase versus cathedral). Standardization in the questions asked, and in the vocabulary of the responses, expedites later retrieval and use of the information. We propose to use inventory standards based on the question typologies in my "Manual on Systems of Inventorying Immovable Cultural Property" (prepared for UN-ESCO/ICOMOS).

Information Handling Advice

The rapid evolution of microcomputing will mean that cultural resource managers will face increasingly difficult purchasing decisions. Thus, the project's training material will give the first time user comparative and explanatory information to aid in purchasing. It might describe, for example, not only what a 200 Kb disk drive is, but what it means in terms of maintaining a data base of 1,100 Moroccan mud-brick ksar forts, or 400 Punic amphorae.

The underpinning of efficient curatorial management is the office; the micro-computer will be an enormous aid in this area, and its potential contributions will form a part of the training package. We wish to prepare the training package, make it available in English, French and Spanish, and obtain its distribution to the curators who can make most use of this needed means to preserving the cultural heritage found in our buildings and artifacts.

Solar Energy and Ecology to Protect a Fragile Himalayan Environment

Louise Helena Isabel Norberg-Hodge
Henbury Lodge, Henbury, Bristol BS10 7QQ, U.K.

British, born January 10, 1946. Volunteer environmental worker in the Ladakh, India area. Educated in Sweden, West Germany, Austria, U.K.; Ph.D. candidate at University of London.

Ladakh is a high-altitude desert located in Jammu and Kashmir, in India. Lying on the Tibetan plateau, at altitudes ranging from 10,000 to 26,000 feet, the climate is extreme; rainfall averages only 4 inches per annum, and temperatures in winter drop as low as −40 degrees. Only 100,000 people inhabit Ladakh's 40,000 square miles. Most live in widely-scattered villages dotted among the mountains, supporting themselves through agriculture and animal husbandry. Communities — and individual households within communities — are almost totally self-sufficient. Few, if any, cultures in the entire world have adapted so well to such a harsh environment. Over the centuries, the population has been controlled, and a perfect equilibrium maintained between Man and the land. The very scarce resources have been used to the absolute utmost, but never overused. The concept of waste does not exist. As a result, the Ladakhis have managed not only to survive — a remarkable achievement in itself — but to prosper.

Tragically, however, this culture is now severely threatened. In 1962, when the Chinese entered neigbouring Tibet, a road was built by the Indian army, in order to bring troops to the Tibetan border; and ever since, the influence of the outside world has steadily increased.

The greatest threats stem from tourism, which has boomed in the last five years. In catering for the 15,000 tourists annually, traditional practices and values have been replaced—almost overnight—by quite alien, and sometimes destructive Western ways. Fossil fuels have begun to replace the renewable sources of energy always used in the past. Water is being wasted and polluted, urbanisation is spreading out, and the Ladakhis are beginning to believe that the modern way of life is better. Fortunately, traditional Ladakhi culture remains 90% intact; what is more, the interest in a more ecological approach to development is increasingly significantly every year.

The main reason for this growing interest has been the impact of solar energy. For the last five years, I have coordinated a series of projects to demonstrate a low-cost method of solar space heating; answering a crucial need without environmentally destructive side-effects. Winter cold is the most serious problem Ladakhis face; discomforting for more than half the year, and

420

responsible, directly or indirectly, for most of the major diseases found in Ladakh. Tuberculosis, skin, lung and eye diseases are all the result of the smoky, unhygienic living conditions made inevitable by the cold. Traditional animal dung and scrub vegetation fuels have been overstretched by the recent increases in population. Burning imported kerosene or coal creates as many problems as it solves, as people must leave for the capital to earn the money needed for this cash-based energy.

Solar energy, on the other hand, makes sense in every way. In fact, Ladakh is probably the best place on the globe for making use of the sun. It receives 325 days of sun per year, and the high altitude causes solar radiation to be particularly intense. The Trombe Wall system, which we have been demonstrating over the last five years, is inexpensive, environmentally sound and marvellously effective. Easily built by local craftsmen from locally available materials, it has demonstrated that living standards can be improved dramatically without the need for great economic, social or environmental dislocation.

The potential for other solar applications is enormous; water heaters (we built a very effective prototype in 1982), greenhouses for extending the short growing season (these, too, have already been tested), crop dryers and cookers. All these technologies could significantly improve the living standards of every Ladakhi family. A further value, shown in the work to date, is that solar energy can play a crucial role as a symbol of ecological development.

The project will continue to develop and demonstrate the Trombe Wall system, along with other technologies, with the goal of establishing solar energy as the number one energy source used throughout Ladakh.

The main thrust of the project, however, is the construction of the Solar Energy and Ecology Centre.

Work over the last five years has created a solid base of interest in the environment and in ecological policies of development. Witness to this interest is the fact that the Government has donated the land for the Centre. Built by local craftsmen, the Centre will have a small office, a meeting room, an extensive library and be, itself, a example of solar energy design functioning. Trombe walls, greenhouses, water heaters and cookers will all be demonstrated, along with an improved version of the traditional composting toilet, simple pumps for raising water, and workshops for the use of local carpenters and blacksmiths. An attractive restaurant will be included to encourage people to come to the Centre and to help cover costs. Locally written plays, emphasizing ecological themes, will be presented to take advantage of the drama that is a favourite Ladakhi pastime. The Centre will be administered by the Ladakh Development Group. Based on firm foundations, the Centre can serve as a focus for ecological activity and education, and thereby lead Ladakh toward an environmentally secure future.

The saddler's workshop *(top)* and a traditional home *(insert)* in the 'living museum' village of Milies, in Greece.

Preserving a Greek Village's Heritage in a Live Museum

♛ Helen Fay Stamatis
Honourable Mention — The Rolex Awards for Enterprise — 1984
58 Souidias Street, Athens 140, Greece.

Greek, born January 14, 1939. Lecturer, photographer, working on this project. Educated in Greece; French Baccalaureat (1st/2nd parts) from French Institute of Athens (1959/60).

Our project has a double goal concerning the Greek village of Milies, located on Mount Pelion of Thessaly. On one hand, we aim at protecting and preserving the historical heritage, traditions, customs, arts and crafts of the area, while on the other hand, we aspire to turn the whole village into a live museum.

The project was first conceived in 1979, after a 2-month stay in Milies, during which period my husband and I realized that the local people were either ignorant of their past, or else badly informed, and that most of the younger generation was unwilling to keep up with local customs, traditions and crafts. What is more, many churches and monasteries, frescoes, icons, old homes and fountains were badly in need of care and repair. Visitors received but little information about the village, and many passed by without even stopping.

Yet, the historical heritage of the area includes the following highlights:

1. Ruins of an ancient Greek temple, dedicated to the god Apollo.
2. Roman tombs.
3. Early Christian tombs.
4. An 18th century monastery built in the Byzantine style, richly ornamented with frescoes of the same period.
5. A church located in the village square, dating back to the early 18th century, with a gilded woodcarved iconostase, many icons and frescoes.
6. The village library, containing some 3,000 volumes and manuscripts, editions of the 15th century, books by leading European scientists and philosophers of the time, as well as the first dictionary and grammar book of the modern Greek language, written and published by a Milies scholar. These volumes were used in the Academy of Greek Letters and Science, established in Milies during the Ottoman rule. At this point, may we stress the fact that 95% of the inhabitants knew how to read and write, while most of Greece was plunged in illiteracy.
7. It should also be noted that Milies joined the Greek war of independence on the 6th of May, 1821. The flag hoisted on that occasion is to be seen in the village library.
8. During the Second World War, the village was set on fire by the

occupation troops. In 1953, terrible earthquakes destroyed most of what was left standing. A few well-built homes, the village library and some churches were spared.

9. Despite all these dreadful experiences, life seems to have blossomed again.

It was nevertheless obvious, during our 1979 visit, that the inhabitants of Milies needed guidance, support and encouragement, in order to keep remembering their past with pride, and to pass on their heritage to future generations and to all visitors. As it was, historical data, century old traditions and customs, old crafts and ways of life were rapidly vanishing, swept away by modern life. My husband and I decided to help. After giving it much thought, we decided that our first move should aim at creating a small local museum to help protect, preserve and show the history and traditions of Milies. To accomplish this, we took the following steps, starting in August 1979 and ending in September 1982:

1. We purchased an old abandoned farrier's workshop, close to the main square.

2. We carefully restored it, respecting the local architectural tradition.

3. At the same time, we started gathering information concerning the historical background of the village, by interviewing mainly elderly people. We then checked these data with history book references.

4. Endless friendly talks were registered on tape, giving useful details of present everyday life, education, health and old age problems, etc.

5. We entered practically all the workshops and watched men at work, always insisting on talking about their problems, aspirations and difficulties.

6. Women welcomed us in their homes and chatted about everyday chores and the upbringing of children. Some women showed us their kitchens, gave away recipes, or else let us gaze at their handmade trousseaux.

7. We went out into the fields and olive groves, saw the farmers at work, visited an old olive press and talked with both men and women about the difficulties they faced working on the land.

8. We visited nearly every church and monastery of the area, making notes and pictures showing the condition of the buildings and that of the icons and frescoes.

9. We attended religious ceremonies, a baptism and a marriage, the local saint's day celebration, Christmas Day, Holy Easter week and Virgin Mary's day. Celebration of Greek Independence Day was also registered. Information about each event was carefully noted.

10. Photographs and slides were taken of every subject of interest, and files were formed. After nearly three years of work, we had some 450 photographs and more than 350 slides, out of which to choose those that would help us go ahead with our project. Ninety two pictures were finally chosen, all of which were enlarged and put in frames, ready to hang up. The slides will be used to illustrate our lectures on Milies.

11. At the same time, all our notes and tapes were being carefully studied, information being double-checked whenever possible. We then grouped the subjects into chapters, forming a coherent, easy to read text, giving as many details as possible to the reader about the history, traditions and customs of Milies. The text was then published (2,500 copies), and now serves as a catalogue to our tiny museum. With ninety two pictures ready to hang up, and

our catalogue in print, we decided we were ready to open our museum.

12.　Furniture was chosen, respecting the local village style. The village carpenter did most of it, and we bought hand-made bedspreads, curtains, cushions, rugs and carpets, all woven and embroidered by the women of Milies.

13.　Our museum thus ready, we hung the photographs in five groups; a) pictures helping visitors to learn about the history of Milies, b) pictures showing every site of interest in the village itself, c) aspects of everyday life in the area, d) workshops and people at work, and e) traditions and customs still in existence.

14.　Last, but not least, we found someone eager to look after the place.

On opening day, 17 September 1982, all the village was invited, and each family received a catalogue free to take home. Surprise, response and enthusiasm shown by the villagers — bringing gifts — and later on by many of the visitors was, and is, our great reward.

During the three-day inauguration festivities, visitors warmly suggested that we print postcards from some of our photographs. Though we were practically at the end of our budget possibilities, we asked them to pick out the subjects they preferred. These were later printed, and are now to be found in our museum, together with a poster of one of the most beautiful sites of the village. The eight postcards and the poster will be offered to each villager as an Easter present.

Our museum is now open to the public free of charge, and is now considered as "their museum" by the villagers. Teachers bring schoolchildren, and visitors from all over Greece and even abroad have written encouraging notes in our visitors book, while the Greek press has given encouraging reviews.

Still full of enthusiasm, we are eager to go on with the second part of the project which includes the following:

1.　Translating the museum's catalog into French and English to help foreign visitors get all the necessary information.

2.　Publishing a book on Milies, containing most of our 92 pictures and a more detailed text.

3.　Most importantly, printing a map of the village, showing all the places of interest; churches, chapels, monasteries, the main library with its rare books, fountains, workshops where one can see the village people at work (among them the saddler, blacksmith, cobbler, farrier, shoemaker, carpenter) and homes where women can be seen baking their own bread, preparing their tomato sauce, grinding wheat, weaving on looms or spinning wool in their yards. The map will show all the easily accessible beautiful spots for the visitor.

We want visitors to enjoy the delight of Milies, and to show that it can become a live museum. We are eager to bring competent people to Milies to study and restore its art and books. We would like to have artists visit for exhibitions, to encourage the promotion of crafts among the villagers.

We could enumerate endless other plans, having found that our work could go on forever. New ideas keep coming as the days go by, most of which could be accomplished, as long as we keep our enthusiasm and our love for the village and its people.

Utilizing Peat to Retard Aquatic Pollution and Recover Metals

A. A. Leslie Gunatilaka

Department of Chemistry, University of Peradeniya, Peradeniya, Sri Lanka.

Sri Lankan, born February 22, 1945. Associate Professor of Chemistry, University of Peradeniya. Educated in Sri Lanka and U.K.; Ph.D. (Organic Chemistry) from University of London in 1974.

The easy availability of peat — an inexpensive and effective material — for the trapping of metals from metal-rich industrial effluents before disposal could be of great benefit to mankind. Our proposed project offers an innovative idea encompassing a relatively simple and economical technique for scavenging metals and recovering them. This will help solve problems associated with pollution both in developed and developing countries; an advantage is the possible recycling of the scavenger used, namely peat.

"Since the Industrial Revolution, the effort of removing man-made pollutants from the natural environment has been unable to keep pace with the increasing amount of waste materials and a growing population that further aggravates the situation. This has often resulted in the transformation of lakes, rivers and coastal waters into sewage depots where the natural biologic balance is severely upset and in some cases totally disrupted," state Forstner and Whitman in their book, *Metal Pollution in the Aquatic Environment*. Commenting upon the solutions to this problem, Forstner and Whitman further state, "Recovered minerals or raw materials which have been processed into materials do not necessarily have to go to waste. In many instances, the processed products can be recycled. Unfortunately, such conversion processes require technology and the availability of energy, and cannot be performed without emissions of waste products. In consequence, they may not be economically viable and environmentally acceptable.... By means of more advanced technological developments and future innovations, it would appear that problems associated with recovery and recycling can be overcome."

Our studies, together with similar investigations, have indicated that metals can be scavenged extremely effectively and almost completely from highly polluted aquatic systems by peat. Moreover, we have shown that by the use of relatively simple and inexpensive chemical processing, the adsorbed metals can be recovered and the peat made available for recycling. Using these two valuable properties of peat and its possible conversion into a transportable form (e.g., blocks), it is proposed to: a) clean the polluted aquatic environment, and b) to recover metals, some of which are extremely valuable (e.g., gold and platinum) and which may be found in economic concentration.

Peat is an earthy carbonaceous material presently used almost exclusively as a low grade fuel. The remarkable capacity of peat as a metal scavenger can be exemplified by the fact that certain metals show a total sorption capacity of 10,000. This means that under natural conditions the metal concentration is 10,000 higher than in water. Furthermore, this metal scavenging capacity of peat holds true for a large number of elements in the periodic table, thereby covering the entire spectrum of elements associated with industrial pollution.

The project involves the following stages:

1. Preparation of 'peat blocks'

Naturally sourced peat must first be prepared for the process by cleaning it to remove unwanted materials. This can be done easily by subjecting the peat to simple washing and flotation techniques.

Once the peat has been 'cleaned', it must be put through an acidification process. This is done by treating the cleaned peat with a dilute hydrochloric acid solution (pH 4.5) in order to activate the metal binding sites.

The resulting acidified peat slurry is then put into moulds of specific size, pressed and sun-dried.

2. Cleansing of the Environment

The first step in this process is to determine the metal scavenging capacity of 'peat blocks' of specific size. Experiments will be conducted to ascertain the optimum size of the blocks for scavenging the maximum quantity of metal per unit of time for water flow-through. As part of this study, the variation in metal sorption capacity related to flow rate of the effluent will be determined.

Based on this information, the following step is to determine the best arrangement of peat blocks within a porous casing to provide maximum efficiency in metal trapping.

3. Recovery and Separation of Metals

This is accomplished by dipping the casing containing 'peat blocks' in hydrochloric acid solution. This enables the metal to be released from the binding sites in the peat and to be brought into solution. Our studies have shown that the optimum pH of the acid solution used in this step should be 3.0.

The acid solution is then removed, and separation of the metal ions is accomplished using standard ion exchange procedures.

4. Recycling of the 'Peat Blocks'

The blocks are then washed in cold water in order to remove excess acid, adjusting the pH back to 4.5, and thus re-activating the metal binding sites. The 'peat blocks' are then ready for re-use.

I anticipate that, with the aid of funds for certain needed equipment, that the duration of this project would be 3 years.

Shelter, Settlement, and Shaping the Third World City

Mario Augusto Noriega
Apartado Aéreo 18991, Bogota, Colombia.

Colombian, born April 22, 1949. Professor of Urban Design at National University, partner in architecture/urban design firm. Educated in Colombia and U.S.A.; M.A. (Architecture) from Rice University, Houston, Texas in 1976.

Solving the problem of shelter in the expanding and poverty-ridden cities of the Third World cannot be viewed as simply putting thousands of low-cost roofs over the heads of the homeless. The solution must be based not only on discovering and analyzing the special needs inherent in this type of housing before providing it, but on the fact that the shape of the city must be controlled, that the need for quality in the urban environment is as pressing a problem as that of furnishing statistically required housing units.

This project, the result of twelve years of work in investigation, design and planning, is based on the search for tools, methods and principles aimed at helping to solve this problem of shelter and the shape of the Third World city.

Mere observation suffices to show that one of the outstanding characteristics of Latin American cities (and Third World cities in general) is their state of permanent change. From literally one day to the next, the urban physiognomy varies: streets change, buildings disappear, neighborhoods spring up while others are swallowed. The lack of resources and infrastructure necessary to withstand this permanent pressure of rapid change has resulted in widespread deterioration of the urban fabric.

In Bogota, Colombia, the city on which this work is based, the situation of change is critical. The majority of the urban population lives in poverty, and the housing shortage is acute. Responses to this problem have been inadequate as much in quality as in quantity, with the result that statistics show more than half the population living in 'spontaneous' (squatter or pirate) illegal settlements.

This project is concerned with detecting and helping to control the processes of architectural and urban change in low-income settlements and housing projects. This specific area was chosen for the following reasons. First, because 'shelter' for families in this sector is not merely a place to live. The house is a vital factor in the family income (rooms are rented out, stores, workshops and other productive facilities are located there, etc.). Solving the problem of the 'house' and its surroundings is thus as high a priority in Third World countries as seeing to the questions of health and welfare. Second, because, despite the fact that the majority of the population lives in this type of housing and settlement, almost nothing is known about their essential characteristics, about the spatial processes of change and the priorities on which their development depends. Such qualititative information is a necessary ingredient in whatever solutions are to be found to the housing problem; its absence in almost all current proposals is alarming.

This project began with the attempt to take such factors into account, and to formulate structured methods for recognizing the essential characteristics and priorities in low-income housing settlements and their process of change, and the transforming of these characteristics into concrete design proposals and models. Its basis is the belief that low-cost housing and high-quality architectural and urban environments are not mutually exclusive.

Between 1971 and 1973, investigations were carried out in several squatter communities on a general level, and in 1974, the Pablo Neruda Settlement was chosen as a basis for detailed investigations. Seven houses were chosen as case studies, recorded in drawings and photographs, with tape recordings of information obtained in conversations with their owners. Very complete data were recorded. In 1979, five years later, a second investigation was made on the same seven case studies, and the state of development described and classified according to three basic aspects: physical development; equipment, furnishing and 'decoration'; and organization and use of space. Analyzed and correlated with other data, we found four stages that emerged clearly as common to the houses in their processes of development. In order of importance they were; 1) definition of territory or the demarcation of the property within the facade, 2) spatial organization, or the creation of one or two multi-use areas; 3) subdivision or specialization of spaces; and 4) the 'extroversion' of the house, or the manifestation of interest in the urban context.

In 1981, our conclusions were translated into design principles used as the basis for prototypical housing units in a low-cost project called Gran Yomasa II. The structured plan, utilizing design principles based on massive data input from squatter/pirate/illegal settlements, provides what we believe to be a solid approach to the better solution of low-income housing problems, for both residents of such settlements and the cities that must handle the needs of rapidly growing populations. At present, equipped with a very large data base of relevant information on such communities, we are building rapidly toward increased knowledge of how best to meet the challenge of reversing urban blight and improving living conditions. We are now investigating the aspects of change in public spaces and the services of low-cost housing settlements.

The Barbados Green Monkey (Cercopithecus aethiops sabeus), once a major agricultural pest, is now becoming a valuable resource for the islanders, under an innovative management control program.

Optimal Utilization, Production and Conservation of Non-Human Primates

Jean Baulu

Barbados Primate Research Centre, "Hillcrest", Bathsheba, St. Joseph, Barbados, West Indies

Canadian, born September 9, 1944. Primatologist, founder and programme leader of Barbados Primate Research Centre. Educated in Canada and U.S.A.; M.Sc. (Primatology/Psychology) from University of Georgia in 1973.

Farmers in Barbados, W.I. have been plagued by an increasing population of pest monkeys introduced from Africa over 300 years ago. The Barbados Green Monkeys have been responsible for at least ten million dollars worth of crop damage a year. Traditional methods of control (shooting, poisoning, use of dogs, etc.) have been totally unsuccessful at decreasing the monkey population and agricultural losses; furthermore, these approaches were cruel as well as wasteful.

As a concerned primatologist, and realizing that this "monkey problem" is common in many developing countries harbouring non-human primates, I came to Barbados in an attempt to solve the farmers' plight, to find ways to manage the pest population, and also to try to change the image of monkeys from that of a destructive nuisance to be exterminated into that of a valuable and renewable natural research resource.

I am happy to report, after two years' effort, the establishment of a successful and on-going Monkey Crop Damage Control Programme which has not only arrested the increase in the monkey population, but is also responsible for the humane capture of at least 20% of the estimated monkey population. Consistent with public acceptability, a market has also been created whereby the captured monkeys are exported for worthwhile and essential bio-medical purposes (i.e., the production and testing of human polio vaccine). With sustained yield cropping, the monkey population is now well under control and the programme has become self-financing. This has meant that, in addition to creating employment and increasing food production, the project is generating a significant amount of foreign exchange on the island.

I also founded the Barbados Primate Research Centre, and developed various programs to try to help sectors of the economy other than agriculture. Among these are: Education (by the establishment of permanent field and laboratory research facilities), Health Science (by the production of captive bred monkeys for specialized research purposes), and Tourism (by the creation of comprehensive monkey displays). As a result of these efforts, Barbados is now considered a model in the optimal utilization, production and conservation of its non-human primate resource.

Green Monkeys, *Cercopithecus aethiops sabaeus,* were introduced in Barbados during the 17th century from Africa, and were by 1684 already considered to be agricultural pests, with a bounty put on their heads. Even today, the bounty exists, but the monkey population has increased over the last 25-30 years, causing losses of millions of dollars in agriculture each year. After preliminary visits between 1977 and 1979, my effort at solving this problem finally got underway in 1980, in the form of researching the various facets of the problem and defining potential solutions.

Barbados is a small island (166 square miles) with a relatively large population (250,000), making it one of the most densely populated countries in the world. The island's gullies are natural boundaries that separate plantations and other agricultural holdings, and monkeys have been allowed to proliferate in these forested gully habitats. They have become dependent upon the produce grown by men, and this parasitic relationship threatens both man and monkey as they compete for the same limited food sources.

By census techniques, combined with intensive trapping efforts in all eleven parishes of Barbados, it was determined that the monkey pest population is approximately 5,000 individuals, or 400 separate groups (ranging from 3-40 individuals, with an average of 13 individuals per group).

Crop damage characteristics were developed through a twelve-month long survey of 110 representative farms throughout Barbados, using a standard questionnaire with over 100 specific questions. Information on monkey sightings, group size, time(s) of day observed, behaviour, location, etc., was also included with the crop-by-crop analysis of the extent of monkey predation. Nearly all crops were raided, and damage exceeded consumption in most cases.

Public acceptance of control methods available was important to the project. Most people interviewed wanted the monkeys exterminated one way or another. Even though all agreed that humane methods of capture would be ideal, very few believed that it was possible, as all previous efforts had failed. Of interest, public opinion was unanimous that the monkeys should be used for worthwhile bio-medical purposes.

Development of successful, humane monkey trapping techniques was the next step. Much research, trial and error went into development of this aspect of the project. At the very outset, one needed to invent successful capture methods which would minimize stress and injury to the animals caught. Once captured, there still remained the problems of restraining, holding, caretaking, conditioning, marketing and shipping, to name a few. Typically, in Asian, African or South American countries that harbour non-human primates, their collection involves selling the animals to middle-men exporters who are not concerned with proper care of the monkeys, large percentages of which die before (or a few days following) their arrival at intended destinations.

Therefore, establishing a central organisation to be responsible for the proper control and management of the primate resource was a first priority. It would be responsible for the provisioning, trapping, holding, marketing, etc., of "pest" monkeys. The second priority was to devise a scheme in which farmers and other residents with a "monkey problem" would be educated and trained in the humane capture of monkeys, would be given appropriate tools that worked, and monetary compensation for their cooperation. The third priority was to earmark areas of intensive agriculture for trapping, and defining others that would be considered protected, in which the monkeys would not be trapped. These latter areas included existing parks or natural ecosystems removed from human agriculture; their protection would ensure the genetic viability and variability of the monkey population which has been separated from its African parent population for over 300 years. Once these standards and general policies were impressed upon local authorities, research on developing techniques began.

Two trapping methods were evaluated; multiple cages and a shooting net.

The multiple cage system is used where monkeys cannot be attracted into open flat terrain. Many cages of varying specifications were tested before finally settling upon one made of wire mesh and wood measuring 60cm x 60cm x90cm, with a sliding door that is left fully open (untriggered) most of the time. After testing a location with baiting (ripe bananas, paw-paws, etc.) for a week to determine if the monkeys will take the bait, a dozen traps are located in a small area, with food place inside. After the monkeys become accustomed to feeding in the cages, the traps are triggered so the doors will shut and lock automatically. Trapped monkeys are quickly removed, sedated with 5mg/kg of Ketamine hydrochloride and transferred to the holding facility.

The shooting net system is used in open areas where monkeys can be induced to raid crops or take bait in the open. A folded net (30m x 30m, of 57mm mesh) attached to four rocket-type cannons is fired by remote control from an observer's hiding place. Up to 14 monkeys at a time have been caught by this method, and it has been used successfully up to four times at the same site to capture remaining group members.

Farmer cooperation has been secured by purchasing bait directly from cooperating farmers, and paying them $50.00 per monkey captured live and unharmed. In the first year, 263 monkeys were captured. In less than two years since then, over 800 more monkeys have been culled from areas sustaining heavy crop losses, representing 20% of the monkey population and over $2 million savings to crops per year minimum.

The Centre has developed a careful and successful program for the handling of the monkeys that has allowed the marketing of them to become successful. Of interest, users are now utilizing Green Monkeys for the testing of polio vaccine, thus substituting for other species traditionally used. The project's work has resulted in publications concerning management of the program and a manual for the trapping techniques. "Pest control" is now viewed as managing a renewable natural resource.

Captive Breeding the Oldest Extant True Mammals

Walter Poduschka

Rettichgasse 12, A-1140 Wien, Austria

Austrian, born August 11, 1922. Free-lance scientist/zoologist. Educated in Austria; Ph.D. (Zoology, Anthropology) from University of Vienna in 1955.

Insectivores (hedgehogs, shrews, moles, tenrecs, otter shrews, golden moles and solenodons) are the oldest still extant true mammals; because of their insectivorous diet, they are considered as beneficial and important to mankind. Furthermore, they are considered man's oldest relatives. Their study reveals many items of our phylogeny, and indications of our own future problems.

Many of these species are endangered in the wild in their native countries, and are in clear danger of becoming extinct. Captive breeding should be tried, which is the objective of this project.

In agreement with the IUCN, I began trying to breed the *Solenodon paradoxus* of Hispaniola, a species with no hope of survival in its native wild. Already extremely rare, these insectivora will have disappeared in a few years if nothing is done. (Some still exist in Cuba, but they have disappeared in Haiti, and are losing the struggle for survival in the Dominican Republic.) To continue our breeding program for *Solenodon*, we urgently need one more room in our facility to provide space for depositing of their pheromones, a vital step in the breeding process.

In Madagascar, one of the two most important mammal groups are the ancient *Tenrecidae*, considered to be true 'living fossils'. Their group consists of more than 20 species. Ten years ago, we began our breeding efforts with five species. Due to old age of most specimens, we were lucky enough only to succeed with two (26 litters since then), one of which was captive bred for the first time. We now need to have new specimens for genetic reasons, and to make the attempt with new species.

In South East Asia, there exists the *Echinosoricini*, another very ancient group of insectivora. Practically nothing is known of their biology, needs, food requirement, propagation rates, etc. They have vanished from some of their habitat in Western Malaysia and Java.

In Mindanao, Philippines, only on two mountain ranges at certain elevations, is an insectivore, *Podogymnura truei*, known only from dead specimens. It has never been studied, and nothing is known of its chances for survival.

In Southern China, Northern Burma and North Vietnam lives (still?) another unstudied insectivore, *Neotetracus sinensis*, again known only from dead specimens.

With the exception of the Chinese species, for which we have little hope to learn about, there are possibilities for saving these creatures.

In Malaysia, with the permission and assistance of the Governmental Wildlife and National Parks Departments, we will be able to obtain a few specimens of the *Echinosoricidae*, which we will have to fetch personally, to prevent loss on the flight.

Another unknown group, the golden moles of Southern and Middle Africa, appear to be doomed, with their native range now reduced to not more than 9,000 hectares. We wish to obtain a few of these for study and breeding, if at all possible.

Techniques of keeping and breeding *Echinosoricidae* and *Tenrecidae* have been developed by us over the last ten years, and have led to successful breeding not only of *Tenrecidae*, but also (for the first time, and now 16 litters) of rare hedgehogs. It can be said, therefore, that probably nobody has more experience with the greater insectivora than we have. We live with them day and night in our small house in the outskirts of Vienna. Breeding facilities, however, are severely limited. In my study/working room, we keep more than 30 specimens, caged and free. As insectivora need, for propagation, ample room to depose their pheromones, we must adapt two more rooms for living quarters for these animals.

To avoid achieving simply "living museum specimens", we wish to continue with our breeding program and be able to send surplus specimens to furnish satellite groups in institutions where we can be sure they will be used for the sake of propagation. Ultimately, it would be the objective to take the surplus from the network of satellite breeding groups and release the animals within protected and appropriate habitat areas where they would be able to survive on their own. We would also seek the assistance of the conservation institutions in determining appropriate re-location of the surplus insectivores. In this project, my wife (a professional zoological assistant) and I have reached the point where we wish to add to our team the very expert aid of Dr. Gilbert L. Dryden, Professor of Biology at Slippery Rock State College, Pennsylvania, U.S.A., who helped me in organizing and leading the first International Insectivore Symposium, in Helsinki, August 1982. Our joint goal is to obtain the knowledge necessary to ensure the survival needs of these animals, which are so important to science and our environment.

The Study of Keratopathy in Salt Pan Workers in the Tuticorin Circle

Francis Joseph Rayen

22 First Street, Chidambaranagar, Tuticorin 628 003, India

Indian, born February 24, 1936. Practicing Ophthalmologist, with own clinic and nursing home. Educated in India and U.K.; Diploma in Ophthalmology from Madras University in 1964.

Nearly five percent of the approximate 10,000 workers in the salt industry in the Tuticorin Circle are afflicted by keratopathy of the cornea of the eye. This keratopathy, if unchecked, leads to total loss of vision. I am trying to study the cause of this "Salt Pan Keratopathy" so that it can be prevented, and thus help the salt pan workers to a safer working environment.

Tuticorin is a town in the South Indian state of Tamil Nadu, on the eastern coast. The 'Tuticorin Circle' of this project extends from Vaipar to Kayalpattanam; the main sources of work in this area are fishing and the salt industry. Of an overall population of 2,100,00 on this coast, some 10,000 men and women work on the 13,406 acres of the salt industry's pans. The climate and conditions in and around Tuticorin are conducive to the development and manufacture of salt, with low rainfall and heavy winds leading to the rapid evaporation of surface water.

Purity of salt in this region is 96-97% NaCl. It is snow white in color, to the extent of being dazzling to the eyes. It is this property that I feel is the cause of damage to the cornea of the eye.

It is well documented that the most damaging solar radiation is ultra-violet light, especially that reflected from water, sand and snow. Usually, exposure to such radiation is temporary, and superficial epithelial damage to the cornea is reversible. The salt pan worker, however, is exposed to the dazzling white light of the Tuticorin salt for 8 hours a day about 200-250 days per year. This prolonged exposure, in high velocity wind and tropical sun, causes the tear film to dry in the interpalpebral area, thus causing damage to the central part of the cornea in the lower pupillary zone. Initial symptoms complained of by the salt pan worker are photophobia, tearing, gritty sensation, lowered vision, and redness of eyes.

Routine examination at my clinic includes visual acuity measurement, examination of the anterior segment of the eyes with focal illumination, slit lamp examination, fluorescein staining, lacrimal duct patency, direct ophthalmoscopy, Schirmers test and tonometry. Signs on examination are lacrimation, blepharospasm, conjunctival and circumciliary congestion, and corneal opacity varying from mild erosion of the epithelium to dense thick white opacity across the pupillary zone. Depending on the progression of the corneal degration, it is possible to classify the cases into one of four 'Grades'. I have done this empirically as a personal assessment permitting identification, follow-up, prognostication and tabulation of salt pan keratopathy patients. No other ocular signs, such as xerosis, dystrophy, trauma or eye disease were present in the keratophy patients.

Different salt pan workers have been given tinted glasses to be worn while working, whithout proper prescriptions for tint. Tuticorin opticians have confirmed providing 1,700 pairs of such glasses, in tints such as yellow, blue, green and brown, on direction of the proprietors of the salt pans.

This project seeks the causative factors and prevention of Salt Pan Keratopathy in the following ways:

I.   Identification of the wavelength of the radiating rays from the salt pans by means of a spectrophotometer. As the cornea is permeable to ultraviolet rays of 3,000-4,000 Angstrom units, this will help in; a) prescribing the correct tint of glass to be worn by the workers to avoid noxious effects, and b) changing the working hours of the salt pan workers, as ultraviolet permeability increases to its zenith at noon and decreases toward evening. Earlier starting, a longer lunch break, and later working would alleviate the worst of the problem.

II.   The chelating action of Disodium Ethylene Diamine Tetraacetate (EDTA), which has proven useful as an ocular wash, must be studied in greater detail, as it definitely helps in the relief of symptoms.

III.   Histopathological study of recipient corneal tissue obtained by penetrating keratoplasty of, say, Grade III or Grade IV salt pan keratopathy patients having no complications or associated degenerative changes. This should be done after proper assessment of visual acuity, description of the cornea, anterior chamber, iris, conjunctiva, lens and intra-ocular pressure, and fundus examination. The recipient corneal button must be preserved in a proper fixative and mailed to a laboratory where an ocular pathologist could give a verdict on the specimen. Tuticorin does not have the facilities for keratoplasty. This could be done in Madurai City at the Aravind Eye Hospital, which is only 100 miles from Tuticorin. Begun in February 1982, I anticipate that the collection of data will be completed in April 1985.

The dangerous and deadly Box Jellyfish, scourge of Australian swimmers, shown here taking a pilchard.

Finding Natural Predators for the Deadly Box Jellyfish

Ben Cropp
Shipwreck Museum, Port Douglas, Queensland 4871, Australia

Australian, born January 1, 1936. Film producer, marine life photographer, museum owner/operator. Educated in Australia; Teachers Certificate from Teachers College, Brisbane, in 1954.

The box jelly fish, *Chironex Fleckeri*, is believed to be the most venomous creature on earth, and yet very little is known of its behaviour, life cycle and ecology. This jelly fish kills at least three times the number of persons taken by sharks in Australia's tropical north.

In 1981, I began a two-year project to film a one-hour TV Special on this creature, entitled "The Deadliest Creature on Earth". I was funded by Seven Network Australia to make the film. Additionally, the National Geographic Society provided support funds for still photos of my field work.

Since so little was known about this creature, much of my field work and underwater observations whilst filming became an added source of knowledge for scientists researching this creature.

On my regular, almost daily, observations and collecting of specimens on Four Mile Beach, I noted that a continuously large population of juvenile box jelly fish remained in that vicinity from December 30, 1981 to March 1982. They ranged in size from 1" to 3" across the bell, with an average diameter of 2". One morning's estimate of population over a 400 meter section of beach, on January 30, exceeded 1,000 juvenile box jelly fish. None were over 3.5" across the bell; in other words, there were no fully developed adults. On every day I saw and caught specimens, they were always there, being regularly replenished by newcomers, presumably from out of the nearby Mowbray River.

I concluded that a significant predation of juveniles exists whilst they are along the beaches. Otherwise, from all these juveniles sighted, we would have to expect a similar population of adults. Instead, the adult population is quite small. If heavy predation or some other death cause were not so, then human fatalities would also be higher. But who or what was the 'predator'?

I increased my field work and filming in this area of behaviour, and discovered four different predators that definitely relished eating the box jelly fish, in both juvenile and adult stage. These species are:

1. Hawksbill Turtle — *Eretmochely's imbricata*
2. Southern Butterfish — *Selenotoca multifasciato*
3. Round Faced Batfish — *Platax teira*
4. Gold Spotted Spinefish — *Siganus chrysospilos*

I had been told that turtles had been seen eating box jelly fish; logical, yes, because they do eat other types of jelly fish. Since they are not frequent inshore visitors, I doubted that they could play a significant role. To carry out my predation experiments, therefore, I selected Reef World at Cairns, because of the wide variety of both estuary and reef fish in its tanks. I could both watch and film under controlled conditions, and experiment with the fish outside their normal feedings times.

When set up, I released a number of juvenile box jelly fish, one by one, into a large tank containing barramundi, grouper, cod, striped spinefoot, bream, butterfish, hawksbill and green and loggerhead turtles. I was greatly surprise to see two Southern Butterfish excitedly and voraciously attack each jellyfish, first nipping off and eating the tentacles, and then eating the body. This predation was filmed and photographed at least 10 times. The butterfish is a very common fish in the estuaries and inshore coastline. Repeating the experiment a week later, the juvenile box jelly fish were this time attacked and eaten by the two hawksbill turtles, who crowded out the butterfish.

In another tank, I released juveniles one by one into a fish group containing striped spinefoot, spotted spinefoot, toadfish, white tip reef sharks, whaler sharks, leopard sharks, rays, one batfish, one spotted butterfish and numerous other smaller reef and estuary fish. The single Round Faced Batfish aggressively attacked and ate the box jelly fish, taking large chunks with each sucking bite. This fish is also common in the estuaries. Several spinefoot of the spotted variety also attacked the jelly fish, first nipping off the tentacles and spitting them out, then eating the bell section. Their predation was not as voracious as the butterfish, but still significant. These spinefoot are perhaps the commonest resident large fish in the mangroves.

In repeated filming, releasing some 20 juveniles one by one, the same excited feeding pattern was observed on the part of the above species. The hawksbill turtles, in fact, became so excited they bit my legs and lamp cable and tried to bite my hands in their rush for the food. They were so aggressive I had to take them out of the tank to film the butterfish predation.

I had been able to observe and photograph repeated predation by the four species, determining beyond doubt that they were true predators of the box jelly fish, and all common to areas frequented by the jelly fish. I also observed six other fish species which had a nibble at the box jelly fish, and these and perhaps other species may also prove to be significant predators when further tests are carried out.

When my film and photo work was completed in May 1982, so also was my funding exhausted. My discoveries and work on predation came to a halt. There is now, however, very strong interest for me to continue this work, because of its importance in learning all we can to combat this deadly creature. When the box jelly fish reach adult size, with a bell of 12cm-15cm in diameter, they possess tentacles up to 3 meters long, and are easily capable of killing humans with their stings.

Thus, Dr. J.T. Baker of Sir George Fisher Centre for Tropical Marine Studies and the Hon. David Thomson, Minister for Science and Technology, are both trying to organise a grant for me. This is difficult, and doubtful of success, because I am not a qualified scientist, although I am recognized as a expert in marine life, with thirty years of field work as part of my diving and film activities.

My project is to conduct a two-year study over two summer seasons of the box jelly fish appearance along our beaches. I will employ full time a qualified marine biologist to assist me, preferably an Honours student working to obtain a Ph.D. on this predation project.

I believe that predation holds the key to the population control of the box jelly fish, and the following steps will be undertaken in my future research:

1. Listing all the predator species through repeated tests and underwater observations and through analysis of the stomach contents of fish species.

2. Determining to what degree the box jelly fish constitutes the regular diet of these predators in the season when the jelly fish are prevalent.

3. Determining what level of destruction these predator's suffer through man's fishing, and to instigate protection and stock build-up for the significant predators in areas popular for bathing.

Developing Local Technology to Protect Nepal's Environment

W Akkal Man Nakarmi

Honourable Mention – The Rolex Awards for Enterprise – 1984
Block No. 12/514, Nagal/Quadon, Kathmandu, Nepal

Nepali, born January 1, 1946. Mechanical Engineer, Product Designer. Educated in Nepal; Civil Engineering Diploma from Engineering School, His Majesty's Government, in 1967.

Introduction. Nepal is a landlocked, hilly country encompassing the world's highest mountains. It is situated between India and China and is about 145,000 square kilometers in area. The population is concentrated in the southern and mid-hilly belt. Most of the 15 million inhabitants live in rural areas and are engaged in agriculture.

Until 1951, the country was closed to foreigners. There were no schools and even today many people cannot read or write. In times gone by, the people could meet their daily requirements from their nearby environment.

At present, however, the environment is greatly endangered. The fast growing population is very often overloading nature's capacity. The forests are cut down to get fuel for cooking. Where wood is not available, the valuable organic fertilizer — animal dung — is burnt to ashes, thus depriving the soil of needed nutrition. Cattle and goats, often the only 'wealth' of the rural farmer, are over-grazing the pastures, leaving dry, semi-desert landscape behind. Often, for time saving and business reasons, modern driving power is introduced. Diesel motors are being installed for efficient agro-processing: all of the fossil fuel has to be imported in our country, and it has to be paid for in "hard currency".

The demands of the population are ever increasing. Modern times require adjustments of standards, and technology does play an important role. The population lives scattered over the country. Therefore, non-centralized facilities, independent of central power supplies, are of great value. There is still a system of payment "in kind" and cash flow is still quite limited — often cash is simply not available.

Improvements. Technology capable of serving the rural population must not only meet the real needs of people, but must be welcomed by them, understood by them, and most definitely, they must not only be able to afford it, but must be able to repair it.

For most of our people, the technology has to be "simple", e.g., made from locally-available materials, made with local skills, and be small enough that even small farmers could — if necessary with a little assistance — afford it.

It is recognized that quite a bit can be achieved by improving existing traditional technologies. For example, improving the over 30,000 water mills estimated to be still in operation in Nepal. Though some 140 water turbines of U.S.A. or European design are now installed, these do not reach all areas where improvements could be significant.

In 1980, we initiated another turbine improvement attempt, based on the existing, local technology — the traditional horizontal water mills. By using one of the devices we developed, a miller can quite easily improve his grinding capacity by four to ten times. More important is the introduction of a pulley, fixed on the main shaft. This system can drive several small scale machines, such as rice hullers, oil expellers, a small dynamo, etc.

The welcome given to this new, locally developed concept is quite encouraging: within less than two years, over 50 units have been produced, with exports to neighbouring India and Bhutan. To buy them, farmers can get loans from the Agricultural Development Bank (Nepal), as the bank has approved their usefulness from the very beginning, and endorses this device which aids low-income farmers.

Prospects and Limitations. The development of traditional technology has a large potential. As with traditional mills and improved stoves, the needs of a good majority of the population are being met. Other solutions are not readily available from the shelf, however, and care must be taken in the use of imported technologies. Constraints for local developments are often our own limitations, but also the low, or lack of, return from innovation means the developers become the artists of technology; they have a job from which little bread is earned.

Personal Situation. The present situation is rather limiting for technology development. There are numerous ideas, but fewer means for implementating them: our workshop is too small and financial constraints limit further developments. Just after the construction of our successful mill operation, I had to take a job as an employee and produce my device in another's workshop. Encouraged by the success of the water mill, however, it is believed that many more things are feasible. The water mill is only one of the many devices developed at our family's workshop. The aim is, and was, to find solutions for our own country's needs; therefore our developments were always at lowest cost for the low-income customers. Work is now continuing on diversification of low-head, low-cost water power utilization for rural applications. Plans include developing pedal power, animal power, and privately owned agro-processing machines (rice flatteners, grinders), fuel saving and smokeless cooking places, and small scale cottage industry machines.

Our achievements are made available to interested people as a product, by photographs, booklets, drawings, etc.

Conclusions. These developments are assisting the rural people in their dire need of ecologically and environmentally sound technology. Financial support would greatly encourage our efforts and widen the path by which we could contribute to improvements in rural living.

Pre-fabricated, Self-fixing Masonry Elements

Luigi Mori

Via Gerolamo Ratton. 24/9, c.a.p. 16157 Genova (Pra), Italy

Italian, born July 7, 1930. Superintendent of construction company. Educated in Italy; Engineering Diploma from Technical Institute, La Spezia in 1950.

This invention is an alternative to traditional masonry in that the pre-fabricated blocks, of unique inside/outside design, require no mortar for their assembly in rapid fashion. With the scarcity of skilled masons, these modules (5kg for half-block, 10 and 16 kg for full blocks) can be made easily on site, and suit multiple uses. Special corner and design elements blocks are also available. Speed and ease of assembly make them particularly appropriate for the needs of developing countries. A further advantage is the ability to remove them easily, if a building is to be taken down or modified.

Landscape Conservation and Job Creation

Heinz Hans Hellin

Bramley Manor, Bramley, Surrey GU5 OHS, U.K.

British, born July 20, 1930. Management and Marketing Consultant. Educated in U.K.; M.A. from University of Oxford in 1954.

The charitable National Trust needs continuing funds for maintenance of large sections of the Lake District. This project seeks to organize conservation 'supporters clubs' in local factories, whose workers benefit from the Lakes. Funds collected would pay for conservation work to be done under National Trust auspices by employees who have been made redundant, thus aiding the environment and helping to reduce unemployment.

Tourism and Community Development in Torrotillo, Costa Rica

Alfred I. Fiks

Hotel Posada Pegasus, P.O. Box 370, 1250 Escazu, Costa Rica

American, born June 2, 1931. Director, Hotel Posada Pegasus. Educated in U.S.A.; Ph.D. from Purdue University in 1962.

This project intends to demonstrate the feasibility of a small tourist hotel in the Third World, and how it can act as; 1) a catalyst for the socio-economic development of the immediate community, and 2) a propellant for grass roots environmental protection/improvement. Starting with a small hotel, a program of purchasing from local suppliers, creation of a local environment education center, and the organizing of necessary local financing, the project is designed to become self-supporting. A report and guidebook on how to develop such projects is a key objective.

Saving the Manatees by Using Them for Pollution Control

Robin Christopher Best

Division of Aquatic Mammal Biology, Instituto Nacional de Pesquisas da Amazônia (INPA), Caixa Postal 478, 69,000 Manaus, Amazonas, Brazil

Canadian, born August 16, 1949. Research biologist at INPA. Educated in Canada; M.Sc. (Zoology) from University of Guelph in 1976.

Amazonian manatees, *Trichechus inunguis*, even though legally protected, have long been killed for their meat and hides. In an innovative program, using 42 of the animals for aquatic plant control at a large hydro-electric dam, this project is collecting valuable data on the physiology of the manatee. Each animal is radio-marked, and daily information is recorded on habits, ecological needs, and reproduction requirements.

Creating a Protection Trust for Chelonians in Antwerp Zoo

Vincent Lucien Paul Bels

31 Rue Sainte Véronique, B-4000 Liège, Belgium

Belgian, born August 4, 1956. Scientific researcher. Educated in Belgium; License in Zoology from University of Liege in 1981.

With the cooperation of the Antwerp Zoo, this project involves the building and equipping of new breeding units for tortoises and turtles, in an effort to create and sustain a breeding stock of some of the world's endangered chelonians. Working in cooperation with other zoos, the program is embarking on a difficult and little-understood process. Highly controlled environmental factors (temperature, light, etc.) are a critical part of successful Chelonian reproduction, and thus key to this project's implementation.

A Study of Reintroduction of Przewalski Horses into the Wild

Claudia Feh

Tour du Valat, Le Sambuc, 13200 Arles, France

Swiss, born September 19, 1951. Graduate student working on Ph.D. Educated in Switzerland; Zoology studies at Zurich University in 1974.

The only true 'wild horse' in the world, the Przewalski now lives only in captivity. Before releasing large numbers of them to the wild, a thorough knowledge of their social behaviour is required to avoid injuries. This three-year study aims to acquire needed knowledge on aggressivity and adult partner bonding, based on research conducted on the animals in semi-reserves in Holland and France. Such data will provide better prospects for successful herd breeding when the animals are returned to appropriate wild habitats on their own.

Establishing a Conservation Policy for Opencast Mines in India

Seeta Lakshmi Indurti

37 Venture, Lakshminagar colony, Saidabad, Hyderabad 500659, India

Indian, born February 13, 1958. Landscape architect. Educated in India; Master of Landscape Architecture from School of Planning and Architecture, New Delhi, in 1981.

Due to the particular geological peculiarities of the Neyveli lignite field in Tamil Nadu, Ms. Indurti has prepared a strategy for an environmentally correct exploitation and reparation of the opencast mine area. It is being undertaken in three parts; Conservation of the artesian aquifer under the lignite bed, implementation of the reclamation policy, and elaboration of the matrix of parameters needed for most effective later use of the land.

Prolonging Lactation by a Safe Family Planning Method

Jouni Valter Tapani Luukkainen

Steroid Research Laboratory, Dept. of Medical Chemistry, University of Helsinki, Siltavuorenpenger 10 A, SF-00170 Helsinki 17, Finland

Finnish, born February 28, 1929. Head of Department, The State Maternity Hospital. Educated in Finland; Dr. of Medical Science from University of Helsinki in 1958.

This project has developed a levonorgestrel-releasing intracervical device (ICD) which has shown good performance in normally menstruating women, and needs no pelvic examination prior to insertion. Its use could be a major benefit in Third World family planning as it allows continued lactation, which helps reduce infant mortality. An extensive, detailed study in the State Maternity Hospital (5,000 births annually) is designed.

Restoring the Dying Wells in Pakistan

Arshad Saud Khosa

P.O. Batil, Bahadargarh, District Dera Ghazi Khan, Pakistan

Pakistani, born December 19, 1955. Agriculturist. Educated in Lahore and Punjab; M.A. (English) from Punjab University in 1978.

"The Dying Wells" project is attempting to restore the traditional old wells of Pakistan to their former importance and usefulness for a population that is in need of water. This is being done by locating the old wells, paying local people to repair them, establishing a pot making factory for producing water buckets, and restoring the ancient rope-making skills that are needed in order to use the wells. Lifting the water is done with the help of bullocks. The project is on-going, and is bringing immediate benefits to local areas. Once again, land is coming back into its traditional uses.

Mitigating Environmental Impact of a Borneo Dam

Anton Lelek
Neugartenstrasse 32 A, 6231 Sulzbach, West Germany

West Germany, born October 11, 1933, Scientist, University of Göttingen Ichthyology II Department Head. Educated in Czechoslovakia; Ph. D. from Czechoslovakia; Academy of Sciences in 1961.

The decision to build a large hydro-electric plant in northern Borneo means that a vast area of land will be inundated, including valuable forest. To reduce the potential ecological harm, this project undertook a major study of the function of the riverine system. Proposals were made to mitigate and/or avoid hazardous consequences involving the very useful dam. The objective is to make this kind of knowledge available to other areas of SE Asia, before such dams cause irreparable damage.

Polycompetent Latrine: Low Cost Solution to Restore Environment

Shridhar Ramchandra Sathe
418 Narayan Peth, Pune 411 030, India

Indian, born November 1, 1909. Manufacturer of cotton mattresses and cushioning. Educated in India; Matriculate of University of Bombay in 1930.

Lack of toilet facilities in urban slums and rural areas in India poses a major health and cultural problem. This project has designed an easily and inexpensively built latrine that combines with a bio-gas production unit to achieve several advantages; 1) Low-cost, 2) Minimum water consumption, 3) Disposes excreta on the spot, avoiding drainage system needs, 4) Harnesses inherent energy of the human sewage, and 5) Provides people in rural areas with the best, cheapest manure. Planning calls for 'adoption' of a slum or village as a demonstration model for others.

Investigating Lung Disease Links with Third World Manufacturing

Christopher Gunapala Uragoda
78/5 Old Road, Nawala, Rajagiriya, Sri Lanka

Sri Lankan, born September 22, 1928. Physician in Charge, Central Chest Clinic. Educated in Ceylon; M.D. from University of Ceylon in 1963.

Although European industries have taken steps to prevent health hazards to workers in many fields, such regulations and efforts are not widely spread in many areas of Third World manufacturing. This project has already studied occupational lung disease in workers employed in such industries as tea, coir, graphite, kapok, ilmenite, chilli, granite, rice, wheat and cinnamon. The present objective is to determine hazards in working with asbestos, cardamon, cloves and papain, in order to propose means of providing appropriate safety measures for workers.

Automated Remote Tracking of the Florida Manatee

Robert Carroll Michelson

1941 Trophy Drive, Marietta, Georgia 30062, U.S.A.

American, born April 24, 1951. A Senior Research Engineer, Georgia Institute of Technology. Educated in U.S.A.; M.S.E.E. from Georgia Institute of Technology in 1974.

With less than 800 manatees estimated to be alive within the U.S.A., a key need for managing the survival of this species is the collection of data about their lives, habitat needs, etc. This project intends to place appropriate equipment on individual manatees that allows a state-of-the-art VHF time-of-arrival, time-segmented automatic tracking and analysis system to be implemented that will provide very large amounts of information for recording and processing. It is a high-technology solution to a challenging ecological problem.

Reconstituting a Population of Terns on the River Rhone

Denis Francis Landenbergue

19 Rue du Vieux-Moulin, 1213 Onex (Geneva), Switzerland

Swiss, born June 5, 1959. Self-taught ornithologist. Educated in Switzerland and U.K.; Diploma of Superior School of Commerce, St.-Jean, in 1978.

This project is designed to rebuild, by artificial means, a breeding population of two "sea-birds" previously widespread in Europe, the Common Tern, *Sterna hirundo*, and the Little Tern, *Sterna albifrons*. Using a large floating platform of unique design (anti-rat fringing, etc.), a habitat for terns was made, and has been successful in attracting these birds to breed once again in Verbois. Other locations have been emplaced, and the birds are now beginning to establish breeding colonies in former habitats.

Captive Breeding of the Endangered Philippine Eagle

Jim Young Abelita

c/o Philippine Oriental Builders, Inc., Don Alejandro Building Shop, J.P. Laurel, Trade School Drive, Davao City, Philippines

Filipino, born February 11, 1956. Manager in a marketing company. Educated in Philippines; Economics student at Ateneo de Davao University.

The great Philippine eagle, *Pithecophaga Jefferji*, is the second largest and most endangered eagle in the world. With a population of between 300-500, and found only on certain of the Philippine islands, its chances of survival are very small unless captive breeding ensures its future. This project involves the expansion of a program that has already produced results in egg-laying by captive Philippine eagles. More caging facilities are needed to build a larger population for future release to the wild.

Effects of Human Use of a Barrier Beach on Bird Populations

Anthony Frank Amos

606 East Street, P.O. Box 466, Port Aransas, Texas 78373, U.S.A.

British, born September 2, 1937. Research Associate, Physical Oceanography, The University of Texas at Austin. Educated in U.K.; GCE from Glynn School, Epsom, Surrey.

This project has counted 500,000 individual birds of over 160 species on Mustang Island, Texas, one of the world's longest (for its width) barrier islands, and computerized all data. The island's commercial development and use for human recreation is directly affecting its long-standing use as a refuge for both local and migratory birds. The project seeks to analyze the degree to which bird populations are suffering, and what steps may be taken to mitigate the effect of further impending construction.

Islands of Hope for Mexico's Ecological Deterioration

Eric William Gustafson

Arbol No. 182, Colonia Santa Engracia, Garza García, N.L. 66220, Mexico

Mexican, born February 12, 1945. International Vice-President, VISA Conglomerate. Educated in Mexico and U.S.A.; Ph.D. (Education) from University of Massachusetts in 1975.

Mexico's rapidly increasing population has placed great pressure on natural habitat land for birds and animals, as well as on the creatures themselves through killing for food or trapping for sale. The goal of this project is the creation of special artificial islands or the preparation of existing islands according to strict parameters in certain areas of Mexico as refuges. In wetland areas, the islands will have natural barriers against intruders, and be designed to support very large populations of birds.

A Fluorescent Electro-Optical Test System for Microorganisms

Morris Aaron Benjaminson

198 Broadway, Suite 1203, New York, N.Y. 10038, U.S.A.

American, born August 6, 1930. Director of Research, North Star Research, Inc. Educated in U.S.A.; Ph.D. (Biology) from New York University in 1967.

This project is developing a fluorescent electro-optical test system that will be employed, initially, to determine the feasibility of automated detection, identification and quantification of viruses. Fluorescently labelled DNase and/or RNase can be introduced and will adsorb to the nucleic acid core of viruses. Passing under a video-linked microscope with ultra-violet exciting light, the nuclease will emit at a specific wavelength. Emissions will be detected, measured and counted. Data will be used in the design of advanced instruments for specific microbial groups.

S.O.S. Guiana: Saving the Caimans

Daniel Rodolphe Louis Berger

21/27 Avenue Garros, 993150 Le Blanc-Mesnil, France

French, born March 5, 1950. Office Clerk in Ministry of Economy and Finance in Paris. Educated in France; School certificate in 1965.

Following studies made in Guiana on three separate expeditions, this project has identified key elements in the systematic destruction of the caimans, which are rapidly diminishing in population. As a means of counteracting the intensive and indiscriminate poaching of the caimans, the project involves the delineation of a reserve area wherein caimans, in a protected habitat, can be encouraged to proliferate rapidly. Key to the success of the reserve will be a system of regular survey by humans to prevent poaching, and developing public awareness of the potential loss if caimans become extinct.

A "First-Look" at Doubled Atmospheric CO_2 in the 21st Century

Sherwood Burtrum Idso

631 East Laguna Drive, Tempe, Arizona 85282, U.S.A.

American, born June 12, 1942. Research Physicist, U.S. Water Conservation Laboratory. Educated in U.S.A.; Ph. D. from University of Minnesota in 1967.

Though the rapid increase of CO_2 in the earth's atmosphere is being documented, along with worries about climatic dangers from the 'greenhouse effect', empirical data discount this hazard, and suggest that there might be numerous agricultural benefits resulting from higher CO_2 levels. This project is designed to demonstrate the ramifications for 21st-century agriculture in terms of increased plant productivity and water use efficiency, as a planning document for governments.

A Massive Case Study of Phenyl Mercury Acetate Contamination

Carlos Alberto Gotelli

San Pedrito 220, 1406 Buenos Aires, Argentina

Argentine, born January 27, 1941. Head of Toxicology Laboratory, School of Medicine, University of Buenos Aires. Educated in Argentina; Bio-Chemistry Doctorate from University of Buenos Aires in 1965.

A massive exposure of infants to phenyl mercury fungicide used by three diaper services in Buenos Aires in the fall of 1980 is providing quantitative information in a benchmark study of the effects of this compound on humans. Between 6,000-12,000 infants were exposed in the situation; this study is providing follow-up data on the development and psychological maturity of these infants for several years, as well as information regarding effects on the central nervous system and other organs, and mercury presence in tissues.

Ice Cave Cooling: A Passive Means

Benjamin Talbot Rogers

P.O. Box 2, Embudo, New Mexico 87531, U.S.A.

American, born October 4, 1920. Consulting Professional Engineer. Educated in U.S.A.; B.S. (Mechanical Engineering) from University of Wisconsin in 1944.

One anomaly of nature is the existence of two types of ice caves in areas where the average yearly temperature is above freezing. The second type is complex, found in high, dry climates, may or may not have a water source, is exposed to blistering summer heat and still produces ice. This project seeks to understand how these caves 'work', and to reproduce them by conventional building means. The objective is architecturally useful cooling, not necessarily the production of ice.

"A Bridge to the Future" — Sugarcane Growing in South Africa

Alfred Anson Lloyd

"Mostyn", 12 Humber Crescent, Durban North 4051, South Africa

South African, born February 27, 1914. Retired business consultant. Educated in South Africa; Ph.D. (Economics, Honoris Causa) from University of Natal in 1981.

This on-going project, supported by the sugar-cane industry, is aimed at providing a means to bring prosperity and ever-improving living standards to thousands of Zulu, Indian and Coloured sugarcane growers and their families. It is a classic example of how private enterprise in any country can assist under-developed communities through providing new job opportunities, the education needed to further community prospects, and the establishement of self-help cooperatives that benefit both community and environment.

Finding Uses for Excess Banana and Plantain Produce

María Cecilia Calderón de Montero

Ap. 5-245 Zapote, San José, Costa Rica

Costa Rican, born February 22, 1930. Manager of a bridge construction company. Educated in Costa Rica; Graduate in Microbiology and Chemistry in 1953.

After all the work of growing bananas and plantains for the market, farmers in Costa Rica must still face rejection of sizable portions of their produce as being unsatisfactory for export. This project is working on; 1) Means of preventing discoloration in the fruits, possibly through action upon the discoloring enzymes, to enhance their marketability, and 2) Methods of utilizing the perfectly nutritious and valuable product now being wasted, by converting it into other foodstuffs at the local level.

Coral Growth — Index of Change in the Thousand Islands, Indonesia

Barbara Elizabeth Brown

Department of Zoology, University of Newcastle upon Tyne, Newcastle upon Tyne NE1 7RU, U.K.

British, born September 23, 1947. University Lecturer in Marine Biology. Educated in U.K.; Ph.D. from University of London in 1972.

The coral reefs of Indonesia provide resources of fish, crustacea and molluscs that directly support many human communities, while their spectacular scenery attracts a rapidly developing tourist industry. Currently threatened by the encroachments of man, the reefs are the subject of the first attempt in Indonesia to assess the effects of pollution on coral reefs using a sensitive growth assay in dominant coral species. Using x-ray technology and sophisticated dyeing techniques, pollution dangers will be quantified and the results published.

Solar-Powered Video for Third World Environmental Education

Elizabeth Irene Combier

315 East 65th Street, No. 4C, New York, New York 10021, U.S.A.

American, born July 11, 1949. Television producer/director. Educated in U.S.A.; M.P.S. from New York University in 1983.

After four years of work in Egypt, this project is now preparing to demonstrate the usefulness of solar-powered video playback machines for environmental and ecological research in local self-help programs. With approval and support from Egyptian authorities, the project seeks to use these machines to allow local people to demonstrate improved agricultural equipment, use of biogas digesters, and language preservation. By letting local people teach their neighbors, via the video machines, faster and better understanding is achieved.

Detoxifying Polluted Air and Water by Coal Sorption Technology

Samia M. Fadl

Environmental/Occupational Science Program, Dept. of Kinesiology, Simon Fraser University, Burnaby, British Columbia V5A 1S6, Canada

Canadian, born July 12, 1944. Assistant Professor at Simon Fraser University. Educated in Egypt and Canada; Ph.D. (Applied Science) from University of British Columbia in 1978.

This project advocates the utilization of a lignite coal sorption technique for the detoxification of polluted air and water. Compared with conventional detoxification systems, this technique has the advantages of low cost, resource availability, broad-spectrum action, simplicity, versatility and recyclability, along with promising energy and reclamation uses for the spent material. A scaled-up field pilot installation is the next step.

A Simple, Easily-built Windmill for Self-help Projects

John Hereford
Route 2, Box 154, New Haven, Missouri 63068, U.S.A.

American, born May 5, 1932. Independent inventor. Educated in U.S.A.; Physics studies at St. Louis University, 1958.

A vertical axis, sail-type wind turbine, easily constructed from a wide variety of materials without unusual skills or tools, is presented as a low-cost alternative to commercial windmills used primarily for production of electricity. This machine, although lower in efficiency than propeller type windmills, is ideal for a variety of tasks such as water pumping, grain grinding and others. Based on an old Chinese design, the 30' high machine was built for $100.80 in 1982. The mechanical output is about 300 watts in a fifteen mile per hour wind. Plans are available.

Lake Tahoe Research: A Model for Preventing Deterioration of Lakes

Charles Remington Goldman
Division of Environmental Studies, University of California, Davis, California 95616 U.S.A.

American born November 9, 1930. Professor of Limnology, University of California at Davis. Educated in U.S.A.; Ph. D. (Limnology and Fisheries) from University of Michigan in 1958.

Lake Tahoe, one of the world's clearest and most beautiful lakes, is located on the California-Nevada border at 1,900 meters in the Sierra Nevada mountains. Faced with very high lake-basin development, the lake is the subject of this project, which is developing an interdisciplinary research effort to prevent its further deterioration. The successful and widely applicable strategy for saving the lake can be used as a model for similar programs elsewhere in the world.

Operation San Antonio — Restoring an Ancient Brazilian Fort

Armando Luiz Gonzaga
Rodovia Haroldo Glavan 759, Florianopolis, S.C. 88,000, Brazil

Brazilian, born April 28, 1936. Executive Director of building company in Brazil.

With a team of volunteer helpers, the Fort of Santo Antonio de Ratones Grande, first built in 1740, is being restored in a unique effort to; 1) preserve an important landmark, 2) make the local population aware of its cultural-historic heritage, and 3) serve as a focal point in calling the attention of the government to helping in the task and implementing an Ecological Control Station in the region. Activation of the community in support of the project has resulted in the cleaning of the fort, which is now in need of restoration work to prevent the deterioration process from beginning once again.

Remote Thermal Imaging: New Tool for Wildlife Management

Lloyd Sherwood Gray

10 Farmstead Road, Storrs, Connecticut 06028, U.S.A.

American, born November 7, 1950. Research Fellow, Sloan Kettering Institute for Cancer Research. Educated in U.S.A.; M.D. from University of Connecticut School of Medicine in 1978.

This project intends to thermographically study healthy animals species in order to collect sufficient data to prove the validity of using remote imaging thermography for the assessment of animal health in the wild. Most heat production occurs in the muscle mass, which is reduced in conditions of malnutrition. By establishing muscle mass data for healthy animals, it may be possible to produce an index consisting of whole body heat as a function of dimensions. Wildlife management and conservation techniques would be aided by such technology.

The Street Kids of Bombay

George William Hall Clarkson

40 Sherwood Drive, Clacton on Sea, Essex, U.K.

British, born June 16th, 1912. Partner in small chain of retail shoe shops. Educated in U.K.; Self-educated since age 14.

Having for many years worked to improve the lives of people suffering from leprosy, for which a unique and effective means of moulding shoes has been accomplished, this project has led to an attempt to tackle the problems of the 'street kids of Bombay'; homeless, orphaned or abandoned, and with little chances for improved lives. Working with local charity organizations, the aim is to provide expanded housing facilities and, more importantly, human affection and caring. While funds are critically needed, the success of the program will depend on human involvement with these youngsters.

A De-centralized Approach to Providing Third World Fuel

Alan Janvier Voelkel

c/o Food For The Hungry, 7729 East Greenway Road, Scottsdale, Arizona 85260, U.S.A.

*American, born April 30, 1958. Volunteer with **Food for the Hungry International**. Educated in Costa Rica, Peru, Colombia and U.S.A.; B.A. (Anthropology) from Wheaton College in 1980.*

This project is designed to provide a locally produced fuel industry in the Dominican Republic, and to serve as a model for similar situations elsewhere in the Third World. The project's aim is to convert agricultural wastes and molasses into 160-190 proof ethanol, and market this ethanol, along with appropriate technology stoves, heating units, etc., via an Ethanol Producers Cooperative. Able to start with minimum investment, the project will grow as rapidly as profits can be plowed back into further expansion.

NAME INDEX

COUNTRY INDEX